According to the syllabus for BE, Visveswaraiah Technological University

Engineering Physics

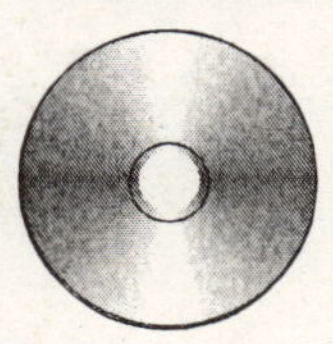

with

CD for **Current Trends in Engineering**

The cover of the book represents

Physics is the study of Nature and Natural phenomenon, nothing other than the study of *Panchabhoota*. The composition enunciates it with a different approach and a rational attitude.

***Bhuvan* (*Pritvi*–Earth)**	: Geology, geography, earth science, etc.
***Gagan* (*Akash*–Ether)**	: Space study, astronomy, astrology, etc.
***Anala* (*Agni*–Fire)**	: Heat, thermodynamics, thermal physics, etc.
***Vayu* (*Vata*–Air)**	: Fluid dynamics, aerospace, aerology, atmospheric study, etc.
***Varuna* (*Jal*–Water)**	: Ocean studies, marine engineering, aerosols, etc.

The surroundings are the earth and the four small boxes are the technological loom of mother environment like *water* is represented by the reservoir, *fire* by rocket, *air* by wind mill, and the *ether* by the planets.

According to the syllabus for BE, Visveswaraiah Technological University

Engineering Physics

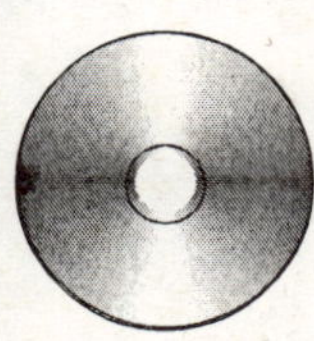

with

CD for **Current Trends in Engineering**

Sathish LA PhD

Head, Department of Physics
Sir M. Visvesvaraya Institute of Technology
Bangalore 562157

CBS PUBLISHERS & DISTRIBUTORS

New Delhi • Bangalore

Engineering Physics

First Edition 2007

ISBN: 81-239-1440-7

Published by :
Satish Kumar Jain for CBS Publishers & Distributors,
4596/1-A, 11 Darya Ganj, New Delhi - 110 002 (India)
E-mail : cbspubs@vsnl.com
Website : www.cbspd.com

Production Director : Vinod K. Jain

Branch Office :
Seema House, 2975, 17th Cross, K.R. Road,
Bansankari 2nd Stage, Bangalore - 560070
Fax : 080-26771680 • E-mail : cbsbng@vsnl.net

Printed at :
India Binding House, Noida, UP (India)

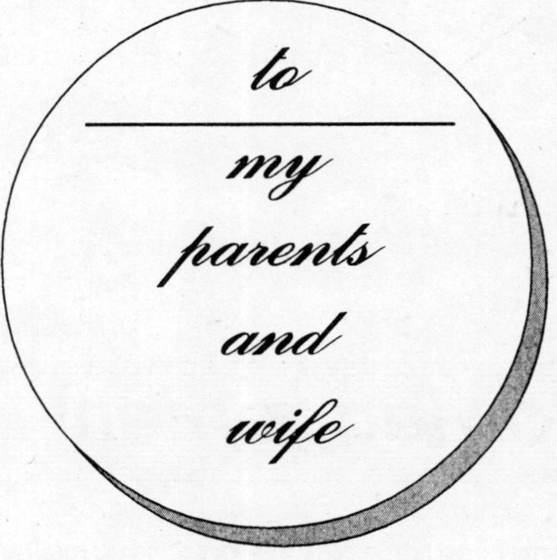
to
my
parents
and
wife

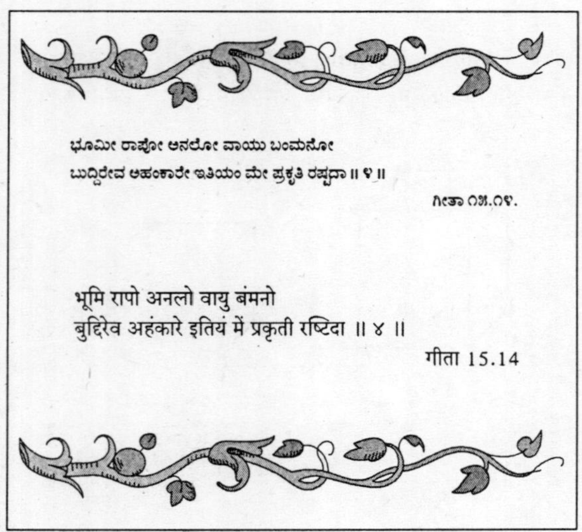

Acknowledgement

At the outset I would wish to bow my head to the Almighty for providing me the intellectual strength and power to write this book.

It is my pleasure to express my deep sense of gratitude to Sri Krishnadevaraya Educational Trust for providing me an opportunity to serve at Sir M Visvesvaraya Institute of Technology, Bangalore, without which this book would not have come.

The development of such a project requires the contributions of many people. I am specially thankful to Mr YN Arjuna, Publishing Director, and Mr Deepak Rao, General Manager (Bangalore branch), CBS Publishers and Distributors, who had enough confidence in me as an author to write a book for their esteemed publishing house.

To my dismay, I discovered that it is impossible at the end of two-year project laboured upon not only at Sir MVIT but also during the extended stays in JSSATE, Noida, SVC, Mysore, and NFGC, Bangalore, to recall all the occasions when students, colleagues and friends gave me an invaluable criticism, advice and instructions among others. I am indebted to Dr Viresh K Basalalli, Principal, Sir MVIT, who is the source of inspiration.

I would wish to express my sincere gratitude to Dr BV Sadashivamurthy, Professor and Head, Department of Physics, SJCE, Mysore, and Dr HS Badrinath, Head, Department of Civil Engineering, Sir MVIT, for their suggestions and critical readings at different stages.

I am most grateful to the MD of MAP System Mr Visvanath and his team members Mr Lokanath, Mr Prem and Mr Vinod for their generous support in completing the DTP and drawing work of this book.

Thanks are due to my colleagues Mr Sasi Kumar, Miss Poornima and Anindita Roy Choudhary for their critical readings of the manuscript.

Sincere thanks and indebtedness are due to one and all who have helped me directly or indirectly in bringing out this book.

Words fail to express my sincere gratitude to my wife Smt Meenakshi B Sathish for her spirit of accommodation at every instance shown to me.

Dr. Sathish LA

Preface

The book *Engineering Physics* is written for the students of I and II semesters of I year BE students of Visvesvaraya Technological University, Belgaum, Karnataka.

The *primary objectives* of the present book are:

(*i*) To present the concept and principles clearly.

(*ii*) To provide concise answers to avoid the bulkness of the book and this approach applies to both the language and the mathematical presentation

(*iii*) To write the content in a simple, lucid and precise manner.

(*iv*) Focused on central issues and highlights.

To create curiosity among the readers, each chapter is provided with an ample number of exercises and unsolved numerical problems, the problems that follow each chapter are tied rather closely to the text and are always thought-provoking. For solutions of these numerical problems, the reader may refer to the other publication ***Problems and Solutions in Engineering Physics*** by the same author.

Since the CURRENT TRENDS IN ENGINEERING is the special topic which has been introduced in the syllabus in different technical universities, the content pertaining to this section is provided on an accomanying CD in the form of power point presentation consisting of over **1300 slides**. In addition to this, latest research in the fields of nanotechnology, quantum computing, smart materials and biotechnology in different countries is also included.

The author has drawn heavily from the existing literature on the subject for which he acknowledges gratefully.

Straight-forward exercises and unsolved numerical problems associated with each chapter are aimed at testing the student's grasp of the preceding material which should encourage active involvement of the reader.

I have tried to make this edition error-free but some mistakes might have crept in in spite of our best efforts. I cordially invite constructive criticism and authentic suggestions from the readers for further improvement of this book.

Dr. Sathish LA

Physics is the study of Nature or Natural phenomena

Physics is the study of Panchabhoota

Prithvi–Earth : Geology, geography, earth sciences, ... , etc.
ಪೃಥ್ವಿ ಭೂಮಿ

Agni –Fire : Heat, thermodynamics, ... , etc.
ಅಗ್ನಿ ಶಾಖ

Vayu–Air : Aerodynamics, atmospheric studies, aerosols, ... , etc.
ವಾಯು ಗಾಳಿ

Varuna–Water : Hydrodynamics, fluid dynamics, ... , etc.
ವರುಣ ನೀರು

Akasha– Ether : Heavenly bodies, space studies, ... , etc.
ಗಗನ ಆಕಾಶದ್ರವ್ಯ

Contents

Introduction

The term *science* originates from Latin remark *scientia* connotation to know.

About four hundred years back various populace whispered that they live on stationary earth which itself is positioned at the centre of the universe. The world further than the solar system was absolutely unidentified at that time. The submicroscopic realm of atoms and molecules was anonymity. Not even a single rule of nature was unerringly invented. There was no science base technology and there was scarcely any progress that could be called as science.

Man has continually been snooping to have knowledge about nature and the natural phenomena happening around him. These four hundred years have elapsed just like a flash of lightning in the era of human race. Previously, which appear as complexity of nature; the man with his advocate to be aware of has exposed them only to be a few type of easy pattern of nature. Footing on the shoulder of physical science, the medical and biological sciences have developed to a commanding height. In parallel with science, mathematics has also developed. The science-based technology has touched every facade of existence.

Antediluvian knowledge was largely qualitative and tentative. The piece of information which distinguishes modern science from prehistoric knowledge is the close up between hypothesis and experimentation. In no other branch of science this relation is further intimate than in physics. With the expansion of science, instruments, experimental and measuring techniques, are becoming more multifarious and expensive. For learning physics it is essential to perform a large deal with easy equipment and apparatus, provided it is used astutely and carefully.

Our understanding of natural phenomena is based on the consequences of massive amount of experiment and interpretation. It is well acknowledged that when experiments are to be performed; one cannot rely in excess upon the human senses of touch, hearing, sight, etc. to make precise scrutiny. The method of measurements which depend upon human senses entirely, termed as subjective method. Methods which make use of orderly apparatus and tools for captivating observations are called objective methods. Information obtained by subjective methods by means of human wits is a little deceptive and also imprecise. The subsequent simple activity illustrates the fake imitation so effortlessly arrived at by employing subjective methods of explanation and also demonstrate the need for using objective methods of inspection for the steps forward and innovation in any branch of science. As an example in summer at midday, the gateway has been exposed to the sun for an adequately long time. Consider the initial touch of the metallic bolt of the gate and any wooden part of the gate and examine the comparative feelings one has in the two cases. One would disembark at a sensation that the bolt is hotter than the wooden part of the door. Of course, this is the conclusion which every body would depict on the basis of the interpretation made by means of "sense of touch" . The information which has been arrived at by the use of subjective method is pretty indefinite because the bolt and the wooden part of the gate are at the same temperature. The exploit of thermometer for measuring temperature, analyzing the internal structure and kind of the material is an objective method and the annotations made by this process do not depend upon the frame of mind, temperament and the mood of the person.

Until the end of eighteenth century, information gathered about various phenomena taking place in nature, by cautious surveillance and close manifestation, was classified into a particular plot of learn called natural philosophy. As comprehension enlarged it was found essential to split the study into two chief twigs, viz. physical sciences and biological sciences.

Physics is largely concerned with the study of those attention-grabbing topics which deal with the laws according to which material react with matter or energy, together with the related concepts of force and motion.

Thus, physics deals with the elementary property of matter and energy. As such physics may be called the science of matter and energy. Even though physics is one whole and cannot be compartmentalized into different parts but for the sake of expediency, the study of physics has been grouped under such heading and technicalities as properties of matter, optics, wave motion, magnetism, electricity, etc. this list is by no means complete. Physics has always been an exceptionally thought-provoking subject, but it has by no means been so thrilling as it is these days. All proceedings and incidence that we stumble upon in life are found to be governed by one or the other laws of physics. A foremost accomplishment of physics has been in illuminating the detail that all form of matters consist of diminutive particles called atoms. It has been further shown that atom is still not the inseparable part of matter. But it has further subdivisions. The study of the ingredient of atom has lead to the finding of many fresh particles called fundamental particles. With the understanding of the fundamental particles, a lot of intricate mysteries of nature have been outspread.

A methodical speculation is a set of connections that help to elucidate a natural occurrence or the activities of a normal structure on the foundation of conventional laws of nature. The enthusiastic observer of nature, who may be called a scientist, was talented to ascertain laws of nature through his inflexible and enduring effort stretch over these four extended centuries. In the present day, no matter what we discern in the meadow of physics can be described with the aid of the hypothesis.

1. *Mechanics:* The theory of motion of material objects at low speeds.
2. *Quantum mechanics:* The theory of mechanical behaviour of sub-microscopic particles.
3. *Electromagnetism:* The theory of electricity, magnetism and electromagnetic radiation.
4. *Thermodynamics:* The theory of heat, temperature and the behaviour of a system of a large number of particles.
5. *Relativity:* The theory of invariance in nature and the theory covering the motion of particle moving at high speed.

In physics, there is a secure relation between theory and experiment. In the past, the method of measurement in physics was subjective in nature. In fact many fabulous and stupendous work in the field of physics have been done with rudimentary apparatus. Since the conclusions obtained by subjective methods can be inaccurate and sometimes misleading the use of objective methods like scientific apparatus is made for taking observations in present day physics. It is because of the accuracy and correctness in the measurement of physical quantities that physics is called an exact science.

The present day physics is the consequence of the revolutionary dexterity of ostentatious scientists, namely Sir Isaac Newton, Albert Einstein, Max Planck, Neils Bohr, JJ Thomson, Michael Faraday, Enrico Fermi, PAM Dirac, Heisenberg, de-Broglie, Schrödinger, Pauli, Archimedes, Galileo, Keplar and others. Indian scientists, CV Raman, S Chandra Sekhar, SN Bose, Meghnad Saha and Homi J Bhabha have continued to the enlargement of physics.

Range and Exhilaration of Physics

The mankind has benefited themselves from the discovery made in the field of physics and technology by an assortment of scientists, these sciatica too had exhilarating moment during the itinerary of invention made by them.

Once Alexander Graham Bell while he was performing an experiment in his laboratory happened to obtain a sachet lying slam to him, fortuitously he was smack with the hard metal shield which was deceitful in the same laboratory and fell down by uproar deafeningly, the co-researchers rushed to his room and took him to the hospital, while he was lying on the bed in the hospital he formulate the law of buzzing of telephone bell.

Archimedes was asked by his king to tell about the purity of gold crown as such without meting it! He got the evidence to explain this problem, while he was in his bathing tub. He was so thrilled that completely unmindfull of his cloth, he came to the street dirging Eureka! Eureka! and formulates the Law of Floation.

While Faraday moved a magnet close to a coil, a galvanometer connected to it showed a deflection representing a current through the coil in the nonexistence of source. An e.m.f. was in fact stirring but Faraday interpreted this watching on a scientific traction and it was recognized as electromagnetic induction. Its reading led to the development of electric generators, motors, etc. Further we know that to shift a body, we have to apply a force on it. It is actually stimulating that earth can exert a pull on a body from a distance, a magnet can draw an iron entity from a distance and an electric charge can cause an another charge to move from a distance. How do forces crop up in these cases, is rather amazing. However, it has been discovered that origin of such forces is due to switch over of particles between two bodies and such forces are termed as exchange forces. A tiny drop of liquid is always spherical in shape. It never assumes a cubical or any other geometrical form. The study of property of liquids exposed that due to surface tension, liquids try to acquire least amount of surface area for a certain volume, the sphere has lowest amount of surface area. There are numerous other such physical phenomena, which initially appear quite electrifying but their study evicted to be of immense significance.

The range of length, mass and time intervals are reasonably gigantic. For example, the length can be as minute as 10^{-14} m and a length may be as big as 10^{25} m. The mass of an entity may differ from 10^{-30} to 10^{55} kg. The time interval may diverge from 10^{-22} s, *i.e.* time taken by the electromagnetic radiation to traverse a nuclear distance to 10^{18} s, *i.e.* lifetime of sun. In the face of such a broad range of length, mass and time interval the method in physics made possible us to gauge all of them. It is fairly awe-inspiring that such a gargantuan number of measurements of mass, length and time interval can be made by methods which are only a few in numbers. It is for the reason that

(*i*) Qualitative lessons of whatever we study in nature inform that these explanations can be elucidated and inferred in terms of a small number of verdict.

(*ii*) Despite the fact that the measurements of length, mass and time vary over an ample array, yet there is moderately a small number of ideology which can be functional to all of them to compute and comprehend them.

(*iii*) By separating an extra imperative attribute of a physical phenomenon from fewer attribute ones, the physical phenomenon can effortlessly be couched and the concealed intricacy become obvious.

Precision of Measurement in Physics: Physics is an investigational science. The experimental data are quantitative and the scale of modification of the measurements, *i.e.* their exactitude is frequently very imperative. Our modern concept of quantum mechanics, relativity and composition of atoms are based upon our capability to quantify a variety of quantity with great meticulousness.

In the early stage, physics was concerned largely with the law illuminating the bulky behaviour of nature. For these comparatively coarse measurements were good enough. Towards the turn of the ninth century, a number of physicists started sensing that the man has educated virtually all that was to be known about the nature and the universe. But finer experiments and more exact measurements, made promising by the upgrading of technology have publicized that there are lots of belongings in nature still unidentified to man. Example: speculation of an element say heavier masses can decompose into lighter but not vice versa. The mass of the neutron is precisely known to be 1.6748×10^{-27} kg; that of proton is 1.6725×10^{-27} kg. Neutron is slightly heavier than a proton, so a neutron can decay into a proton but not vice versa. Had our dimensions been correct only up to two places of decimals, and then we could not put in the picture which way the reaction would go.

Physics is the most primitive of all the sciences and it has played a crucial part in the progress of other sciences.

Physics in Relation to Mathematics: Physics is quantitative discipline. The theories in physics make use of various mathematical concepts and it has proved a most authoritative contrivance in the improvement of theoretical physics.

Physics in Relation to Chemistry: The study of structure of atom, radioactivity, X-ray diffraction, etc. has brought about an upheaval in the present century chemistry. These studies have lead to reshuffle of elements in periodic table, to identify even traces of substances in a sample, to recognize the nature of valence and chemical bonding and to realize multifaceted chemical structures.

Physics in Relation to Biological Sciences: Life sciences have also been benefited by the progress in physics. Optical microscope is established to be massive utilize in the study of biology. The electron microscope has made it feasible to see even the composition of a cell. The X-ray and neutron diffraction have lent a hand in understanding the constitution of nucleic acids, which have assisted to manage the vital process of life activity. X-ray has been used in medical sciences for diagnostic purpose. Radioisotopes are used in radiotherapy for the cure of lethal disease like cancer and other skin diseases. The irradiation of fungi and bacteria produce novel yield which have been used to generate enzymes, vitamins and antibiotics.

Physics and Astronomy: Physics has also helped in the expansion of astronomy. Galileo with the help of his crudely deliberated telescope was talented to view the moon of Jupiter. Sir William Herschel was capable to find out a innovative planet with the assist of the optical telescope. Thus, he was clever to add one more planet to the record of five planets previously revealed by the ancients. The progresses in picturesque technique in physics proved a fantastic aid in making several elemental researchers in the corral of astronomy. The employ of radio telescope has enable astronomers to witness up to the uttermost restrictions of the space and find out fresh stuff in the sky like quasars, pulsars, etc. These new discoveries in astronomy have new glow on the theme of cosmology.

Physics in Relation to other Sciences: The law of physics when appropriately interprets apply to living organism also and it has lead to the growth of latest discipline called *biophysics*. The mechanical, nuclear, gravimetric and acoustic methods are extensively used in sciences such as geology, meteorology, oceanography and seismology.

Physics is making terrific involvement to the societal advancement of mankind and scarcely any bough of science can prolong and progress without making use of modern physics modus operandi.

Physics in Relation to Technology: Machinery is the appliance of science in general and physics in fastidious for practical purpose. The sighting of electromagnetic induction by Faraday can be regarded as one of those grand methodical detection, which have not only benefited a common man but have formed the foundation of the technology. Physicist, to resolve a difficulty, usually measures convinced property of matter and then examines the consequences to perceive if there is any mathematical relationship between them. Mathematical equations used in physics have their corporeal meaning and it is a suitable means to articulate the law prevail the actions of matter and energy in the universe. Unerringly this is the major aspire of science in general and physics in particular, *i.e.* to seek out the laws of the universe and to express them in precise mathematical form. Technologists apply this in sequence, *i.e.* the law of the universe for such purpose as for scheming electric motors, dynamos, radio, television, radar, artificial satellites, computer space crafts, nuclear power generators, dams and so on, for improving the standard of existence of the entire mankind. It is also supposed that a precise discovery is imperfect and undeveloped until it has sensible appliance. Encroachments made in the technology on the basis of physics are:

(*i*) The cram of lever system has assisted to propose a large number of useful machines.

(*ii*) The reading of satellite has prepared it to be able to forecast weather correctly.

(*iii*) Conversion of heat energy into mechanical work has lead us to blueprint heat engines.

(*iv*) Discovery of vacuum tubes have lead to wireless communications and to the design of radio and televisions. Now the evolution in the field of semiconductors has provided altogether new magnitude to the technology in this pasture.

(*v*) Discovery of X-ray have been originated to be colossal use in medical sciences. Whereas it is used to diagnose the assorted types of diseases, X-ray is also used as radiotherapy. X-ray has proved of vast weight in industries and research.

(*vi*) The discoveries of nuclear fission have proved a fabulous source of energy. In nuclear reactors and nuclear bombs, a large amount of energy is obtainable due to the conversion of mass into energy.

Science and Technology as a Companion: Engineering and technology make use of the scientific information to advance the material well being of the mankind.

(*i*) Electric generators are the outcome of the research made by Faraday.

(*ii*) Radio and television are the out come of the discovery of the intimate relationship between electricity and magnetism made by James Clark Maxwell.

(*iii*) Space vehicles and artificial satellites are based upon Newton's laws of motions and Keplar's law of planetary motions.

(*iv*) Nuclear power stations and nuclear energy generations are possible only because of Rutherford's ideas about the structures of matter.

(*v*) Electronics came into being from JJ Thomson's discovery of the electron and it has many applications in communications, computers systems, domestic products, etc. Modern telephone exchanges now make extensive use of electronics and have brought about a revolution in the life of man.

Science and Technology as an Antagonist: Nevertheless if science and technology are not used in a conscientious way it may establish blight for the society. Environmental problems such as pollution, nuclear weapon, etc. are the outcome of the irresponsible use of technology.

The communication between physics and technology is two means interchange. The discoveries of physics provide ascend to new-fangled technology and sequential advances in technology are used for more research in physics. Enhanced technologies provide innovative utensil and technique which assist for the extra intensification of physics.

Physics is a Global Activity: In very old era the movement of physics or science generally was slightly sluggish for the reason that the scientist occupied in technical investigation in different countries could not without difficulty communicate with each other. With the spread in the means of transport and communication the scientist from different countries are in a position to get in touch with each other and they are capable to converse mutually their findings. With this kind of interface of scientists functioning in different country, the cadence of scientific growth in latest occasion has noticeably gone up and the scientific fact has been doubling itself approximately after every decade. Modern physics in general is an worldwide action. The atomic energy could be harnessed only as a consequence of the combined hard work and shared communication of scientists like Einstein, Fermi, Bohr, etc. who put together their experience and knowledge. Thus, the advancement in physics these days is an outcome of the combined global commotion. It cannot be restricted inside the boundaries of thin nationalism.

Physics and Society: The record of science and technology can be traced back to the civilization of primitive man. But the alert use of science for social economic advance is moderately a latest happening. It is in progress only with the methodical uprising. These days marvelous transform is being brought about by science in common and physics in exacting the standard of living and thinking of populace all over the globe. The invention of radio, television, computer, rocket, reactor, integrated circuits, transistor, semiconductors, etc. wherein physics has played a foremost position, have totally revalorized not only the existence but the incredibly accepted wisdom of the man on this sphere. The entire milieu of life has undergone a reflective amend with the escalating exercise of science and technology. The things which were absurd earlier have been made veracity recently. What urbanized countries encompass and the emergent country do be deficient in the modern science and technology. The modern sciences own the potentiality to eliminate starvation and deficiency entirely from the exterior of the orb and to shepherd in an epoch of profusion and leisure.

On the other hand incorrect use of science may end in catastrophe nuclear weapon like atom bombs and hydrogen bombs, which can obliterate the whole of humanity in no time also the result of the progression of modern physics and engineering. The science and technology with their astounding proceed have put into the hand of man of grand comprehension that can be functional alike, to the formation of almost paradise on earth or to the total destruction, if not of life, at least of all civilization. Dropping of atom bombs on two cities of Japan—Hiroshima and Nagasaki during World War II, has opened the eyes of scientists as well as of political leaders of the world in respect of the devastation and destruction that can be caused by the wrong use of the scientific knowledge.

1 Modern Physics

INTRODUCTION

The branch of physics which deals about the study of light like nature, properties, sources, etc. are called optics.

Optics is basically divided into two branches.

(*i*) Linear optics.

(*ii*) Non-linear optics.

The linear optics is further classified into two groups.

(*i*) Geometrical optics.

(*ii*) Physical optics.

In geometrical optics light is considered as a beam.

Ex: Reflection, refraction, aberretion and study about optical instrument.

In physical optics the theories of light is studied and is further divided into two groups.

(*i*) Wave optics **Ex:** Interference, diffraction and polarization.

(*ii*) Quantum optics **Ex:** Photoelectric effect and compton effect.

Light is a form of energy and can be transferred from one point to another either by particle motion or by means of wave motion.

Different theories have been proposed for the explanation of nature of light. The important theories of light and their salient feature are discussed.

1. Newton's Corpuscular Theory

In the later half of the 17th century, *i.e.* in the year 1672, Sir Issac Newton put forth the corpuscular theory of light. According to this theory, light consists of a stream of very minute particles called corpuscules. They are short in size and travel in a straight line with very high speed. When they fall on the retina of the eye they produces the sensation of vision. Newton explained different colours of light by attributing different size to the corpuscule.

On the basis of corpuscular theory of light the phenomenon of reflection, refraction, simultaneous reflection and refraction, etc. were explained successfully. This theory fails to explain the phenomenon like interference, diffraction, polarization, double refraction, photoelectric effect, etc.

The following conclusion of corpuscular theory are against to the observed fact.

1. When a source of light emit corpuscules it mass should decrease, but no such change in mass is ever detected.
2. According to corpuscular theory the velocity of light in a denser medium is higher than that in air. The reverse is observed by Focult's experiment.
3. According to this theory, greater is the temperature of the source, the greater is the agitation of its molecules and hence higher should be the velocities of the corpuscules coming out from it. Experimental observation show that velocity of light is independent of temperature.

To overcome the difficulties of corpuscular theory. In the year 1679 the Dutch physicist, Christian Huygens proposed wave theory of light.

2. Wave Theory of Light

According to this theory each point of a luminous source of light sends out waves in all directions in an hypothetical medium called ether. Ether was suppose to be continuous medium which pervades all space. Due to the tremendous velocity of light, it was assumed that ether has large elasticity and extreemly low density.

Wave theory could explain reflection and refraction satisfactorily. According to this theory light travels faster in a rarer medium than in a denser medium. This prediction is in agreement with experimental results. The theory also explained interference and diffraction satisfactorily. How ever it failed to account for polarization of light. This difficulty was overcome by Fresnel. He suggested that light waves are not longitudinal as assumed by Huygen but transverse in nature.

Even though the wave theory could satisfactorily explain several optical phenomena, the presence of ether could not be detected by any experiment. In fact disproved the existence of ether.

3. Maxwell's Electromagnetic Theory

Electromagnetic theory of light was put forward by James Clerk Maxwell in the year 1873. According to this theory light consists of fluctuating electric and magnetic field propagating in the form of an electromagnetic wave. In an electromagnetic wave the electric and magnetic fields are perperndicular to each other and to the direction of propagation of the wave.

The alternatively expanding and collapsing electric and magnetic fields travel in space with the velocity of light.

In the year 1885 German Scientist, Heinrich Rudolf Hertz experimentally confirmed the electromagnetic theory. The experiments of Hertz together with those which followed lent support to Maxwell's electromagnetic theory. But with the discovery of photoelectric effect a new problem arose, and electromagnetic theory failed to explain the results of photoemission.

According to the electromagnetic theory of light, a substance exposed to light is subjected to an oscillatory electric field, the intensity of light being proportional to the square of the amplitude of electric vector. The electrons within the substance interact with the field, gain kinetic energy, and leave the substance as soon as the kinetic energy exceeds the binding energy of the electron. The electromagnetic energy gained by the electron depends upon.

(*a*) intensity of incident light

(*b*) area of the surface over which light falls

(*c*) effective area of the electrons exposed to light

(*d*) time for which the surface is illuminated

If we consider the light falling on a sodium surface, a detectable photoelectric current will be obtained when 10^{-6} W/m^2 of electromagnetic energy is absorbed by the surface. A layer of sodium of thickness 1×10^{-10} m and 1 m^2 in area has about 10^{19} atoms. If we suppose that the incident light is absorbed by the upper most layer of sodium atoms, each atom will receive an average amount of energy equal to 10^{-25} W so that 1.6×10^{6} sec will be required by an atom to have 1 eV of energy and an electron in sodium will gain sufficient energy to be liberated in about 2 months.

Hence, the electromagnetic theory fails to explain why there is no time lag between the instant light falls and that of the emission of photoelectrons.

Secondly, according to classical electromagnetic theory, the ejection of an electron should depend upon the incident energy, *i.e.* intensity of light and not on frequency. In other words, there should be no threshold frequency for any material. This is contrary to the observed facts.

Again, there is no limit on the maximum energy to be transferred to the electron, *i.e.* the electron can have any value of maximum energy. This is also contrary to observed facts.

Hence, electromagnetic theory of light fails to explain the basic facts of photoelectric effect.

We know that classical mechanics can successfully explain the motion of astronomical bodies such as planets, satellites, comets, etc. as well as macroscopic bodies such as the motion under gravity, rotational motion on an inclined plane, oscillatory motion on either side of equilibrium position. Apart from this, the motion of charged particles in electromagnetic fields, vibrations in stretched strings, elastic vibrations in solids propagation of sound waves in gases, flow of fluids, molecular motion in gases, etc. can also be successfully explained by the classical mechanics. In the beginning of twentieth century, some experimental facts were discovered from which it was concluded that the phenomenon such as black body radiation, photoelectric effect, Compton effect, specific heat of solids at low temperature, emission and absorption of light, etc. in which there is mutual interaction of radiation with matter, could not be explained by the classical mechanics. To explain these phenomenon successfully, a new principle was propounded which is called the *quantum principle.*

Followings are some important experimental and theoretical results which could not be explained by the classical mechanics but could be explained by the quantum principle.

1. Black Body Radiation

A body which absorbs all the radiations incident on its surface whatever may be their wavelengths is called a *black body.* In practice, an isothermal cavity in which radiations from outside can enter in through a small aperture in it, can be considered to be a black body. When radiations inside the cavity are in equilibrium with its walls, the emissivity in a unit wavelength range at wavelength λ be $E\lambda$, then the emissivity in the wavelength range λ and $\lambda + d\lambda$ will be $E\lambda d\lambda$.

In the year 1899, Lummer and Pringshein experimentally studied the distribution of radiant energy emmitted from a black body in different wavelengths by keeping the body at different temperatures. Figure 1.1 shows the experimentally obtained energy distribution curves at different temperature.

Following are the conclusions drawn from these curves:

(*i*) At each temperature, radiant energy is continuously distributed in all wavelengths from zero to infinity. The value of $E\lambda$ is zero at wavelength $\lambda = 0$. The value of $E\lambda$ increases with increase in wavelength λ and at a particular wavelength λ_m, the value of $E\lambda$ is maximum. As the wavelength λ further increases beyond λ_m, the value of $E\lambda$ decreases and at $\lambda = \infty$, the value of $E\lambda$ becomes zero. Thus, at each temperature, $E\lambda - \lambda$ curve is continuous. In other words, *the black body spectrum is continuous at each temperature.*

(*ii*) At each temperature, the area enclosed by the curve with λ-along x-axis represents the total energy emitted by the black body. Higher the temperature of the black body, greater is the total radiant energy emitted by it. The total energy emitted by a hot body, *i.e.* the area enclosed by the curve with the λ-along x-axis is directly proportional to the fourth power of absolute temperature of the body.

(*iii*) As the temperature of the body increases, the amount of energy $E\lambda$ emitted for wavelength λ, increases rapidly with temperature.

(*iv*) As the temperature of body increases, the wavelength λ_m corresponding to the maximum radiant energy, decreases, *i.e.* the peak of the curve shifts towards the shorter wavelength region.

(*v*) The wavelength λ_m corresponding to the maximum emissivity, is inversely proportional to the absolute temperature T of the body, *i.e.*

$$\lambda_m \propto \frac{1}{T} \quad \text{or} \quad \lambda_m T = b$$

Where b is a constant, whose value is 2.88×10^{-3} mk.

From this expression it is clear that when the body is heated to an ordinary temperature, it emits the radiation waves only in the longer wavelength region. As the temperature of body increases, in the emitted radiations the energy of radiation waves of short wavelengths increases. This is why an iron ball on heating first appears red at nearly 500° C then yellow at 800° C and white at a temperature above 1000° C.

To explain the spectrum distribution of black body radiation following laws were propounded:

(*i*) **Stefan's Law:** According to the Stefan's law of black body radiation, "The total energy radiated per second from unit surface area of a black body is directly proportional to the fourth power of absolute temperature of the body", *i.e.*

$$E \propto T^4 \quad \text{or} \quad E = \sigma T^4$$

Where σ is a constant, called as Stefan's constant.

(*ii*) **Wien's Law:** According to Wien law.

(*a*) The wavelength of the radiation emitted by a black body at any temperature is inversely proportional to the absolute temperature of the body.

$$i.e.\ \lambda \propto \frac{1}{T} \quad \text{or } \lambda T = \text{constant.}$$

(*b*) At any temperature, the energy of radiations of wavelength λ emitted by the black body is directly proportional to the fifth power of its absolute temperature, *i.e.* $E\lambda \propto T^5$.

Hence, the emissivity of radiations in the wavelength range λ and $\lambda + d\lambda$ emitted by a black body at a temperature T is given as

$$E_\lambda d\lambda = \frac{A}{\lambda^5} f(\lambda T)\, d\lambda$$

where $f(\lambda T)$ is an unknown function and A is a constant.

Later Wein considered the unknown function $f(\lambda T)$ to be exponential. Then according to Wein's law.

$$E_\lambda d\lambda = \frac{c_1}{\lambda^5} e^{-c_2/\lambda T}\, d\lambda \quad \text{(where } c_1 \text{ and } c_2 \text{ are constants)}$$

This law could explain only the lower wavelength part, *i.e.* from 0 to λ_m of the experimental curve at any temperature as shown in Fig. 1.1 because in this range of wavelength, the exponential function $e^{-c_2/\lambda T}$ is more effective as compared to the other function $\frac{1}{\lambda^5}$, hence the value of $E\lambda$ increases with the increase in the wavelength λ.

(*iii*) **Rayleigh-Jean's Law:** To explain the spectral distribution of black body radiation, Rayleigh and Jean assumed that inside the cavity, due to superposition of radiation waves incident on the wall the waves are reflected from it, and they form stationary waves. Hence, the energy density $v\lambda\, d\lambda$ in the wavelength range λ and $\lambda + d\lambda$ is equal to the number of modes of vibrations in unit volume in the wavelength range λ and $\lambda + d\lambda$ mean energy per mode of vibration.

According to classical theory, mean energy per mode of vibration at an absolute temperature T is kT, where k is Boltzmann's constant. Hence, according to Rayleigh-Jean's law, it is clear

$$v_\lambda d\lambda = \left(\frac{8\pi}{\lambda^4}\, d\lambda\right) \times kT \quad \text{or} \quad v_\lambda d\lambda = \frac{8\pi kT}{\lambda^4}\, d\lambda$$

That, this law could explain only the higher wavelength part of the experimental energy distribution curve as shown in Fig. 1.1 because from the above relation we find that $\upsilon\lambda$ (or $E\lambda$) decreases with the increase in λ.

(*iv*) **Planck's Law:** To explain the black body radiation, Planck in the year 1900, propounced the quantum theory of radiation. According to this theory, the emission and absorption of radiation is not continuous, but the is in the form of packets of definite and discrete set of energy. Each packet is called the *quanta or photon.* Each photon has a definite energy and definite momentum whose value is proportional to the frequency of radiation. If ν is the frequency of radiation, the energy of each photon is given by,

$$E \propto \nu \quad \text{or} \quad E = h\nu$$

and momentum of each photon $p = \dfrac{h\nu}{c}$

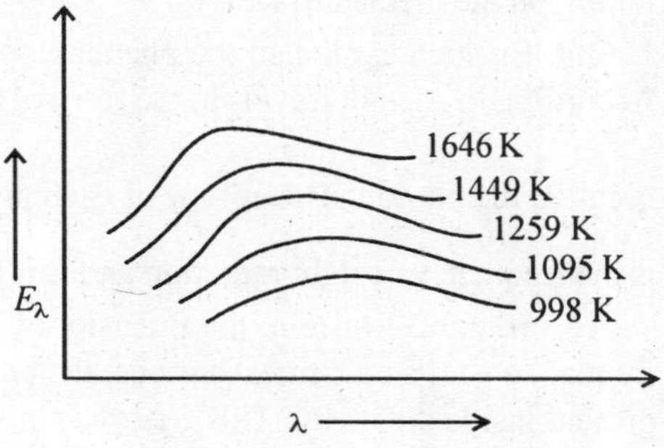

where h is a universal constant, called the Planck's constant. Its value is 6.625×10^{-34} JS, c is the speed of light. Thus, the energy emitted or absorbed by a body is an integral multiple of $h\nu$ such as $h\nu$, $2h\nu$, $3h\nu$.

1646 K
1449 K
1259 K
1095 K
998 K
E_λ
λ

Fig. 1.1

On the basis of above hypothesis, following relation was given by Planck for the energy distribution in the black body radiation. Which could successfully explain the experimental curves of Lummer and Pringsheim.

$$v_\lambda d\lambda = \frac{8\pi hc d\lambda}{\lambda^5 \left(e^{hc/\lambda kT} - 1\right)} \quad \text{or} \quad u_\nu d\nu = \frac{8\pi h\nu^3 d\nu}{c^3 \left(e^{h\nu/kT} - 1\right)}$$

Quantum theory successfully explained the phenomenon of photoelectric effect and Compton effect. However this theory could not explain interference diffraction and polarization.

2. Photoelectric Effect

In the year 1888, Hallwacks experimentally found that when an ultraviolet light or X-rays are made to fall on any metal surface, electrons get ejected from the metal surface and they begin to emit from it. This phenomenon is called the *photoelectric effect*.

(*i*) The photoelectric emission occurs only when the frequency ν of the light incident on the metal surface is greater than a definite minimum value ν_o. This minimum frequency ν_o is called the threshold frequency of the given metal surface.

(*ii*) The maximum kinetic energy of the electron emitted depends on the frequency of the incident light. It does not depend on the intensity of incident light.

(*iii*) The number of electrons emitted depends on the intensity of incident light and it does not depend on its frequency.

(*iv*) Electrons are emitted from the metal surface as soon as the light falls on it.

The above experimental facts could not be explained by the Maxwell's electromagnetic wave theory because (*a*) according to wave theory, on increasing the intensity of incident light, the amplitude of waves will increase and hence the energy propagated by the waves will increase. Consequently, the free electrons will get more energy from the metal and hence the energy of photoelectrons emitted must increase. But it is against the experimental fact. (*b*) According to wave theory, if the intensity of incident light is such that it can provide sufficient energy to bring out electrons from the metal, then whatever be the frequency of incident light, the electrons should get emitted from the metal surface. But experimentally it is found that for the electron emission, the frequency of incident light must be more than a definite minimum frequency ν_o. (*c*) Since the energy carried by the light waves will get distributed amongst all the electrons present in the area of the metal surface on which the light falls, therefore, it must take some time to acquire the energy needed for the emission of electron. But experimentally it is found that there is no time lag between the emission of electron and the light falling on the metal surface.

But, Einstein explained the phenomenon of photoelectric effect on the basis of Planck's quantum theory. According to it, light travels in the form of small packets of energy with the speed of light. This packet of light is called the photon. The energy of each photon is $h\nu$ and its momentum is $\frac{h}{\lambda}$ where h is Planck's constant, ν is the frequency of light radiation and λ is its wavelength.

According to Einstein, the intensity of light depends on the number of these photons. When radiation of frequency ν falls on a metal surface, the energy of each incident photon is $h\nu$, which is gained by the electron present inside the metal. This energy is used up in two ways:

1. A part of the energy is used in making the electron free from the metal surface. This energy is called the work functiion or threshold energy of the metal and is denoted as ϕ.
2. The remaining part of the energy is used in providing kinetic energy to the free electron to emit out. If m is the mass of the emitted electron and v is its velocity, then maximum kinetic energy of the emitted electron is given by

$$\frac{1}{2} m v^2 = h\nu - \phi$$

This equation explains all the experimental observations. Obviously, the maximum kinetic energy of the photoelectrons emimtted depends on the frequency ν of the incident light and on the nature of metal surface, *i.e.* work function. It does not depend on the intensity of incident light. When the intensity of incident light is changed, the number of photons falling on the metal surface changes. Since each photon has its separate identity, hence greater the number of photons falling on the metal surface, greater is the number of electrons emitted from it. Thus, the number of electrons emitted depends on the intensity of incident light.

3. Compton's Effect

In the year 1923, Compton observed that when a beam of monochromatic X-rays is made to incident on a substance such as carbon, there are two components in the scattered radiation

(*i*) Unmodified radiation whose wavelength is same as that of the incident radiation and which is obtained in the direction of incident radiation.

(*ii*) Modified radiation whose wavelength is longer than that of the wavelength of incident radiation. The difference in wavelength increases with the increase in scattering angle and it does not depend on the wavelength of incident radiation and the nature of scatterer. This phenomenon is called the Compton's effect, and the difference in wavelength is called Compton's shift. As the angle of scattering increases, the intensity of the modified radiation also increases.

Compton's effect could not be explained by the wave theory. On the other hand, this phenomenon could be successfully explained by the quantum theory according to which light or radiation is in the form of packets of a definite set of energy and a definite momentum, which are called the photons. If ν is the frequency of incident radiation, the energy of each photon is $h\nu$ and momentum of each photon is $\frac{h}{\lambda}$, where h is Planck's constant and c is the speed of light.

Compton assumed that in the process of scattering, there is an elastic collision between a photon and a free electron of the scatterer, in which energy and momentum are conserved. When a photon falls on the scatterer, it imparts some of its energy and momenum to emit out free electron from the scatterer. As a result, the energy and momentum of the photon decreases, *i.e.* the wavelength of the scattered photon increases. The free electron so emitted is called the recoiled electron and is as shown in Fig. 1.2. If λ is the wavelength of the scattered photon, then Compton's shift, is given by

$$\lambda' - \lambda = \frac{h}{m_0 c}(1 - \cos\theta)$$

where m_0 is the rest mass of electron, h the Planck's constant, c the speed of light and θ the angle of scattering. The above relation has been experimentally verified.

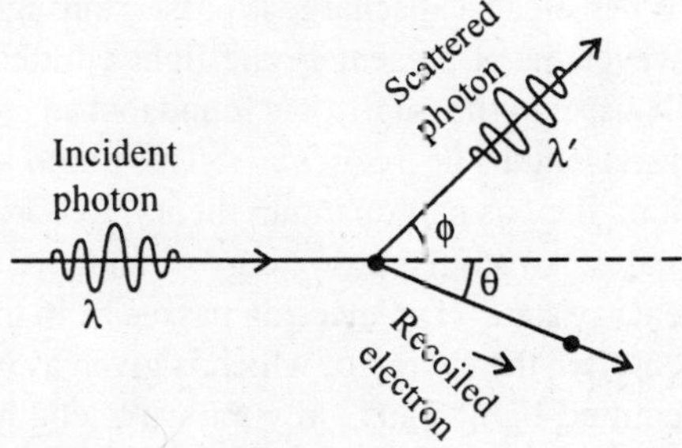

Fig. 1.2

4. Specific Heat of Solid at Low Temperatures

According to Dulong and Petit's law, the atomic specific hea t of each solid is constant and is given by $c_v = 3R$, where R is the universal gas constant, the atomic specific heat of a solid does not depend on its temperature. But experimentally, it is found that this law is valid only at ordinary and high temperatures. At a low temperature, the atomic specific heat of solid decreases with the decrease in temperature and it becomes zero at an absolute zero. This experimental fact could not be explained by the classical theory because, according to classical law of equipartition of energy, atomic specific heat of solid $Cv = 3R$. In the year 1907, Einstein explained the variation of specific heat of solid with temperature on the basis of quantum theory. He assumed that each atom of the solid as a three dimensional periodic oscillator of a definite frequency v, whose energy is an integral multiple of (quanta) hv. On the basis of this, the mean energy of an oscillator an at an absolute temperature T is not kT, but it is $\frac{hv}{\left(e^{hv/kT} - 1\right)}$ and the atomic specific heat is given by,

$$C_v = 3R\left(\frac{hv}{kT}\right)^2 \frac{e^{hv/kT}}{\left(e^{hv/kT} - 1\right)^2}$$

This relation states that at a low temperature, the atomic specific heat varies with the temperature. It may be mentioned that experimentally it is found that at a very low temperature, *i.e.* near absolute zero, $C \cdot \propto T^3$. This fact was explained by Debye in 1912 after modifying the Einstein's quantum theory.

5. Frank-Hertz Experiment

In the year 1913, Frank and Hertz experimentaly established that an atom can absorb only discrete energies. In other words, an atom has only some discrete energy states. If v is the frequency of radiation emitted or absorded, the energy that can be emitted or absorbed by the atom is hv, $2hv$, $3hv$, ····· On the other hand, according to classical electrodynamics, the absorption or emission of energy by an atom is continuous.

6. Stern and Gerlach Experiment

In the year 1922, Stern and Gerlach in their experiment applied a non-homogeneous magnetic field on atoms and found that in this magnetic field, only the discrete directions of magnetic moment vector of atom are possible. On the other hand, according to classical theory there can be any direction of the magnetic moment vector.

7. Davisson-Germer Experiment

In the year 1925, Davisson and Germer allowed an electron beam to fall on a metal target and they obtained a diffraction pattern similar to that obtained by grating with light. This observation could not be explained by the classical theory since electron is a particle.

8. Line Spectra

When electric discharge is passed through a gas enclosed in a tube at a low pressure, there are only the discrete wavelengths present in the light emitted from it. These spectral lines are the characteristic of that atom. Thus, experimentally it is found that an atom absorbs or emits the electromagnetic radiations of some particular wavelengths or frequencies. This could not be explained by the classical theory. But in 1913, Bohr explained it on the basis of quantum theory. According to Bohr, in an atom, electron can remain only in discrete energy states. An atom emits energy when there is a transition of an electron from the higher energy state to the lower energy state. If an electron passes from the energy state E_1 to the energy state E_2 (where $E_1 > E_2$), a photon is emitted, the energy of which is given as $hv = E_1 - E_2$, where h is Planck's constant and ν is the frequency of an emitted light. Thus, only the wavelength or frequency of the spectral line could be explained by the Bohr's hypothesis, but it could not provide any explanation regarding the intensities of spectral lines.

Thus, all the theoretical and experimental facts described above establish the quantum theory propounded by Planck to explain the black body radiation and they could not be explained by the classical theory.

1.1 DUAL NATURE OF LIGHT

From all these theories it is clear that certain phenomenon can be explained only by regarding light is made up of particles while certain other phenomenon can be explained only by regarding light as a wave. To explain this we have to agree that light appears to have a dual nature sometimes it behaves like waves and at other time as particles or photons.

Louis de-Broglie proposed a theory where in certain microparticles such as electrons behaves as waves. Let a particle of mass m moving with a velocity v actually behaves as a wave of wavelength λ such that

$$\lambda = \frac{h}{mv} \quad \text{or} \quad \lambda = \frac{h}{P}$$

This is de-Broglie's hypothesis of matter waves and was verified by electron diffraction experiments.

Thus, the nature of light is not unique. Light behaves as waves or as particles. The best analysis is that when light is emitted or absorbed it exhibits particle nature and when light is propagated it is always an electromagnetic wave travelling with a speed nearly equal to 3×10^8 ms^{-1} in air or vacuum.

1.2 WAVE-PARTICLE DUALISM

To understand the wave and particle duality, it is necessary to know what is a particle and wave.

1.2.1 Concept of Wave

A wave is spread out over relatively large region of space, it cannot be said to be located here or there it is hard to think of mass being associated with a wave. A wave is specified by its frequenly (v), wavelength (λ), phase (ϕ), amplitude (A), and intensity (I).

1.2.2 Concept of Particle

A particle has mass, it is located at some definite point, it can move from one place to another it gives energy when slowed down. Thus, a particle is specified by mass (m), velocity (v), momentum (p) and energy (E).

Radiation including visible light, infrared, ultraviolet, X-rays, etc. behave like wave in experiments based on interference, diffraction, polarization, etc. this is due to the fact that these phenomena require the presence of two waves at the same position at the same time. Obviously it is difficult for two particles to occupy the same position at the same time, thus, we can conclude that radiation behave like wave.

Planck's quantum theory was successful in explaining block body radiation, the photoelectric effect, the Compton effect, etc. had clearly established that radiant energy, in its interaction with matter behaves as though it consists of corpuscles. Here, radiation interacts with matter in the form of photons or quanta. Thus, we can conclude that radiations behave like particle.

Therefore radiation sometimes behave as a wave and at some other time as a particle, *i.e.* it has a wave-particle qualism.

Note: It should be remembered that radiation cannot exhibit its particle and wave properties simultaneously.

1.3 DE-BROGLIE'S HYPOTHESIS

According to Max Planck's quantum theory of radiation, the electromagnetic radiation (light) behaves as 'quanta'. On the basis of this theory de-Broglie in the year 1924 proposed the novel statement that "A moving body what ever be its nature has wave property associated with it". So a fast moving electron, proton or a neutron is not a simple particle, it can be treated as "Matter waves" this concept was agreed with the experimental results and moving particles (body) do exhibits wave property and corpuscular property.

According to Planck's quantum theory of radiation light is a form of energy, it is emitted in the form of discrete set of energy packets. Each energy packet is called photon. The energy of each photon is given by,

$$E = h\nu \quad (1)$$

Where, h is a constant called Planck's constant and ν is the frequency of the radiation.

If m is the mass of photon and c be its velocity then according to theory of relativity the energy of moving photon is given by

$$E = mc^2 \quad (2)$$

On comparing Eq. (1) and (2), we get

$$h\nu = mc^2$$

$$m = \frac{h\nu}{c^2} \quad (3)$$

If a particle of mass m moving with a velocity c then its momentum is given by,

$$p = mc \quad (4)$$

substituting this value in the above expression (3), we get

$$mc = \frac{h\nu}{c}$$

$$\therefore \quad p = \frac{h\nu}{c} \quad (5)$$

this is the expression for momentum of a photon in terms of frequency ν

we know that the velocity, wavelength and frequency are related as $c = \nu\lambda$

then, $$p = \frac{h\nu}{\nu\lambda}$$

$$\therefore \quad p = \frac{h}{\lambda} \quad (6)$$

This is the expression for momentum of a photon in terms of wavelength λ

$$\therefore \quad \lambda = \frac{h}{mc} \qquad | \because p = mc$$

or $$\lambda = \frac{h}{p}$$

This is the de-Broglie's wavelength of a photon. The same concept is also applicable for material particles such as electron, proton and neutron.

If a material particle of mass m is moving with a velocity v, then its momentum p is given by

$$p = mv.$$

Therefore, de-Broglie wavelength of matter wave (electron) or material particle is given by,

$$\lambda = \frac{h}{mv} \quad \text{or} \quad \lambda = \frac{h}{p}$$

This equation combines the concept of both particle and wave nature, because λ has a meaning associated with waves and p is associated with particles.

This is known as de-Broglie wave equation and λ is called de-Broglie wavelength. The associated wave is termed as matter wave or guide or pilot or de-Broglie wave.

The phase velocity of the wave is given by

$$v_p = v\lambda = \frac{mc^2}{h}, \frac{h}{mv}$$

$$\because \quad \lambda = \frac{h}{mv}$$

$$V_p = \frac{c^2}{v}$$

Like other waves the de-Broglie wave can also be represented by a function $\psi\left(\vec{r}t\right)$. It is a function of position $\vec{r}$ and time t. It is also known as wave function representing a matter wave guiding the particle. The wave function can be positive, negative or a complex quantity.

The wave vector $\vec{K}$ for de-Broglie wave is given by

$$\left|\vec{K}\right| = K = \frac{2\pi}{\lambda} = \frac{2\pi}{\frac{h}{p}} = \frac{p}{\frac{h}{2\pi}} = \frac{p}{\hbar}$$

Where $$\hbar = \frac{h}{2\pi}$$

$$\therefore \quad \left|\vec{K}\right| = \frac{P}{\hbar}$$

1.4 DE-BROGLIE WAVELENGTH IN TERMS OF ENERGY AND TEMPERATURE (HELIUM ATOM)

The de-Broglie wavelength λ is given by the relation

$$\frac{h}{p} = \frac{h}{mv}$$

If E is the kinetic energy of the particle, then

$$E = \frac{1}{2}mv^2 = \frac{1}{2}\frac{m^2v^2}{m} = \frac{p^2}{2m}$$

or $$p = \sqrt{2mE}$$

hence, de-Broglie wavelength

$$\lambda = \frac{h}{p} = \frac{h}{\sqrt{2mE}}$$

according to kinetic theory of gases, the average kinetic energy of the material particle on the basis of kinetic theory of gases given by $E = \frac{1}{2}mv^2 = \frac{3}{2}\ TK.$

where, k is Boltzmann's constant = 1.38×10^{-23} JK^{-1} and T is the absolute temperature.

$$\therefore \quad \lambda = \frac{h}{\sqrt{2mE}} = \frac{h}{\sqrt{3mKT}}$$

This is the expression for de-Broglie wavelength in terms of energy and temperature or of the atoms having energy corresponding to the mean thermal energy at a TK.

1.5 CALCULATION OF DE-BROGLIE'S WAVELENGTH

Let us consider an electron of mass m having charge e is accelerated through a potential difference of V volts, the work done on the electron will be converted into kinetic energy and it moves like a matter wave with velocity v, *i.e.*

The energy gained by the electron = its kinetic energy

i.e.
$$Ve = \frac{1}{2}mv^2$$

$$v = \sqrt{\frac{2\ Ve}{m}}$$

$$v = \sqrt{\frac{2\ Ve}{m}}$$

The momentum of the electron is $P = mv$, $P = m.\sqrt{\frac{2\,Ve}{m}} = \sqrt{m}\ \sqrt{m}\ \frac{\sqrt{2\,Ve}}{\sqrt{m}}$

i.e.
$$P = \sqrt{2\ Ve\ m}$$

The de-Broglie's wavelength of matter wave is given by

$\lambda = \frac{h}{p}$ by substituting the value of p, we get

$$\lambda = \frac{h}{\sqrt{2\ Vem}}$$

substituting the values of h, e and m, we get

$$\lambda = \frac{6.625 \times 10^{-34}}{\sqrt{2 \times 1.6 \times 10^{-19} \times 9.1 \times 10^{-31} \times V}}$$

$$\lambda = \frac{12.265}{\sqrt{V}} \times 10^{-10}\ \text{m}$$

$$\lambda = \frac{12.27}{\sqrt{V}}\ \text{Å}$$

This is the expression for de-Broglie wavelength of a wave associated with a material particle like electron.

Note : An electron beam subjected to a potential difference of 150 V will have the wavelength of 10^{-10} m in contrast with X-ray beam which requires a potential difference of 12000 V to have the same wavelength.

1.6 DAVISSON AND GERMER'S EXPERIMENT

Davisson and Germer succeded in confirming de-Broglie's wavelength of matter wave associated with electron beam.

The schematic of the apparatus used by Davisson and Germer is as shown in Fig. 1.3.

The electron beam from the filament *F* is accelerated to different velocity under electric potential of about 30 V – 100 V. The beam of electrons are collimated into a fine beam by allowing them through a pin hole *G* made on metalic block electrode.

The electron beam is directed on nickel crystal capable of sliding over a circular scale. A microammeter/galvanometer is used to measure the ionisation current developed in the ionisation chamber due to the admission of scattered electron beam. The whole apparatus is enclosed in an evacuated chamber.

The ionisation chamber is kept at different positions and in each position the intentisy *I* of the diffracted electron beam is measured. If θ is the angle between the incident electron beam on the crystal and the diffracted electron beam towards the chamber. A graph is plotted between *I* and θ and is as shown in Fig. 1.4.

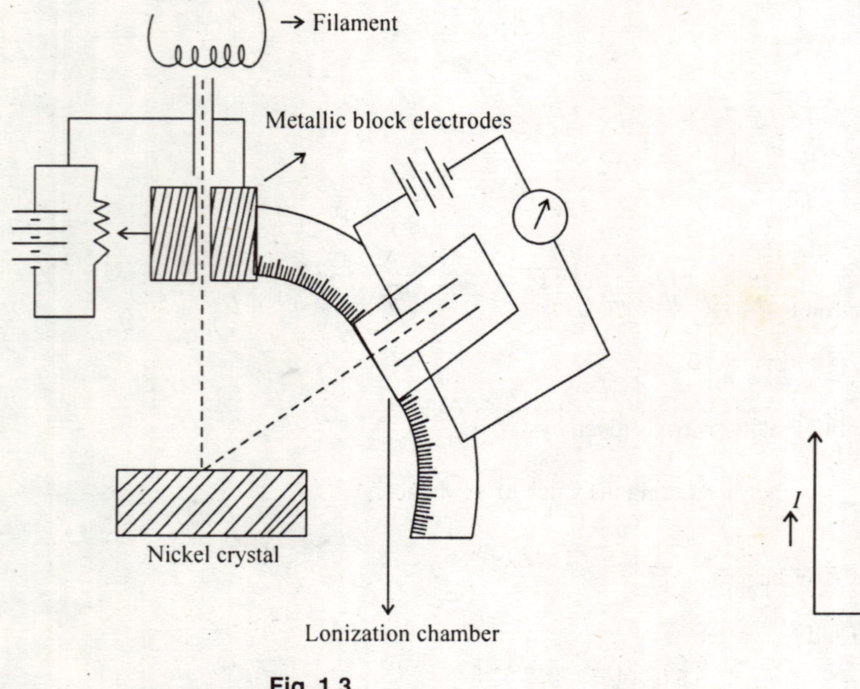

Fig. 1.3

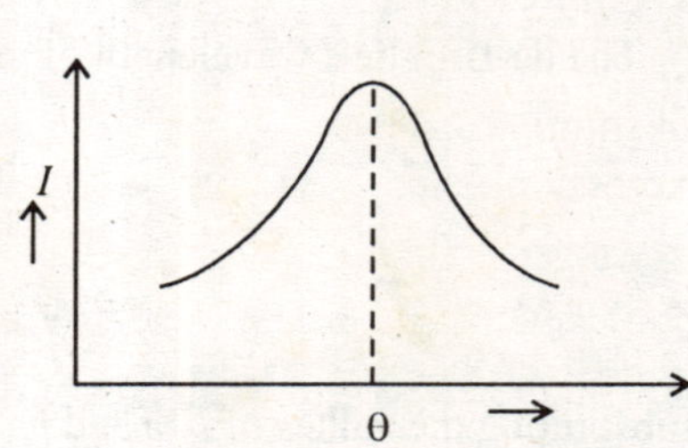

Fig. 1.4

The experiment is repeated by gradually increasing the accelerating voltage V of the collimating aperture *G*, these graphs at 40 V, 44 V, 48 V, 54 V, 64 V and 68 V respectively are shown in Fig. 1.5.

The atomic layers of the nickel crystal act as three dimensional grating for the electron beam, as shown in Fig. 1.6, the angle between incident and diffracted beam be set up for 50° (co-latitude angle). The electric potential of electron beam is increased steadily and the collection current is measured. It is found to be maximum at an accelerated potential of 54 V, which is as shown in the graph in Fig. 1.4.

If the colatitude angle is 50°, the glancing angle θ = 65°, by X-ray diffractometer the lattice space is determined and it is found to be 0.91 Å for nickle crystal.

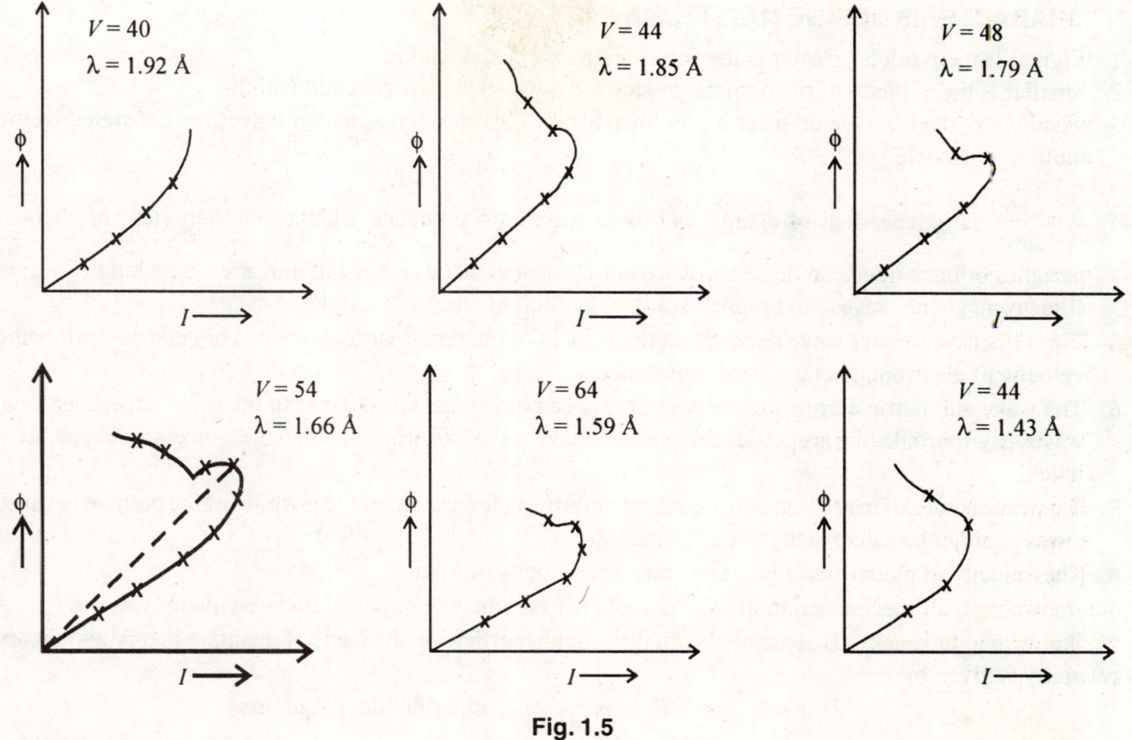

Fig. 1.5

We know that, the Bragg's equation for diffraction is given by

$$2\, d \sin \theta = n\lambda$$

here $n = 1$, for first order spectrum, $\theta = 65°$ and $d = 0.91$ Å substituting these values in the above expression.

We get, $\lambda = 2 \times 0.91 \times 10^{-10} \times \sin 65° = 1.65 \times 10^{-10}$ m

or $\therefore\ \lambda = 1.65$ Å

but the de-Broglie's wavelength is given by

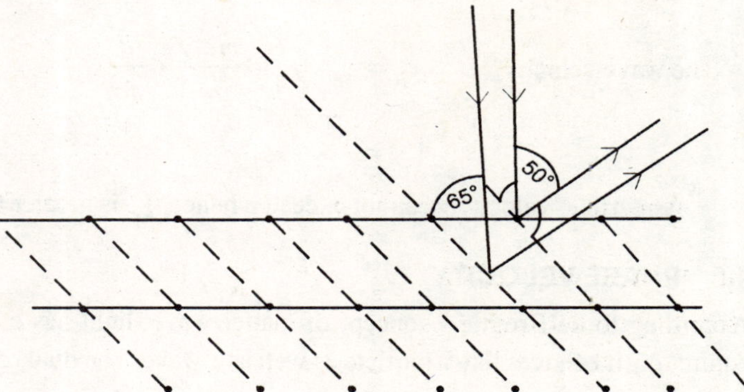

Fig. 1.6

$$\lambda = \frac{12.27 \times 10^{-10}}{\sqrt{V}} \quad \text{here } V = 54 \text{ V}$$

$$\therefore\ \lambda = \frac{12.27 \times 10^{-10}}{\sqrt{54}} = 1.66 \times 10^{-10} \text{ m}$$

$$\therefore\ \lambda = 1.66 \text{ Å}$$

i.e. the wavelength calculated by the above two methods are in good agreement. Thus Davisson and Germer experiment provide a direct verification of de-Broglie hypothesis of matter wave.

1.7 CHARACTERISTICS OF MATTER WAVES

1. Lighter is the particle, greater is the wavelength associated with it.
2. Smaller is the velocity of the particle greater is the wavelength associated with it.
3. when $v = o$, then $\lambda = \infty$ and when $v = \infty$, then $\lambda = 0$. This shows that matter waves are generated by the motion of particle.
4. $\lambda = \dfrac{h}{mv}$ is independent of charge, *i.e.* these waves are produced whether the particles are charged particles or uncharged. So these wave are not electromagnetic waves but they are a new kind of waves (Electromagnetic waves are produced only by motion of charged particles.
5. The velocity of matter wave depends on the velocity of material particle, *i.e.* it is not constant while the velocity of electromagnetic wave is constant.
6. The wave and particle aspect of moving bodies can never appear together in the same experiment, *i.e.* waves have particle like properties and particles have wave like properties and the concepts are separately linked.
7. The wave nature of matter introduces an uncertainty in the location of the position of the particle because a wave cannot be said exactly at any point.
8. The velocity of matter wave is greater than the velocity of light.

Let the velocity of mechanical motion of particle is v and the velocity of associated matter wave is V_p.

We know that, the energy E associated with the moving particle on the basis of quantum theory and theory of relativity is given by

$$E = hv \quad \text{and} \quad E = mc^2 \quad \text{on comparing these equations}$$

We get,

$$hv = mc^2 \quad \text{or} \quad v = \frac{mc^2}{h}$$

The wave velocity

$$v_p = v \times \lambda = \frac{mc^2}{h} \times \frac{h}{mv}$$

$$V_p = \frac{c^2}{v}$$

As particle velocity v cannot exceed c hence, V_p is greater than the velocity of light.

1.8 PHASE VELOCITY

According to de-Broglie's concept of matter wave, light has a dual nature, *i.e.* it behaves like a particle as well as a wave. The dual nature can be explained interms of wave packet.

A wave packet comprises a group of waves of slightly differing in velocity and wavelength with phases and amplitude such that they superimpose constructively over a small region of space where the particle can be located and when they superimpose destructively in the space the amplitude reduces to zero. Such a wave packet which moves with its own velocity is as shown in Fig. 1.7 and is known as the group velocity.

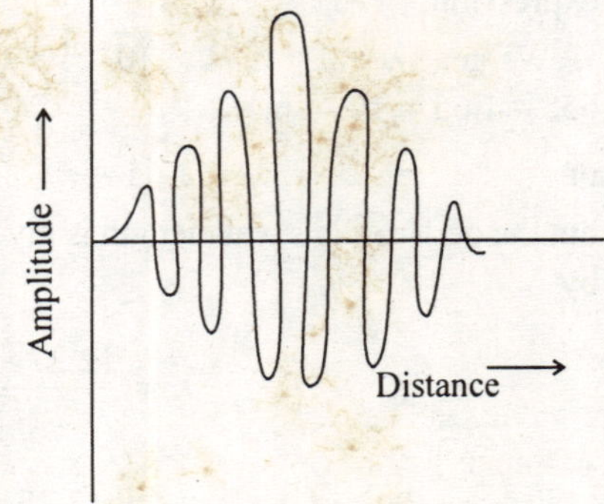

Fig. 1.7

1.9 PHASE VELOCITY OR WAVE VELOCITY

When a wave of single frequency and single wavelength travel through a medium, its velocity of advancement in the medium is called phase velocity or wave velocity.

or

A phase velocity or a wave velocity is the velocity with which planes of constant phase advance through medium and is denoted by V_p.

If λ is the wavelength and ν is the frequency of the wave, then the phase velocity V_p is given by

$$V_p = \nu\lambda \tag{1}$$

But according to Max Planck's quantum theory of radiation, the energy of the wave having frequency, ν is given by

$E = h\nu$ where $h \rightarrow$ Planck's constant
and $\nu \rightarrow$ frequency of radiation

$$\therefore \quad \nu = \frac{E}{h} \tag{2}$$

and the momentum of the wave on the basis of de-Broglie's concept of matter wave is given by $P = \dfrac{h}{\lambda}$

$$\because \quad \lambda = \frac{h}{p} \tag{3}$$

substituting the values of (2) and (3) in (1)

we get $$V_p = \nu\lambda = \frac{E}{h} \cdot \frac{h}{p}$$

$$\therefore \quad V_p = \frac{E}{P}$$

This is the expression for phase velocity interms of energy and momentum. This relation also holds good in the case of de-Broglie wave associated with a moving particle.

1.10 SHOW THAT DE-BROGLIE PHASE VELOCITY EXCEEDS THE VELOCITY OF LIGHT

Let us consider a particle of mass m moving with a velocity v, then the total energy and momentum of a particle is $E = mc^2$ and $p = mv$ respectively. But we know that the phase velocity associated with the de-Broglie wave is given by,

$$\therefore \text{ phase velocity } \quad V_p = \frac{E}{P}$$

substituting the values of E and P

we get, $$v_p = \frac{mc^2}{mv}$$

$$\therefore \quad v_p = \frac{c^2}{v} \tag{1}$$

As $v < c$, therefore the phase velocity of the associated wave is always greater than the velocity of light.

The de-Broglie waves associated with a moving particle of rest mass m_o having a momentum P then the total energy in a relativistic case is given by

$$E = \sqrt{m_o^2 c^4 + p^2 c^2}$$

but phase velocity $v_p = \dfrac{E}{P}$

where P = momentum of the particle

$$\therefore \quad V_p = \frac{E}{P} = \frac{\sqrt{m_o^2c^4 + p^2c^2}}{P}$$

$$= \left(\frac{m_o^2c^4 + p^2c^2}{p^2}\right)^{1/2}$$

$$= c\left(\frac{m_o^2c^2}{p^2} + 1\right)^{1/2}$$

$$v_p = c\left(\frac{m_o^2c^2\lambda^2}{h^2} + 1\right)^{1/2}$$

$$\because \quad \lambda = \frac{h}{p} \qquad \therefore \quad P = \frac{h}{\lambda}$$

As the term $\frac{m_o^2c^2\lambda^2}{h^2}$ is always a positive quantity for all values of m_o and λ (except zero). Therefore, the phase velocity associated with the wave is always greater than the velocity of light in vacuum according to equation (1) phase velocity of the associated wave is always greater than c.

(*i*) Since the speed of a material particle can never be equal to the velocity of light, therefore the phase velocity exceeds the particle velocity.

(*ii*) Further, even in free space the velocity of de-Broglie wave depends upon wavelength. As a result, de-Broglie waves of different wavelengths have different phase velocities. Thus vacuum is a dispersive medium, de-Broglie waves of electromagnetic waves are not dispersive medium, in this case the waves of all wavelengths travel with the same velocity in vacuum.

1.11 GROUP VELOCITY

Let a group consists of a number of waves of different frequencies super imposed upon each other. **Ex:** Let a white light consists of a continuous visible wavelength spectrum ranging from 3000 Å to 7000 Å.

"The velocity with which the slowly varying envelop of the modulated pattern due to a group of waves travels in a medium" is known as the group velocity.

Consider a group of waves consisting of only two components of equal amplitude and having angular frequencies ω_1 and ω_2 differing by a small amount represented by the equation.

$$y_1 = a\cos(\omega_1 t - k_1 x)$$

and

$$y_2 = a\cos(\omega_2 t - k_2 x)$$

where $\frac{\omega_1}{K_1}$ and $\frac{\omega_2}{K_2}$ represent their respective phase velocities.

$$\omega = 2\pi v \quad \text{and} \quad K = \frac{2\pi}{\lambda}$$

It is further supposed that $\frac{\omega_1}{K_1} \neq \frac{\omega_2}{K_2}$, *i.e.* it is a dispersive medium.

The resultant amplitude is given by

$$y = y_1 + y_2$$

$$= a\left[\cos(\omega_1 t - k_1 x) + \cos(\omega_2 t - k_2 x)\right]$$

$$= 2a \cos\left[\frac{(\omega_1+\omega_2)}{2}t - \frac{(k_1+k_2)}{2}x\right]\cos\left[\frac{(\omega_1-\omega_2)}{2}t - \frac{(k_1-k_2)}{2}x\right]$$

$$= 2a \cos(\omega t - kx)\cos\left(\frac{\Delta\omega}{2}t - \frac{\Delta k}{2}x\right)$$

where $\omega = \frac{\omega_1+\omega_2}{2}$; $k = \frac{k_1+k_2}{2}$ and $\Delta\omega = \omega_1 - \omega_2$; $\Delta k = k_1 - k_2$

The resultant wave has two parts.

(*i*) A wave of frequency ω, propagation constant *k* and velocity

$$v_p = \frac{\omega}{k} = \frac{2\pi v}{2\pi/\lambda} = v\lambda$$

$$V_p = v\lambda \tag{1}$$

This is the phase velocity or wave velocity

(*ii*) A wave of frequency $\frac{\Delta\omega}{2}$, propagation constant $\frac{\Delta k}{2}$ and velocity $V_p = \frac{\Delta\omega}{\Delta k}$

It consists of a group of waves of very slowly varying envelop of frequency $\frac{\Delta\omega}{2}$ and propagation constant $\frac{\Delta k}{2}$.

The modulated pattern moves with a velocity $V_g = \frac{\Delta\omega}{\Delta k}$ known as the group velocity as shown in Fig. 1.8.

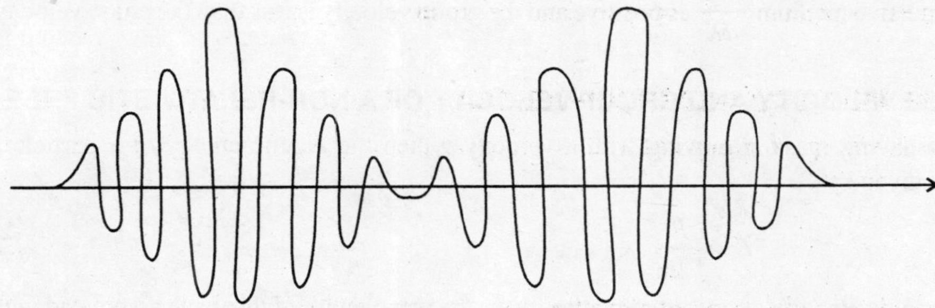

Fig. 1.8

If Δω and Δ*k* are very small, we can put

$$V_g = \frac{d\omega}{dk} \tag{2}$$

also
$$V_g = \frac{2\pi dv}{2\pi d\left(\frac{1}{\lambda}\right)} = -\lambda^2\frac{dv}{d\lambda}$$

1.12 RELATION BETWEEN PHASE VELOCITY AND GROUP VELOCITY

We know that, the phase velocity interms of angular frequency and propagation constant is given by

$$V_p = \frac{\omega}{k} \quad \therefore \quad \omega = V_p k$$

and $$V_p = \frac{d\omega}{dk} = \frac{d(v_p k)}{dk} = V_p + k\,\frac{dv_p}{dk} \quad (1)$$

But $$K = \frac{2\pi}{\lambda}$$

$$\therefore\ dk = -\frac{2\pi}{\lambda^2}\,d\lambda \quad \text{and} \quad \frac{k}{dk} = -\frac{\lambda}{d\lambda} \quad (2)$$

on substituting this in the above expression
we get,

Hence, $$\boxed{v_g = v_p - \lambda\,\frac{dv_p}{d\lambda}}$$

This is the relation between phase velocity and group velocity.

1.13 DIFFERENT CASES

1. If $\dfrac{dv_p}{d\lambda} = O$, *i.e.* the medium is such that in it the phase velocity does not depend on the wavelength, then $v_p = v_g$, *i.e.* the group velocity and the phase velocity are equal. Such a medium is called as non-dispersive medium.
2. In a dispersive medium $\dfrac{dv_p}{d\lambda}$ is positive and the group velocity is less than the phase velocity, *i.e.* $v_g < v_p$.

1.14 PHASE VELOCITY AND GROUP VELOCITY OF A NON-RELATIVISTIC FREE PARTICLE

Let a particle having mass m moving with a velocity v then the kinetic energy of a particle E interms of momentum is given by

$$E = \frac{p^2}{2m} \quad (1)$$

but according to de-Broglie's concept of matter wave, the wavelength of the wave associated with the particle is given by

$$\lambda = \frac{h}{p} \quad \text{or} \quad p = \frac{h}{\lambda} \quad (2)$$

on substituting this in the above equation, we get

$$E = \frac{h^2}{2m\lambda^2} = \frac{h^2k^2}{8\pi^2 m} \quad (3)$$

$$\because\ k = \frac{2\pi}{\lambda} \quad \text{or} \quad \lambda = \frac{2\pi}{k}$$

we also know that, according Max Planck's quantum theory of radiation the energy E of a particle having frequency v is given by

$$E = hv = \frac{h\omega}{2\pi} \quad (4)$$

$$\because\ \omega = 2\pi v \quad \text{or} \quad v = \frac{\omega}{2\pi}$$

on equating (3) and (4), we get

$$\frac{h\omega}{2\pi} = \frac{h^2k^2}{8\pi^2 m}$$

or
$$\omega = \frac{hk^2}{4\pi m} \qquad (5)$$

but the phase velocity v_p is given by

$$v_p = \frac{\omega}{k} = \frac{hk}{4\pi m} = \frac{h}{4\pi m} \times \frac{2\pi}{\lambda}$$

$$v_p = \frac{h}{2m\lambda} \qquad (6)$$

the group velocity
$$v_g = \frac{d\omega}{dk} = \frac{hk}{2\pi m} = \frac{h}{2\pi m} = \frac{h}{2\pi m} \times \frac{2\pi}{\lambda} \qquad \boxed{\because\ k = \frac{2\pi}{\lambda}}$$

$$\therefore\ v_g = \frac{d\omega}{dk} = \frac{h}{m\lambda}$$

$$= \frac{h}{m} \times \frac{mv}{h} \qquad \boxed{\because\ \lambda = \frac{h}{mv}}$$

$$v_g = \frac{d\omega}{dk} = v$$

$$\therefore\ v_g = v \text{ (particle velocity)}$$

Hence, the de-Broglie wave group associated with a moving particle moves with the same velocity as the particle. Thus, a moving particle is equivalent to a wave group.

Note 1: From Eq. (6)

$$v_g = \frac{h}{2m\lambda}$$

$$\therefore\ v_g = \frac{p}{2m} \qquad \boxed{\because\ \lambda = \frac{h}{p}}$$

$$= \frac{mv}{2m}$$

$$\therefore\ v_p = \frac{v}{2} = \text{half the velocity of the particle}$$

$$\therefore\ v_g = v$$

$$\therefore\ v_p = \frac{1}{2} v_g$$

Thus, the phase velocity of a non-relativestic free particle is half of the group velocity.

Note 2: A single electron moving with a uniform velocity v is associated with a group velocity equal to v_g as shown in Fig. 1.9.

A uniform beam of electrons moving with equal velocity v is equivalent to a monochromatic wave having a wavelength equal to $\frac{h}{mv}$ is shown in Fig. 1.10.

Fig. 1.9

1.15 HEISENBERG'S UNCERTAINTY PRINCIPLE

The de-Broglie wavelength associated with a moving particle travelling with a uniform velocity v is given by,

$$\lambda = \frac{h}{mv}$$

This is a monochromatic wave of infinite extent the phase velocity V_p of such a monochromatic de-Broglie wave is given by

$$v_p = \nu\lambda \qquad \text{where, } \nu \rightarrow \text{frequency}$$

$$\lambda \rightarrow \text{wavelength}$$

$$= \frac{h\nu}{\frac{h}{\lambda}} = \frac{E}{P} = \frac{mc^2}{mv}$$

$$v_p = \frac{c^2}{v}$$

Fig. 1.10

and is thus always greater than the velocity of light in vacuum. Therefore, it is theoretically impossible for a monochromatic de-Broglie wave train to transport a particle or carry energy. The phase velocity, merely represents the rate at which a given phase of a monochromatic wave train advances and is purely a mathematical concept. Further, the stability of a material particle like an electron demands that it should be concentrated over a small region of space at any given instant of time. Thus, the mass and charge of a particle like the electron are localised entities whereas the de-Broglie wave with which we represent the moving particle is of infinite extent by virtue of it is being monochromatic.

By taking a group of waves of nearly equal wavelengths centered round the de-Broglie wavelength $\lambda = \frac{h}{p}$.

We obtain a wave packet the amplitude of which is different from zero, only in a certain small region of space having dimensions of the particle. This small region of space is associated with the position of the particle and this wave packet travels with the velocity known as group velocity V_g. Hence, instead of associating a single monochromatic de-Broglie wave with a particle we associate with it a wave packet consisting of a group of waves of nearly equal amplitude centered round de-Broglie wavelength $\lambda = \frac{h}{p}$ of the particle. The wave packet has the dimensions of the localised particle and travels with the same velocity as the particle.

1.16 UNCERTAINTY PRINCIPLE

In classical physics dynamical variables like position and momentum of a particle can be measured accurately, at any instant of measurements. According to quantum mechanics the moving particle is treated as matter wave and there is a limit to the accuracy with which we can measure its position and momentum, the same concept is applied in indeterminacy principle it states that "the simultaneous determination of exact position and momentum of a moving particle is impossible".

If Δx is the error involved in the measurement of the position along x-axis and Δp be the error involved in the measurement of its momentum, then

$$\Delta x \cdot \Delta p \geq h, \text{ A more accurate calculation gives } \Delta x \cdot \Delta p \geq \frac{h}{2\pi} = \hbar$$

suppose $\Delta x = 0$ then $\Delta p = \frac{h}{0}$

i.e. $\Delta p = \infty$

This means to say that, if we know the exact position of the particle, we shall never show its momentum with any certainty.

Similarly, if $\Delta p = 0$ then $\Delta x = \frac{h}{0}$

$$\Delta x = \infty$$

This means to say that if the momentum of the particle is determined precisely its position cannot be known with any certainty.

The above relation shows that the extent of uncertainty involved in simultaneous measurement of both position and momentum can never be less than h. This principle can also be expressed in the following two forms. If uncertainty in the measurement of energy is ΔE and the time interval is Δt.

Then, $\Delta E \cdot \Delta t \geq \hbar$ $\boxed{\hbar = \frac{h}{2\pi}}$

and angular displacement $\Delta\theta$ and angular momentum ΔL.

Then, $\Delta L \cdot \Delta\theta \geq \hbar$ $\boxed{\hbar = \frac{h}{2\pi}}$

where, the notation Δ in association with the respective variables, indicates the minimum uncertainty involved in the measurement of the corresponding variable. The uncertainty principle is applicable to subatoms like positron (e^+), electron (e^-), etc.

If at any instant, the position and momentum of a particle are simultaneously measured and let Δx and Δp be the uncertainty in measurement of position and momentum, then

$$\Delta x \cdot \Delta p = \frac{h}{2\pi} \quad \text{or} \quad \Delta x \cdot \Delta p = \hbar$$

obviously if Δx is small, Δp will be large, in other words, if it is possible to measure the position of particle with certainty, it is not possible to measure the momentum with certainity.

In an exact form, according to Heisenberg's uncertainty principle the product of uncertainities in position and momentum of the particle cannot be less than $\frac{1}{2}\hbar$, *i.e.* $\Delta x \cdot \Delta p \geq \frac{1}{2}\hbar$.

1.17 PHYSICAL SIGNIFICANCE OF HEISENBERG'S UNCERTAINTY PRINCIPLE

The Heisenberg's uncertainty principle is given by

$$\Delta x \cdot \Delta p = \hbar \qquad \boxed{\because \ \hbar = \frac{h}{2\pi} = 1.05 \times 10^{-34} \text{ JS}}$$

(*i*) If the position coordinate x of a particle in motion is accurately determined at some instant, so that $\Delta x = 0$, then at the same instant the uncertainty ΔP_x in the determination of the momentum becomes infinite.

(*ii*) If the momenum P_x of a particle is accurately determined at some instant. So that $\Delta P_x = 0$, then at the same instant the uncertainity Δx in the determination of the position coordinate becomes infinite. Thus, if an experiment is designed to measure x or P_x accurately, the other quantity will become completely uncertain. We can measure both the quantities by means of an experiment, but only within certain limits of accuracy specified by the uncertainty relation.

(*iii*) For a particle of mass m moving with velocity v the product of the uncertainity Δx and the uncertainty Δv in the velocity is given by

$$\Delta x \cdot \Delta v \geq \frac{\hbar}{m}$$

For heavy bodies $\frac{\hbar}{m}$ is very small, and therefore, the product $\Delta x . \Delta v$ of the two uncertainity becomes very small. For such particles both the position x and the velocity v can be determined accurately.

For very heavy bodies if m is such that $\frac{\hbar}{m} = 0$, the uncertainities vanish and all quantities can be determined with perfect accuracy this is the limiting case of classical mechanics. Thus classical mechanics is applicable for heavy bodies and the uncertainities are the characteristic of quantum mechanics which is applicable to light particles such as electron, proton, neutron, etc.

1.18 γ-RAY MICROSCOPE

Heisenberg deviced an hypothetical experiment for the simultaneous measurement of both position and momentum of an electron. Let us imagine an electron travelling along x-axis and the same being observed through a microscope of high resolving power, as shown in Fig. 1.11 when the moving electron is illuminated some of the scattered photon will enter through the lens thus the observer can find out the position and momentum of the electron.

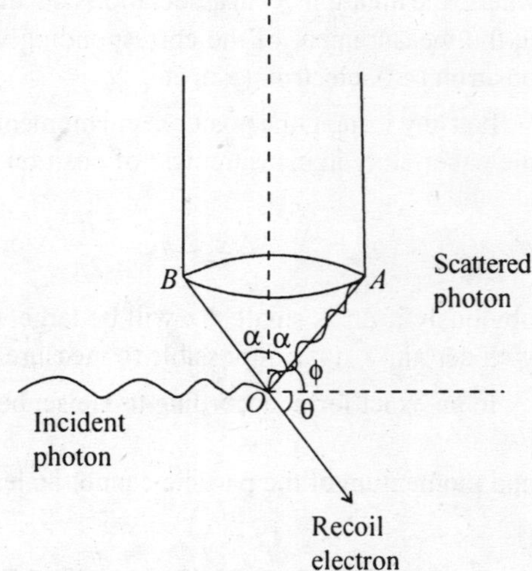

Fig. 1.11

Due to the diffraction effect the two object is said to be just resolvled if they are separated by a distance $\Delta x = \frac{\lambda}{2 \sin \theta}$ which is the resolving power of the microscope.

To make the separation too small the radiation of very small wavelength like γ-ray is used to illuminate the electron, if the position of electron changes by Δx due to the impact of γ-ray photons it can be just seen separated in the field of view. So the uncertainty in the measurement of position of electron is $\Delta x = \frac{\lambda}{2 \sin \theta}$. Where θ = semi solid angle subtended at the electron. Due to the interaction between the γ-ray and the electron there will be change in momentum of electron.

Let $\frac{\hbar}{\lambda'}$ be the momentum of incident photon and $\frac{\hbar}{\lambda}$ be the momentum of scatterd photon.

If the photon enter the microscope along OA its momentum along x-axis is $\frac{\hbar}{\lambda} \sin \theta$

i.e. change in momentum along OA $\quad P_1 = \frac{\hbar}{\lambda'} - \frac{\hbar}{\lambda} \sin\theta$

loss in momentum of along OB $\quad P_2 = \frac{\hbar}{\lambda'} - \left(-\frac{\hbar}{\lambda} \sin\theta\right)$

$$P_2 = \frac{\hbar}{\lambda'} + \frac{\hbar}{\lambda} \sin\theta$$

$\therefore$ Change in momentum of scattered photon $\Delta P_x = P_2 - P_1$

$$= \frac{2\hbar}{\lambda} \sin\theta$$

$\therefore$ It is possible to reduce the uncertainty in the measurement of position by using light of short wavelength, but it increases the uncertainty in its momentum.

$\therefore$ $\Delta x \cdot \Delta p_x$ cannot be minimum and they cannot be determined simultaneously with accuracy.

Now
$$\Delta x \cdot \Delta P_x = \frac{\lambda}{2\sin\theta} \times \frac{2\hbar}{\lambda} \sin\theta$$

$$\Delta x \cdot \Delta P_x = \hbar$$

i.e. from the above result, we can conclude that the position and momentum cannot be determined simmultaneously and the uncertainty will never be less than $\hbar$. However a more rigorous treatment gives

$$\Delta x \cdot \Delta p_x = \frac{\hbar}{2}.$$

The significance of this uncertainty principle is that, the microscopic phenomenon of simultaneous measurement of position and momentum is impossible. At any instant of measurement either position or velocity can be precisly but not both.

In case of macroscopic phenomena the uncertainty in the determination of both position and momentum are so small that they are neglected.

1.19 SHOW THAT THE UNCERTAINTY IN THE VELOCITY IS EQUAL TO PARTICLE VELOCITY, IF UNCERTAINTY IN THE LOCATION (POSITION) IS EQUAL TO DE-BROGLIE WAVELENGTH

Let a particle of mass m moving with velocity v is associated with the wave. Then according to de-Broglie the wavelength of the particle associated with the wave a given by

$$\lambda = \frac{h}{mv}$$

According to uncertainty principle

$$\Delta x \cdot \Delta P_x = h \quad \text{or} \quad \Delta P_x = \frac{h}{\Delta x} \tag{1}$$

If
$$\Delta x = \lambda \quad \textit{or} \quad \Delta x = \frac{h}{mv}$$

then Eq. (1) becomes
$$\Delta P_x = \frac{h}{h/mv}$$

$$\Delta P_x = mv \quad (2)$$

but, we know that, $P = mv$

$$\therefore \quad \Delta P_x = m\Delta v_x$$

Equation (2) becomes

$$m\Delta v_x = mv$$

$$\Delta v_x = v$$

$\therefore$ In general $\Delta v = v$ Hence, proved,

this means that uncertainty in the velocity of the moving particle is equal to its velocity if the uncertainty in the location is equal to de-Broglie wavelength.

1.20 APPLICATION OF UNCERTAINTY PRINCIPLE

1.20.1 Non-Existence of Electrons in Nucleus

We know that, according to Einstein's special theory of relativity the energy of the particle is given by

$$E = mc^2$$

but from relativistic mass-velocity relation, we know that $m = \dfrac{m_0}{\sqrt{1-\dfrac{v^2}{c^2}}}$

where m_o is the rest mass of the particle and m is the mass of the particle moving with a speed v.

On substituting this in the above expression, we get

$$E = \frac{m_o}{\sqrt{1-\dfrac{v^2}{c^2}}} \times c^2$$

squaring on both sides, we get $E^2 = \dfrac{m_o^2 c^4}{c^2 - v^2} \times c^2$ $\qquad \because \left(\sqrt{1-\dfrac{v^2}{c^2}}\right)^2 = \dfrac{c^2 - v^2}{c^2}$

$$E^2 = \frac{m_o^2 c^6}{c^2 - v^2} \quad (1)$$

The momentum of the particle moving with a speed v is given by,

$$P = mv$$

we know that $m = \dfrac{m_o}{\sqrt{1-\dfrac{v^2}{c^2}}}$

on substituting we get, $p = \dfrac{m_o}{\sqrt{1-\dfrac{v^2}{c^2}}} \times v$

multiplying by c on both side and on squaring we get,

$$P^2c^2 = \frac{m_o^2 v^2 c^2}{c^2 - v^2} \times c^2$$

$$P^2c^2 = \frac{m_o^2 v^2 c^4}{c^2 - v^2} \qquad (2)$$

Equation (1) – (2) gives

$$E^2 - P^2c^2 = \frac{m_o^2 c^6}{c^2 - v^2} - \frac{m_o^2 v^2 c^4}{c^2 - v^2} = m_o^2 c^4 \frac{\left(c^2 - v^2\right)}{c^2 - v^2}$$

$$\therefore \quad E^2 - p^2c^2 = m_o^2 c^4$$

or

$$E^2 = p^2c^2 + m_o^2 c^4$$

or

$$E^2 = c^2\left(p^2 + m_o^2 c^2\right)$$

We know that, Heisenberg's uncertainty principle state that $\Delta x \cdot \Delta P_x \geq \dfrac{h}{2\pi}$ (4)

We also know that, the diameter of the nucleus is of the order of 10^{-14} m. If an electron is to exist inside the nucleus, then the uncertainty in its position Δx must not exceed the size of the nucleus, *i.e.* $\Delta x \leq 10^{-14}\ m$

In such an event, the uncertainty in the momentum as per Eq. (4) becomes

$$\Delta P_x \geq \frac{h}{2\pi \Delta x}$$

i.e.

$$\Delta P_x \geq \frac{6.625 \times 10^{-34}}{2\pi \times 10^{-14}}$$

$$\therefore \quad \Delta P_x \geq 1.055 \times 10^{-20}\ Ns \qquad (5)$$

which is the uncertainty in the momentum of the electron. Then the momentum of the electron must be at least equal to the uncertainty in the momentum, *i.e.* 1.055×10^{-20}

$$\therefore \quad P_x \geq 1.055 \times 10^{-20}\ Ns \qquad (6)$$

Now from Eq. (3)

$$E^2 = c^2\left(P^2 + m_o^2 c^2\right)$$

We know $\quad m_o = 9.11 \times 10^{-31}$ kg

Now, by making use of the condition in Eq. (6).

We can say that, in order that an electron may exist within the nucleus its energy E must be such that,

$$E^2 \geq \left(3 \times 10^8\right)^2 \left[\left(1.055 \times 10^{-20}\right)^2 + \left(3 \times 10^8\right)^2 \left(9.1 \times 10^{-31}\right)^2\right]$$

i.e.

$$E^2 \geq \left(3 \times 10^8\right)^2 \left[1.113 \times 10^{-40} + 7.4692 \times 10^{-44}\right]$$

Since the second term in the bracket is smaller by more than 4 orders of magnitude compared to the first term and it is neglected.

Taking the square root on both sides and simplifying we get the condition as

$$E = 3.1649 \times 10^{-12}\ \text{J}$$

expressing in eV, we get $E \geq 20$ Mev

This means to say that, in order that an electron may exist inside the nucleus, its energy must be greater than or equal to 20 Mev. But, the experimental investigation on beta-decay reveal that, the kinetic energy of the beta particles (which are same as electrons) is of the order of 3 to 4 Mev. This clearly indicates that, electrons can not exist within the nucleus.

1.20.2 Existence of Protons and Neutrons in the Nucleus

For protons and neutrons, $m \approx 1.67 \times 10^{-27}$ kg and this is a non-relativistic problem as $v = 3 \times 10^{6}\ \text{ms}^{-1}$. For a proton or neutron to be confined within such a nucleus of radius 5×10^{-15} m, the maximum uncertainty in its position is $\Delta x = 5 \times 10^{-15}$ m using uncertainty relation $\Delta x \cdot \Delta p = \dfrac{h}{4\pi}$, the corresponding (minimum) uncertainty in its momentum is therefore.

$$\Delta p = \frac{h}{4\pi \Delta x} = \frac{6.625 \times 10^{-34}}{4 \times 3.14 \times 5 \times 10^{-15}\ m}, \quad E = \frac{p^2}{2m} = \frac{\left(1.05 \times 10^{-20}\right)^2}{2 \times 1.67 \times 10^{-27}} = 0.5\ \text{Mev}$$

$$= 1.05 \times 10^{-20}\ \text{kg} \cdot \text{ms}^{-1} \qquad \therefore E = 0.5\ \text{Mev}$$

Since, this energy is smaller than the energies carried by these particles emitted by nuclei, both these particles can exist inside the nucleus.

1.21 RADIATION OF LIGHT FROM AN EXCITED ATOM

We know that, the average time period that an atom takes to come to its unexcited state from the excited state is of the order of 10^{-8} *s*. Using uncertainty principle between energy and time is given by

$$\Delta E \cdot \Delta t \geq \frac{h}{4\pi}$$

Thus, the uncertainty in the photon energy is given by

$$\Delta E \geq \frac{h}{4\pi \Delta t}$$

$$\geq \frac{6.625 \times 10^{-34}}{4 \times 3.14 \times 10^{-8}} \approx 10^{-26}\ \text{J}$$

Thus, the uncertainty in the frequency of the light is

$$\Delta v = \frac{\Delta E 4\pi}{h} \geq \frac{10^{-26} \times 4 \times 3.14}{6.625 \times 10^{-34}} = 18.9 \times 10^{7} \text{ Hz}$$

$$\Delta v \approx 10^{7} \text{ Hz}$$

This is the maximum limit to the accuracy with which one can determine the frequency of the radiation emitted by an atom.

1.22 EXPRESSION FOR DE-BROGLIE WAVELENGTH λ INTERMS OF v_g AND v_p

According to de-Broglie's concept of matter wave a fast moving material particle is associated with a wave packet, which comprises the group of waves of slighty varying wavelength and frequencies Therefore, the modulated pattern due to a group of wave is associated with group velocity and phase velocity.

We know that, the group velocity v_g interms of angular frequency and propagation constant is given by,

$$v_g = \frac{d\omega}{dk} \tag{1}$$

where $\omega = 2\pi v \; ; \; k = \dfrac{2\pi}{\lambda}$

and $\omega = 2\pi v \; ; \; k = \dfrac{2\pi}{\lambda}$

on substituting these values in the above equation, we get $v_g = \dfrac{dv}{d\left(\dfrac{1}{\lambda}\right)}$

but we know that the velocity associated with the de-Broglie wave group is equal to particle velocity itself,

i.e. $v_g = v$

then $v = \dfrac{dv}{d\left(\dfrac{1}{\lambda}\right)}$

or $$d\left(\frac{1}{\lambda}\right) = \frac{dv}{v} \quad \text{or} \quad dv = vd\left(\frac{1}{\lambda}\right) \tag{2}$$

Let us consider a particle having mass m moving with a velocity v then the total energy E of the particle is given by the sum of potential energy and kinetic energy.

$$\therefore \; E = \frac{1}{2}mv^2 + V \tag{3}$$

but, we also know that, according to Max Planck's quantum theory of radiation the energy E of a particle of frequency υ is given by

$$E = hv \tag{4}$$

on comparing, we get $$hv = \frac{1}{2}mv^2 + V \tag{5}$$

If the particle is moving with a constant potential then on differentiating Eq. (5), we get

$$hdv = mvdv$$

or
$$dv = \frac{m}{h} v\, dv \tag{6}$$

on comparing with (2), we get

$$vd\left(\frac{1}{\lambda}\right) = \frac{m}{h} v\, dv$$

$$d\left(\frac{1}{\lambda}\right) = \frac{m}{h} dv$$

on integrating, we get

$$\frac{1}{\lambda} = \frac{mv}{h} + \text{constant}$$

where $p = mv$, momentum of the particle and constant of integration is zero.

$$\therefore \quad \frac{1}{\lambda} = \frac{p}{h}$$

or
$$\lambda = \frac{h}{p}$$

$$\therefore \quad \lambda = \frac{h}{mv}$$

This is the required expression for de-Broglie wavelength λ interms of phase velocity and group velocity.

1.23 HEISENBERG'S UNCERTAINTY PRINCIPLE FROM DE-BROGLIE CONCEPT OF MATTER WAVE

Heisenberg's uncertainty principle can be proved with the help of de-Broglie's wave concept that a particle in motion can be taken as a group of waves, and the group velocity being equal to the particle velocity.

Consider a group of waves consisting of only two components of equal amplitude and having angular frequencies ω_1 and ω_2 differing by a small amount and represented by the equations.

$$y_1 = a\cos(\omega_1 t - k_1 x)$$

and
$$y_2 = a\cos(\omega_2 t - k_2 x)$$

where, $\frac{\omega_1}{k_1}$ and $\frac{\omega_2}{k_2}$ represent their respective phase velocities

$$\frac{\omega}{k} = \frac{2\pi v}{\frac{2\pi}{\lambda}} = v\lambda = v_p$$

The resusltant amplitude is given by

$$y = y_1 + y_2 = 2A\cos(\omega t - kx)\cos\left(\frac{\Delta\omega}{2} t - \frac{\Delta k}{2} x\right)$$

where, $\omega = \dfrac{\omega_1 + \omega_2}{2}$; $k = \dfrac{k_1 + k_2}{2}$

$$\Delta\omega = \omega_1 - \omega_2 \quad ; \quad \Delta k = k_1 - k_2$$

The resultant wave is as shown in Fig. 1.12.

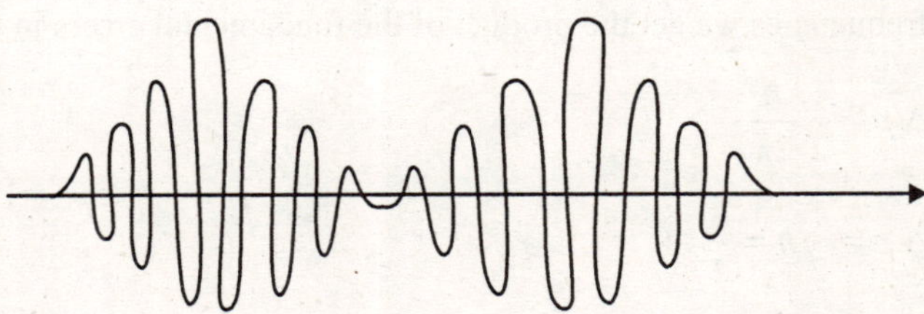

Fig. 1.12

where the loop formed travels with the group velocity $v_g = \dfrac{\Delta\omega}{\Delta k}$

As the group velocity of the de-Broglie wave group associated with the moving particle is equal to the particle velocity, the loop formed is equivalent to the position of the particle. Thus, the position of the particle cannot be given with certainty. It will lie some where between the two consecutive nodes. In other words, the fundamental error in the measurement of the position of the particle is equal to the distance between two consecutive nodes.

The condition for the formation of a node is given by,

$$\cos\left(\frac{\Delta\omega}{2}t - \frac{\Delta k}{2}x\right) = 0$$

$$\therefore \quad \left(\frac{\Delta\omega}{2}t - \frac{\Delta k}{2}x\right) = \frac{\pi}{2}, \frac{3\pi}{2} \cdots\cdots (2n+1)\frac{\pi}{2}$$

Thus, if x_1 and x_2 are the positions of two consecutive nodes, then

$$\frac{\Delta\omega}{2}t - \frac{\Delta k}{2}x_1 = (2n+1)\frac{\pi}{2}$$

and

$$\frac{\Delta\omega}{2}t - \frac{\Delta k}{2}x_2 = (2n+3)\frac{\pi}{2}$$

$$\therefore \quad \frac{\Delta k}{2}(x_1 - x_2) = \pi$$

where $(x_1 - x_2)$ is the error in the measurement of the position of the particle.

$$\therefore \quad \Delta x = \frac{2\pi}{\Delta k}$$

Also $k = \dfrac{2\pi}{\lambda} = \dfrac{2\pi p}{h}$ $\qquad \because \lambda = \dfrac{h}{p}$

$$\therefore \quad \Delta k = \frac{2\pi}{h}\cdot\Delta p$$

Hence, $$\frac{2\pi}{\Delta x} = \frac{2\pi}{h} \cdot \Delta p$$

or $$\Delta x \cdot \Delta p = h$$

where, Δp is the fundamental error in the measurement of momentum P.

The above treatment is very elementary. If we consider a group consisting of a very large number of waves of continuously varying frequencies we get the product of the fundamental errors in the measurement as

$$\Delta x \cdot \Delta p \geq \frac{1}{2} \frac{h}{2\pi}$$

$$\Delta x \cdot \Delta p = \frac{1}{2} \hbar = \frac{\hbar}{2}$$

$$\Delta x \cdot \Delta P_x = \frac{\hbar}{2}$$

This value is used for more accurate calculations the constant $\hbar = \frac{h}{2\pi} = 1.054 \times 10^{-34}$ JS

$$= 1.054 \times 10^{-27} \text{ erg} \cdot \text{s}$$

$$h = 6.58 \times 10^{-16} \text{ eVs}$$

1.24 LIMITATIONS OF CLASSICAL MECHANICS AND DEVELOPMENT OF WAVE MECHANICS

The Bohr's theory of atom is able to account for the hydrogen spectrum and hydrogen isotopes, but is has a number of severe limitations.

It does not explain the fine structure of spectral lines and the intensity variation. And its use is limited to hydrogen and hydrogen like atoms; it cannot explain the spectra of complex atoms. The most important one is it does not explain how electrons interact with atoms, and atoms with atoms.

In classical mechanics Newton's laws describe the motion of the particles. The future history of particle is completely determined by its initial position and momentum together with the forces that act on it.

But due to uncertainty principle the future characteristics of classical mechanics is not possible, because the initial position and momentum of the particle cannot be measured accurately. However, according to de-Broglie's concept of matter wave a material particle is associated with a wave. So a mathematical reformation using a wave function associated with matter waves is needed. Such a mathematical reformation is known as quantum mechanics or wave mechanics which was developed by Schrodinger in the year 1926. This wave mechanics is consistent with the uncertainty principle. The amplitude of matter wave is a complex quantity which is described by Schrodinger and is represented as $\psi(x\ y\ z\ t)$ known as wave function or the state of the system.

1.25 DIFFERENCE BETWEEN CLASSICAL MECHANICS AND WAVE MECHANICS

The fundamental difference between classical mechanics and wave mechanics lies in what is that they describe.

(*i*) Classical mechanics is concerned with the motion of a particle under the influence of applied forces and the quantities such as position, mass, velocity and acceleration can be measured. This assumption is of course completely valied in our every day experience and classical mechanics provides the correct explanation for the behaviour of moving bodies in the sense that the values it predicts for observable magnitudes agrees with the measured values of those magnitude.

(*ii*) Wave mechanics deals with relation between observable magnitudes, but the uncertainity principle radically alters the definition of *observable magnitude* in the atomic realm. According to uncertainty principle the position and momentum of a particle cannot be accurately measured simultaneously while in classical mechanics both are assumed to have definite as certainable values at every instant. The quantities whose relationships wave mechanics explores are probabilities.

At first glance wave mechanics seems to poor substitute for classical mechanics but closer observation reveals a striking fact, that the classical mechanics is nothing but an approximate version of wave mechanics.

1.26 SCHRODINGER WAVE EQUATION

1.26.1 Wave Function

The relation $\lambda = h/p$ between momentum p and wavelength λ which has been experimentally verified for both photon and particles suggests that it is possible to use a concentrated group of waves or a wave packet to describe localised particles and photons of radiation. The associated de-Broglie wave function $\psi(\vec{r}\ t)$ is a function of space variables $(x\ y\ z)$ and t and it can give complete information about the state of a physical system. It is also known as state function.

1.27 SIGNIFICANCE OF $|\psi|^2$

The wave function ψ has three basic properties.

(*i*) ψ can interfere with itself. This property explains the phenomenon of electron diffraction.

(*ii*) ψ is large in magnitude where the particle is small else where

The probability of finding the practicle described by the wave function ψ is proportional to $|\psi|^2$.

Thus $|\psi|^2$, *i.e.* the square of the amplitude of the wave function ψ gives the probability density of finding the physical system at a particular place at a given time.

The wave function is generally a complex quantity and thus consists of two parts, the amplitude and phase.

If we write the complex quantity $\psi = A + iB$ where A and B are real functions, then its complex conjugate

$$\psi^* = A - iB$$

and
$$\psi\psi^* = A^2 - i^2 B^2$$

$$\psi\psi^* = A^2 + B^2$$

which is always a positive real quantity

$\therefore$ Probability density $= |\psi|^2$

i.e. if ψ is a complex quantity, the probability density is given by the product $\psi\psi^*$ of ψ and its complex conjugate ψ^*.

(*iii*) The wave function ψ describes the behaviour of a single particle or photon and not the statistical distribution of a number of such quanta.

1.28 CONDITIONS OF WAVE FUNCTION

Due to the interaction of $|\psi|^2$ as a probability density, the wave function ψ must obey the following conditions.

(*i*) ψ must be finite for all values of x, y and z.

(*ii*) ψ must be single valued *i.e*, for each set of values of x, y and z ψ must have one value only.

(*iii*) ψ must be continuous in all regions except in those regions where the potential energy $V(x\,y\,z) = \infty$.

(*iv*) The partial derivities of ψ, *i.e.* $\frac{\partial\psi}{\partial x}, \frac{\partial\psi}{\partial y}$ and $\frac{\partial\psi}{\partial z}$ must also be continuous.

1.29 NEED FOR DIFFERENTIAL WAVE EQUATION

The motion of a particle of atomic dimensions cannot be described by Newton's law of motion which are applicable to macroscopic particles. The behaviour of microscopic particles are governed by the associated de-Broglie wave or the wave function $\psi(\vec{r}\,t)$. Since ψ is a complex quantity and square of its modulus $|\psi|^2$ or $\psi\psi^*$ gives the probability of finding the particle at the position vector $\vec{r}$, corresponding to cartesion coordinates $x\,y\,z$ and time t.

To describe the motion of the particle in space and time we must know the value of $\psi(\vec{r}\,t)$ at all times and for this purpose it is necessary to find a differential equation which controls the space time behaviour of the wave function $\psi(\vec{r}\,t)$. The solution of such a differential equation will gives the possible motion of the particle. Hence, the necessity of such a differential equation.

1.30 TIME INDEPENDENT SCHRODINGER WAVE EQUATION

According to de-Broglie's concept of matter wave a particle of mass m moving with a velocity v is always associated with a wave whose wavelength is given by $\lambda = \frac{h}{mv}$. If the particle has wave properties, it is expected that there should be a wave equation which describes the behaviour of the particle. Let x, y and z be the coordinates of the particle and ψ the wave displacement for the de-Broglie wave at any instant of time t, thus according to classical differential equation, we know that, equation of a wave motion is given by

$$\frac{\partial^2\psi}{\partial t^2} = v^2\left(\frac{\partial^2\psi}{\partial x^2} + \frac{\partial^2\psi}{\partial y^2} + \frac{\partial^2\psi}{\partial z^2}\right)$$

$$\frac{\partial^2\psi}{\partial t^2} = v^2\,\nabla^2\psi$$

$$\therefore \quad \nabla^2\psi = \frac{1}{v^2}\frac{\partial^2\psi}{\partial t^2} \tag{1}$$

where $\nabla^2 = \frac{\partial^2}{\partial x^2} + \frac{\partial^2}{\partial y^2} + \frac{\partial^2}{\partial z^2}$ and v is the velocity (∇^2 being laplacian operator)

the solution of the Eq. (1) gives ψ as a periodic displacement in terms of time, *i.e.*

$$\psi(x\,y\,z\,t) = \psi(x\,y\,z)\,e^{-i\omega t} \tag{2}$$

where $\psi_o(x\,y\,z)$ is a function of $x\,y\,z$ and gives the amplitude at the point considered and ω is the angular frequency of the wave.

the Eq. (2) can also be expressed as

$$\psi(\vec{r}\ t) = \psi_o(\vec{r})\, e^{-i\omega t} \tag{3}$$

on differentiating Eq. (3) with respect to t twice, we get

$$\frac{\partial^2 \psi}{\partial t^2} = -\,\omega^2 \psi_o(\vec{r})\, v^{-i\omega t}\ \frac{\partial^2 \psi}{\partial t^2} = -\,\omega^2 \psi \qquad \boxed{\because\ \psi = \psi_o}$$

substituting the values of $\frac{\partial^2 \psi}{\partial t^2}$ in Eq. (1), we get

$$\frac{\partial^2 \psi}{\partial x^2} + \frac{\partial^2 \psi}{\partial y^2} + \frac{\partial^2 \psi}{\partial z^2} + \frac{\omega^2}{v^2}\psi = 0 \tag{4}$$

But $\qquad \omega = 2\pi v = 2\pi \frac{v}{\lambda}$

$$\therefore \quad \frac{\omega}{v} = \frac{2\pi}{\lambda}$$

substiting the values of $\frac{\omega}{v}$ in Eq. (4) we get

$$\frac{\partial^2 \psi}{\partial x^2} + \frac{\partial^2 \psi}{\partial y^2} + \frac{\partial^2 \psi}{\partial z^2} + \frac{4\pi^2}{\lambda^2}\psi = 0$$

or

$$\nabla^2 \psi + \frac{4\pi^2}{\lambda^2}\psi = 0 \tag{5}$$

but we know that, according to de-Broglie's relation $\lambda = \frac{h}{mv}$ on substituting these value in the above expression, we get

$$\nabla^2 \psi + \frac{4\pi^2}{h^2} m^2 v^2 \psi = 0 \tag{6}$$

Let E and V be the total and potential energies of the particle respectively, then its kinetic energy is given by

$$\text{Kinetie energy} = \frac{1}{2} mv^2 = E - V$$

or

$$m^2 v^2 = 2m\,(E - V) \tag{7}$$

on comparing Eq. (6) and (7), we get

$$\nabla^2 \psi + \frac{8\pi^2 m}{h^2}(E - V)\,\psi = 0$$

$$\frac{\partial^2 \psi}{\partial x^2} + \frac{\partial^2 \psi}{\partial y^2} + \frac{\partial^2 \psi}{\partial z^2} + \frac{8\pi^2 m}{h^2}(E - V)\,\psi = 0 \tag{8}$$

This is the time independent Schrodinger's wave equation for a particle having mass m, total energy E and potential energy V moving in space.

The time independent Schrodinger wave equation for a particle which is moving along the x-axis become

$$\frac{\partial^2 \psi}{\partial x^2} + \frac{8\pi^2 m}{h^2}(E - V)\psi = 0 \tag{9}$$

Substituting $\hbar = \frac{h}{2\pi}$ in the above expression, we get

$$\nabla^2 \psi + \frac{2m}{\hbar^2}(E - V)\psi = 0 \tag{10}$$

For a free particle $V = 0$, hence, the Schrodinger wave equation can be expressed as

$$\nabla^2 \psi + \frac{2mE}{\hbar^2}\psi = 0$$

1.31 PHYSICAL SIGNIFICANCE OF THE WAVE FUNCTION

The dual nature of matter wave can be appriciated by the interpretation of wave function ψ. Since the electron beam is diffracted like X-ray beam is the one cause of an optical analogue to arrive at the importance of ψ. The intensity of electromagnetic waves at a point in space is represented by ψ, represents the amplitude of the wave. If ψ is the amplitude of matter wave ψ^2 represents the particle density at a point in space. Later on Maxborn slightly twisted the idea and suggested that $|\psi|^2$ represents the probability of finding the particle at a point. More exactly the probability of finding the particle in a volume dv given by the relation $\iiint |\psi^2| \, dv = 1$.

According to this interpretation the wave in space is a simultaneous representation of what we know about the particle. For instance; in phenomenon of electron diffraction the electron behaves as particle when it interacts with the atomic layers. While its probability of being present is controlled by the wave associated with it. Thus, the particle and wave aspect are complementary to each other (Fig. 1.13).

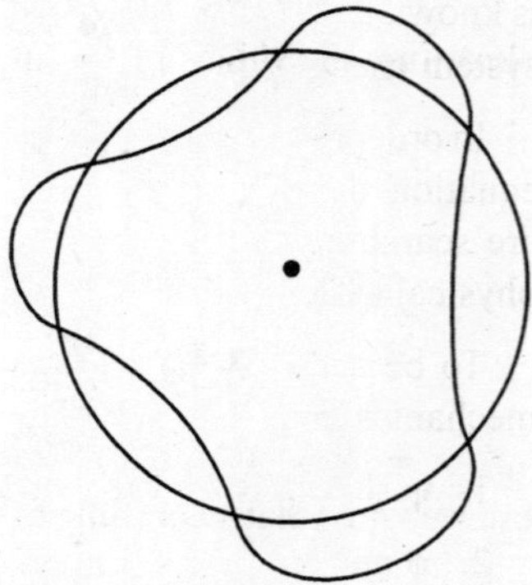

Fig. 1.13

1.32 NORMALISATION

The probability density is given by

$$|\psi^2| = \psi^* \psi$$

where ψ^* is the complex conjugate of ψ

Let us consider a small elemental of volume dv defined by the coordinates $(x, x + dx)$; $(y, y + dy)$ and $(z, z + dz)$, then the probability of finding the particle within the elementary volume dv is given by

$$\psi^* \psi \, dv = \psi^* \psi \, dx \, dy \, dz$$

The probability of finding the particle in a finite volume v is given by

$$\int \psi^* \psi \, dv = \iiint_v \psi^* \psi \, dx \, dy \, dz$$

The particle must always be some where in the space so that extending the integral over all space, the probability becomes certainty, *i.e.* it tends to unity.

$$\therefore \quad \iiint_{\text{All space}} \psi^* \psi \, dx\, dy\, dz = 1$$

or

$$\iiint_{\text{All space}} |\psi^2| \, dx\, dy\, dz = 1$$

This is called normalisation condition.

The process of integration over all possible locations to give unity is called normalisation.

If there are n-particles in the space, then

$$\int_{\text{All space}} \psi^* \psi \, dx\, dy\, dz = n$$

where ψ is unnormalised wave function.

1.33 NATURE OF EIGEN FUNCTIONS AND EIGEN VALUES

A system is defined by its energy, momentum, position, etc. If the wave function ψ corresponding to a system is known then the system can be determined. A wave function corresponding to a system will change when the system changes its state.

In order to find ψ, the Schrodinger's wave equation has to be solved. But since it is a second order differential equation, there is not one but several solutions, and all of them may not be the correct wave functions which we are searching for. So, we have to select those wave functions which would correspond meaningfully to a physical system, such wave functions are said to be acceptable wave functions.

To be accpetable wave function it has to satisfy the following criteria which are postulated in quantum mechanics.

1. ψ should be single valued every where.
2. ψ and its first derivatives with respect to its variables are continuous every where.
3. ψ should be finite every where.

Such acceptable wave functions are named as Eigen functions. Once the Eigen functions are known they could be used in Schrodinger's wave equation to evaluate energy. Since there is only a restricted set of Eigen functions there is also restricted set of energy values those values are called Eigen values.

1.34 APPLICATIONS OF SCHRODINGER'S WAVE EQUATION PARTICLE IN ONE DIMENSIONAL BOX

Let us consider an one dimensional potential box of length (width) L having an infinite height. Let the particle of mass m is bombarding back and forth in a potential box along x direction as shown in Fig. 1.14. The potential of the particle becomes zero. When the particle is with in the box, *i.e.* $V = 0$, when $O < x < L$.

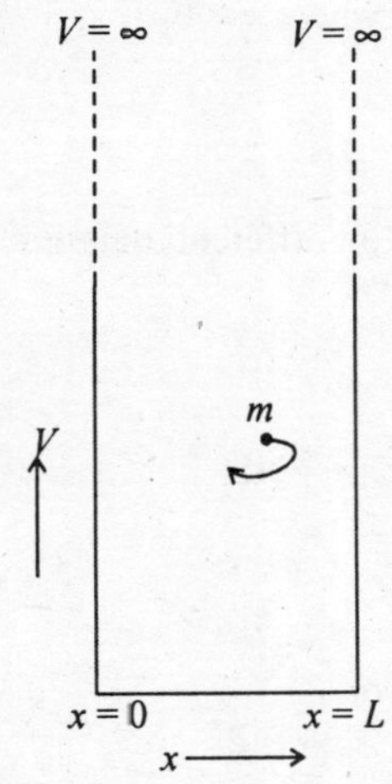

Fig. 1.14

The potential of the particle is infinite when the particle is at the walls of the box.

i.e. $\quad V = \infty$ for $x \leq 0$

and $\quad V = \infty$ for $x \geq L$

$\therefore$ when the partaicle is moving inside the potential box between $x = 0$ and $x = L$, then potential $V = 0$. Hence, the one dimensional Schrodinger equation becomes.

$$\frac{\partial^2\psi}{\partial x^2}+\frac{8\pi^2 m}{h^2}(E-O)\,\psi=0 \qquad \because \quad \frac{\partial^2\psi}{\partial x^2}+\frac{8\pi^2 m}{h^2}(E-V)\,\psi=0, \quad V=0$$

$$\therefore \quad \frac{\partial^2\psi}{\partial x^2}+\frac{8\pi^2 m}{h^2}E\psi=0 \tag{1}$$

or
$$\frac{-h^2}{8\pi^2 m}\frac{\partial^2\psi}{\partial x^2}=E\psi$$

This is of the form $H\psi = E\psi$

putting
$$K^2=\frac{8\pi^2 m}{h^2}E$$

$\therefore$ Equation (1) becomes

$$\frac{\partial^2\psi}{\partial x^2}+k^2\psi=0 \tag{2}$$

Since the Eq. (2) is in the form of II order equation.

The general solution for the above equation is $\psi = A\sin kx + B\cos kx$ (3)

The constants A and B can be evaluated by using boundary conditions.

i.e. $\psi = 0$ for $x = 0$ and hence $B = 0$

$\psi = 0$ for $x = L$ and hence $0 = A\sin kL$

Since $A \neq 0$ $\therefore$ $kL = n\pi$

or
$$k=\frac{n\pi}{L} \tag{4}$$

where $n = 0, 1, 2, 3$ ..... and is called quantum number, by substituting the value of (4) in (3), we get

$$\psi = A\sin\frac{n\pi}{L}x \tag{5}$$

on differentiating Eq. (5) twice with respect to x, we get

$$\frac{\partial\psi}{\partial x}=\frac{n\pi}{L}\cdot A\cos\frac{n\pi}{L}x$$

$$\frac{\partial^2\psi}{\partial x^2}=\frac{-n^2\pi^2}{L^2}\cdot A\sin\frac{n\pi}{L}x$$

or
$$\frac{\partial^2\psi}{\partial x^2}=\frac{-n^2\pi^2}{L^2}\psi \qquad \boxed{\because \quad \psi = A\frac{n\pi}{L}x} \tag{6}$$

substituting the values of $\frac{\partial^2\psi}{\partial x^2}$ in Eq. (1), we get

$$\frac{-n^2\pi^2}{L^2}\psi + \frac{8\pi^2 m}{h^2}E\psi = 0$$

$$\frac{8\pi^2 mE}{h^2} = \frac{n^2\pi^2}{L^2}$$

or
$$E = \frac{n^2h^2}{8mL^2}$$

$$\therefore \quad E_n = \frac{n^2h^2}{8mL^2} \tag{7}$$

This is the expression for energy of the particle in a potential box of length L in an nth state. In Eq. (7) the value $n = 0$ is not accetable because, when $n = 0$, $\psi = 0$ by Eq. (5) which means to say that the electron is not present inside the box which is not true. Hence, the lowest value of n is 1. Consequently the lowest energy corresponds to $n = 1$ is called the zero point energy.

The zero point energy of an electron in a box is given by

$$E_{\text{Zero point}} = \frac{h^2}{8mL^2}$$

The lowest permitted state of energy is referred as the ground state energy. Thus zero point energy is taken as the ground state energy. The states corresponding to $n > 1$ are called excited state.

It is clear from Eq. (7) that inside an infinitely deep potential well, the particle can have only discrete set of values of energy, *i.e.* the energy of the particle is quantised. The discrete energy values are given by

For $\quad n = 1 \quad E_1 = \dfrac{h^2}{8mL^2}$,

For $\quad n = 2 \quad E_2 = \dfrac{4h^2}{8mL^2} = 4E_1$,

For $\quad n = 3 \quad E_3 = \dfrac{9h^2}{8mL^2} = 9E_1$,

For $\quad n = 4 \quad E_4 = \dfrac{16h^2}{8mL^2} = 16E_1$,

and so on. The energy level are shown in Fig. 1.15.

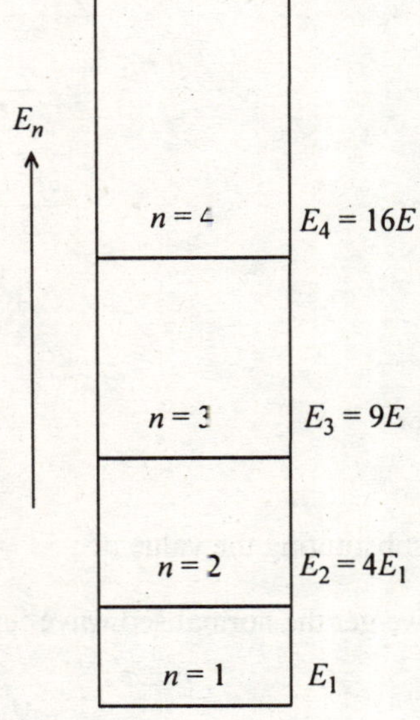

Fig. 1.15

The energy values for $n = 1, 2, 3 \ldots$ are Eigen values of the particle in the well and integer n corresponding to energy level E_n is called its quantum number the zero point energy is inversely proportional to m and square of the length of the box. Hence, if m is very large, *i.e.* for a particle of laboratory size zero point energy is almost zero. If $L \to \infty$ the particle behaves as a free particle with out any box and does not possess any zero point energy.

The lowest permitted state of energy is referred as the ground state energy. Thus zero point energy is taken as the ground state energy. The state corresponding to $n > 1$ are called excited state.

Energy levels of a particle inside an infinite potential well.

1.35 EIGEN VALUES AND EIGEN FUNCTION

We know that the wave function for a particle in a one dimensional potential box is given by

$$\psi = A \sin \frac{n\pi}{L} x$$

Where A is a constant, L is the length (width) of the potential box and x is the direction in which the particle is move with in the potential box. The probability of finding the particle is depending on ψ^2. The probability of finding the particle should be equal to unity when the particle is within the box.

∴ The probability of finding the particle within the box is given by

$$\int_{x=0}^{x=L} \left|\psi_n^2\right| dx = 1$$

substituting the value of ψ_n

$$\psi_n = A \sin \frac{n\pi}{L} x$$

$$\therefore \quad \int_{x=0}^{x=L} A^2 \sin^2 \frac{n\pi}{L} x \, dx = 1$$

$$A^2 \int_{x=0}^{x=L} \frac{1}{2}\left[1 - \cos^2\left(\frac{n\pi}{L}\right) x\right] dx = 1 \qquad \boxed{\because \; \sin^2\theta = \frac{1}{2}(1 - \cos 2\theta)}$$

$$\frac{A^2}{2}\left[X - \frac{L}{2n\pi} \sin\left(\frac{2n\pi}{L}\right) x\right]_o^L = 1$$

$$\frac{A^2}{2}\left[L - \frac{L}{2n\pi} \sin \frac{2n\pi}{L} \times L - (0)\right] = 1$$

or

$$\frac{A^2 L}{2} = 1 \qquad \boxed{\because \; \sin 2n\pi = 0}$$

$$\therefore \quad A^2 = \frac{2}{L}$$

or

$$A = \sqrt{\frac{2}{L}}$$

substituting the value of $A = \sqrt{\frac{2}{L}}$

we get the normalised wave function of a particle in a one dimensional box.

$$\psi_n = A \sin \frac{n\pi}{L}$$

$$\therefore \quad \psi_n = \sqrt{\frac{2}{L}} \sin \frac{n\pi}{L} x$$

This is the expression for wave function or Eigen function in terms of the width of the potential box.

1.36 WAVE FUNCTIONS AND PROBABILITY DENSITIES

In the case of a free particle, in a one dimensional box, the solution for the Schrodinger wave equation is given by

$$\psi(x) = A \sin\left(\frac{n\pi}{L}\right)x \;\therefore\; \psi(x) = \sqrt{\frac{2}{L}} \sin\left(\frac{n\pi}{L}\right)x \tag{1}$$

For $n = 1, 2, 3$ ..... the first three normalised wave function ψ_1, ψ_2, ψ_3 and corresponding probability densities $\left|\psi_1^2\right|, \left|\psi_2^2\right|, \left|\psi_3^2\right|$ are shown in Figs 1.16 and 1.17.

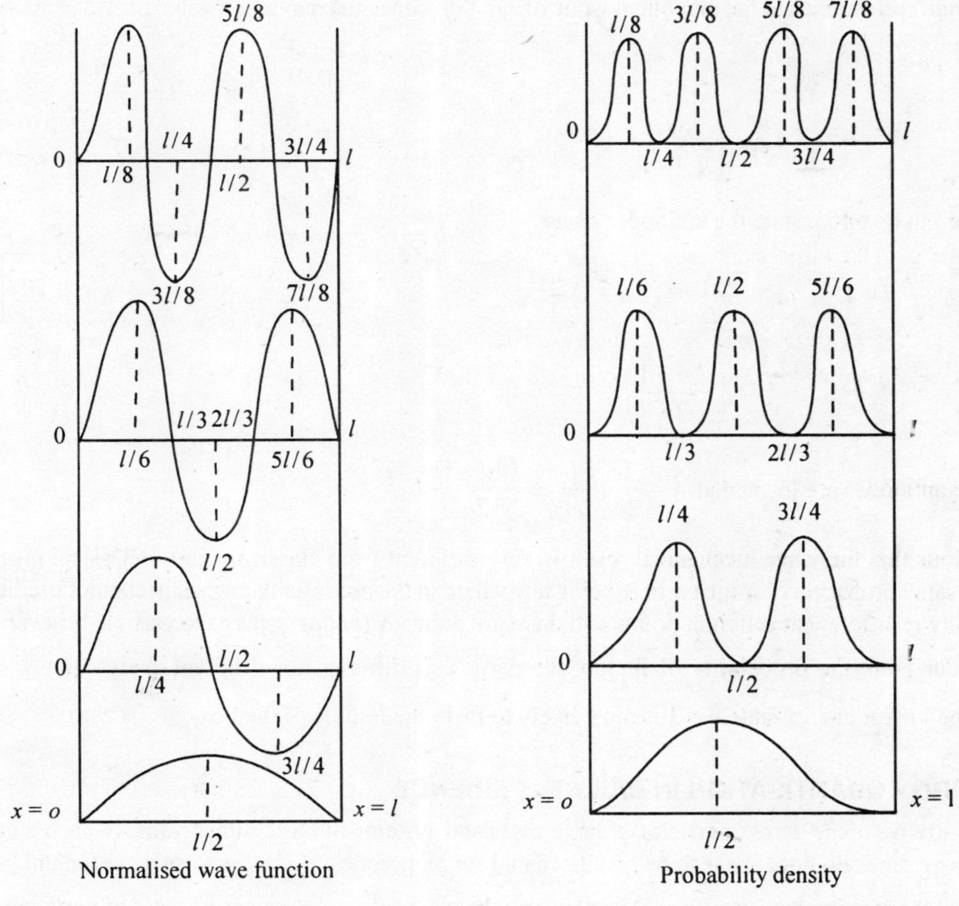

Fig. 1.16 **Fig. 1.17**

It is clear from the Figure that ψ_n has positive as well as negative values where as $\left|\psi_n^2\right|$ is always positive.

A node is the position of minimum or zero displacement and an antinode is the position of maximum displacement. Thus a node corresponds to $\psi_n(x) = 0$.

$\therefore$ For the nth quantum state the nodes occur where

$$\sin\frac{n\pi x}{L} = 0 \text{ or } \frac{n\pi x}{L} = m\pi$$

where $m = 0, 1, 2 \cdots\cdots$

$$\therefore \qquad x = m\frac{L}{n}$$

thus, the nodes are located at $x = 0, \frac{L}{n}, \frac{2L}{n} \cdots\cdots \frac{nL}{n}$.

An antinode corresponds to the maximum value of $\psi_n(x)$. Since the maximum value of $\sin\frac{n\pi}{L}x = 1$, for an antinode.

$$\text{i.e.} \qquad \psi_n(x) = \left(\frac{2}{L}\right)^{1/2}$$

$\therefore$ For the nth quantum state the antinodes occur

where $\sin\frac{n\pi}{L}x = 1$ or $\frac{n\pi x}{L} = (2m+1)\frac{\pi}{2}$

$$\therefore \qquad x = \left(\frac{2m+1}{2n}\right)L$$

Thus, the antinodes are located at $x = \frac{L}{2n}, \frac{3L}{2n} \cdots\cdots \frac{(2n-1)}{2n}L$

It is obvious that the wave mechanical result is very different from classical result. Classical mechanics predicts the same probability for the particle being any where in the box. But the wave mechanics predicts that the probability is different at different points and there are points N (nodes) where the particle is never found. At a particular point the probability of finding the particle is different for different energy states. **E.g.:** a particle in the lowest energy state $n = 1$ is most likely to be in the middle of the box, ψ_2^2 is zero.

1.37 ENERGY QUANTISATION IN DAILY EXPERIENCE

In our daily life we come across particles of large mass and systems of large dimensions. In such a case the allowed energy states do not appear to be quantised and for all practical purposes form a continuum.

Let us find the spacing between two adjacent energy levels for different values of mass of particle moving in boxes of different sizes.

1.37.1 Molecule of a Gas in Vessel

Let m be the mass of a gas molecule $(10^{-26}$ kg$)$ and L be the dimension of vessel (0.1 m), then for the molecule of a gas, the difference in energies of two adjacent level is given by

$$\Delta E_n = E_{n+1} - E_n$$

$$= \frac{(n+1)^2 h^2}{8mL^2} - \frac{n^2 h^2}{8mL^2}$$

$$= \frac{h^2}{8mL^2}(2n+1)$$

$$= \frac{\left(6.625 \times 10^{-34}\right)^2 (2n+1)}{8 \times 10^{-26} \times (0.1)^2} = \frac{5.5 \times 10^{-40} (2n+1)}{1.6 \times 10^{-19}}$$

$$\Delta E_n = 3.4 \times 10^{-21} (2n+1) \text{ eV}$$

The permissible energy levels are so close together that they may be regarded as being practically continuous. Although energy quantisation occurs in principle, it will not affect the motion of the molecules. Hence, in every day experience quantum effects are imperceptible. This accounts for the success of Newtonian mechanics in every day life.

1.37.2 Electron in a Region of Atomic Dimensions

Let m be the mass of an electron $\left(9.1 \times 10^{-31} \text{ kg}\right)$ and L be the atomic dimension $\left(1 \text{ Å} = 10^{-10} \ m\right)$, then the difference in the energies of two adjacent level is given by

$$\Delta E_n = E_{n+1} - E_n$$

$$= \frac{(n+1)^2 h^2}{8mL^2} - \frac{n^2 h^2}{8mL^2}$$

$$= \frac{h^2}{8mL^2}(2n+1)$$

$$= \frac{\left(6.625 \times 10^{-34}\right)^2 (2n+1)}{8 \times \left(9 \times 10^{-31}\right) \times \left(10^{-10}\right)^2} \text{ J}$$

$$= \frac{6 \times 10^{-18}}{1.6 \times 10^{-19}}(2n+1)$$

$$= 38(2n+1) \text{ eV}$$

It is clear that the spacing between the lowest ($n = 0$) and the next level ($n = 1$) is 114 eV, and increases in higher adjacent levels, thus, these energy levels are sufficiently far apart to make the quantisation of energy thus, is passing from atomic to much large dimensions the results of wave mechanics approach to those of classical physics.

UNSOLVED NUMERICAL PROBLEMS

1. Calculate the de-Broglie wavelength associated with the following :
   (*i*) A ball of 50 g moving with a velocity of 20 m/s.
   (*ii*) A proton moving with a velocity of 2200 m/s.
   (*iii*) An electron moving with a kinetic energy of 10 eV.
2. What is the de-Broglie wavelength of an electron which has been accelerated from rest through a potential difference of 100 V.
3. Calculate the de-Broglie wavelength associated with a proton moving with a velocity equal to 1/20th of the velocity of light.
4. Find the energy of the neutron in electron-volt whose de-Broglie wavelength is 1 Å. Given mass of neutron = 1.674×10^{-27} kg and Planck's constant $h = 6.625 \times 10^{-34}$ JS.
5. What voltage must be applied to an electron microscope to produce electrons of wavelength 0.40 Å.
6. Compute the de-Broglie wavelength of a proton whose kinetic energy is equal to the rest energy of an electron. Mass of a proton is 1836 times that of the electron.
7. Calculate the de-Broglie wavelength of an electron accelerated through a potential difference of (*i*) 100 volts, (*ii*) 300 volts, (*iii*) 1 kV.
8. Compute the de-Broglie wavelength of an electron whose kinetic energy is 50 eV. Given: $h = 6.625 \times 10^{-34}$ JS, $m_o = 9.1 \times 10^{-31}$ kg, 1 eV = $1.6 \times 1.6 \times 10^{-19}$ J.
9. Each of a photon and an electron has an energy of 1 keV. Determine their corresponding wavelengths. Given: $h = 6.625 \times 10^{-34}$ JS, $m_o = 9.1 \times 10^{-31}$ kg, 1 eV = 1.6×10^{-19} J.
10. An electron microscope is functioning at 10,000 volts. Compute the wavelength associated with the electron beam.
11. Determine the de-Broglie wavelength of a neutron of energy 12.8 MeV. Mass of neutro = 1.675×10^{-27} kg, $h = 6.625 \times 10^{-24}$ JS, $c = 3 \times 10^8$ m/s and 1 eV = 1.6×10^{-19} JS.
12. Compute the wavelength associated with (*a*) 1 MeV (kinetic energy) electron, (*b*) 1 MV (kinetic energy) proton and (*c*) 1 MeV photon. The rest masses of electron and proton are 9.1×10^{-31} kg, 1.67×10^{-27} kg respectively. Given: $h = 6.625 \times 10^{-34}$ JS, $c = 3 \times 10^8$ m/s, 1 eV = 1.6×10^{-19} J.
13. Find the mass and momentum of a photon of 1 Å. Given $h = 6.625 \times 10^{-34}$ JS and $c = 3 \times 10^8$ ms^{-1}.
14. Calculate de-Broglie's wavelength of a particle of mass 9×10^{-26} kg and velocity 3×10^8 m/s. Given $h = 6.625 \times 10^{-34}$ JS.
15. Find the wavelength for a beam of electrons whose kinetic energy is 100 eV. Mass of electron = 9.1×10^{-31} kg, $h = 6.625 \times 10^{-27}$ erg·s.
16. Calculate the de-Broglie wavelength of a proton of energy one million electron volt, given mass of proton = 1.66×10^{-24} g and $h = 6.625 \times 10^{-27}$ erg·s.
17. What is the wavelength associated with an electron accelerated through 1000 volt. Given mass of electron = 9.1×10^{-28} g and $h = 6.625 \times 10^{-27}$ erg·s.
18. Find the de-Broglie wavelength associated with an electron moving with a velocity of 0.5 c and rest mass 9.1×10^{-28} gm. Given $h = 6.625 \times 10^{-27}$ erg·s and $c = 3 \times 10^{10}$ cm/s.

    Velocity of the electron $v = 0.5 \times 3 \times 10^{10} = 1.5 \times 10^{10}$ cm/s.

19. Calculate the wavelength of thermal neutron kept at 127° C. Given Boltzmann's constant $k = 1.38 \times 10^{-16}$ erg/K and mass of neutron $m = 1.67 \times 10^{-24}$ gm, $h = 6.625 \times 10^{-27}$ erg·s. Temperature $T = 273 + 127 = 400$ K.

20. What voltage must be applied to an electron microscope to produce electrons of wavelength 0.5 Å? Given $e = 4.8 \times 10^{-10}$ e.s.u., $m = 9 \times 10^{-28}$ g, $h = 6.625 \times 10^{-27}$ erg·s.

21. Find the de-Broglie wavelength for an electron having kinetic energy of 1 MeV. Given $h = 6.625 \times 10^{-27}$ erg·s, velocity of light $c = 3 \times 10^{10}$ cm/s.

22. Find the wavelength for a beam of neutrons whose kinetic energy is 100 eV. Given mass of neutron 1.676×10^{-24} g and $h = 6.625 \times 10^{-27}$ erg·s.

23. The de-Broglie wavelength for a beam of electrons is 0.315 Å. Calculate the velocity of the electrons. Given $h = 6.625 \times 10^{-34}$ JS and $m = 9 \times 10^{-31}$ kg.

24. If the uncertainty in position of an electron is 4×10^{-10} m, calculate the uncertainty in its momentum.

25. An electron is confined to a box of length 10^{-9} m, calculate the minimum uncertainty in its velocity. Given, mass of electron $m = 9 \times 10^{-31}$ kg and Planck's constant $h = 6.625 \times 10^{-34}$ JS.

26. An electron has a speed of 6.6×10^{4} m/s within accuracy of 0.01%. Calculate the uncertainty in the position of the electron. (Mass of electron = 9×10^{-31} kg, Planck's constant = 6.625×10^{-34} JS).

27. A nucleon (neutron or proton) is confined to nucleus of radius 5×10^{-15} m. Calculate the minimum uncertainty in the momentum of nucleon. Also, calculate the minimum kinetic energy of the nucleon. Given that, $m = 1.67 \times 10^{-27}$ kg and $h/2\pi = 1.05 \times 10^{-34}$ JS.

28. An excited atom has an average life-time of 10^{-8} s. That is during this period it emmits photon and returns to the ground state. What is the minimum uncertainty in the frequency of this photon?

29. The life-time of an excited state of nucleus is 10^{-12} s. What is the uncertainty in energy of γ-ray photon emmitted? ($h/2\pi = 1.054 \times 10^{-34}$ JS).

30. A microscope using photons is employed to locate an electron in an atom to within a distance of 0.2 Å. What is the uncertainty in the momentum of the electron located in this way? What is the uncertanity in velocity?

31. The velocity of an electron is 3×10^{4} m s^{-1} with 0.01% accuracy. Find the uncertainty in measurement of its position.

32. An electron is inside a box of length 10^{-9} m. Calculate the minimum uncertainty in its velocity. Given: $m_e = 9.1 \times 10^{-31}$ kg, $h = 6.625 \times 10^{-34}$ JS.

33. If the uncertainty in the position of a moving particle is equal to the wavelength associated with it, show that the uncertainty in its velocity is equal to $\left(\frac{1}{2\pi}\right)$ times its velocity.

34. Prove on the basis of uncertainty principle that electrons are not present in nuclei. A typical atomic nucleus is about 5×10^{-15} m in radius and the energy of a β-particle emitted from a nucleus is at most of the order of 1 MeV.

35. Wavelengths can be determined with accuracies of one part in 10^{6}. What is the uncertainty in the position of a 1 Å X-ray photon when its wavelength is simultaneously measured ?

36. If an excited state of hydrogen atom has a life-time of 2.5×10^{-14} sec, what is the minimum error with which the energy of this state can be measured? Given $h = 6.625 \times 10^{-34}$ JS.

37. An electron has a speed of 600 m/s to an accuracy of 0.01%. With what fundamental accuracy we can locate its position. Given: Mass of electron $m = 9.11 \times 10^{-31}$ kg.

38. An electron is confined to a box 10^{-10} m in length. What is the uncertainty in the measurement of its velocity?

39. If we are certain that an electron is in an atom of diameter 3×10^{-10} m, what is the uncerainty in the momentum of the electron? Given $h = 6.625 \times 10^{-34}$ J.

    Uncertainty position $\Delta x = 3 \times 10^{-10}$ m, $h = 6.625 \times 10^{-27}$ erg·s.

40. An electron is confined to move between two rigid walls separated by 10^{-9} m. Determine the de-Broglie wavelengths representing the first three allowed energy states of the electron and the corresponding energies. Given: electron mass $= 9.1 \times 10^{-31}$ kg and $h = 6.625 \times 10^{-34}$ JS.

41. Compute the energy difference between the ground state and the first excited state for an electron in a one-dimensional rigid box of length 10^{-10} m. Given: Mass of electron $= 9.1 \times 10^{-31}$ kg and $h = 6.625 \times 10^{-34}$ JS.

42. Determine the probabilities of finding a particle trapped in a box of length L in the region from 0.45 L to 0.55 L for the ground state.

43. The x-component of the velocity of an electron is known to lie somewhere between 2.35×10^2 m/s and 2.16×10^2 m/sec. What is the minimum uncertainty in its x-component? Given $h = 6.625 \times 10^{-34}$ JS, $m = 9.11 \times 10^{-31}$ kg.

44. The energy of an electron at any instant is equal to 100 eV with a possible error of 0.01 eV. Find the least error with which its position can be determined. Given mass of electron $m = 9.11 \times 10^{-31}$ kg, and charge on it is 4.8×10^{-10} e.s.u.

45. The time interval between excitation of an atom and radiation of a photon from it is 10^{-8} sec. What is the uncertainty in the energy of emitted photon? Given $h = 6.625 \times 10^{-34}$ JS.

46. A particle is in motion along a line between $x = 0$ and $x = L$ with zero potential energy. At points for which $x < 0$ and $x > L$, the potential energy is infinite. The wave function for the particle in the nth state is given by

    $\psi_n = A \sin \dfrac{n\pi x}{L}$. Determine the expression for the normalised wave function.

47. An electron is moving with a velocity 10^6 m/s. Calculate the wavelength associated with electron.

48. Calculate the de-Broglie wavelength of alpha particles accelerated through 2000 V.

49. What potential difference is required to accelerate electrons so that the de-Broglie wavelength is 0.5 Å?

50. Calculate the de-Broglie wavelength associated with an electron of energy, (*i*) 10 eV, (*ii*) 15 keV.

51. A proton is accelerated through a potential difference of 1.0 kilovolt. Calculate the de-Broglie wavelength.

52. Calculate the de-Broglie wavelength of a proton moving with a velocity $\dfrac{1}{20}$th the velocity of light.

53. Calculate the momentum and energy of electron whose de-Broglie wavelength is 2.0 Å.

54. Calculate the de-Broglie wavelength associated with a thermal neutron at 27° C.

55. Find the energy of a neutron in eV whose de-Broglie wavelength is 1 Å.

56. Calculate the de-Broglie wavelength associated with an electron in the ground state of hydrogen atom according to the Bohr's model.

57. A neutron beam of energy 0.04 eV is diffracted from the (100) plane of a crystal. If the glancing angle for the first maxima is 13°10', find the separation between the two consecutive (100) planes of the crystal. (mass of neutron = 1.67×10^{-27} kg, sin 13°10' = 0.2273).
58. Find the wavelength of a 1 kg object whose velocity is 1m/sec.
59. An electron is accelerated between two points of potentials 50 and 70 volt from rest. Calculate the kinetic energy, velocity and de-Broglie wavelength for the electron.
60. What is the frequency of an X-ray photon whose momentum is 1.1×10^{-23} kg-m/s?
61. What voltage must be applied to an electron microscope to produce electrons of wavelength 1.0 Å?
62. What is the wavelength associated with an electron whose speed is half the speed of light?
63. The electrons of an electron microscope accelerated at 60,000 volts. Calculate their de-Broglie wavelength.
64. What is the energy of gamma photon having a wavelength of 1 Å?
65. An electron microscope use 1.25 keV electron. Find its ultimate resolving power on the assumption that this is equal to wavelength of electron.
66. Compute the de-Broglie wavelength of electrons in Bohr's first orbit.
67. Calculate the de-Broglie wavelength of an electron of energy 10 eV.
68. Calculate the de-Broglie wavelength associated with electron accelerated to potential difference of 1 keV.
69. Calculate the momentum of an electron possessing de-Broglie wavelength 6.6×10^{-11} m.
70. Calculate the wavelength of thermal neutrons at 20° C.
71. Calculate the de-Broglie wavelength of an electron whose kinetic energy is 50 eV.
72. A radar signal of 0.2 μ is used to measure the distance between the signal source and a distant aeroplane which reflects the pulse. Calculate the uncertainty in the measurement of its distance. ($c = 3 \times 10^8$ m/s)
73. An electron is bound in a space of 1.0×10^{-14} m. What will be the uncertainty in its momentum?
74. Calculate the minimum uncertainty in the position of an electron of velocity 3×10^7 m/s^{-1}.
75. An electron can be seen within a distance 0.1 Å inside the atom with the help of a light photon using a microscope. What will be the uncertainty in the measurement of the momentum of electron?
76. To what accuracy the velocity of an electron can be determined if its position is known to the accuracy of 10^{-10} m?
77. A proton is inside the nucleus of radius 5×10^{-15} m. Find the minimum uncertainty in its momentum. What should be the minimum kinetic energy of proton?
78. It takes an average time 10^{-8} s in exciting an atom and emitting energy from it. Find the uncertainty in the energy of photon emitted. With what accuracy the frequency of photon emitted can be measured?
79. The uncertainty in velocity is 1 m/s. Calculate the uncertainty in the position of (*i*) a pebble of mass 20 g, (*ii*) electron, (*iii*) proton.
80. An oil drop of mass 10^{-14} g is floating on the free surface of a liquid. At any instant, the position of drop can be determined within the error of 10^{-6} m. What will be error in measurement of its velocity?
81. An electron is accelerated through a potential difference of 40 volt. The error in the measurement of its energy is 1 eV. What is the error in the measurement of its position?
82. A proton is moving in a one-dimensional box of length 10^{-9} m. Calculate: (*i*) the uncertainty in the position of proton, (*ii*) the uncertainty in the velocity of proton.
83. Compare the uncertainties in the velocities of electrons and protons present in a box of dimension 10 Å.
84. What is the uncertainty in the momentum of an electron which is restricted to a region of linear dimension equal to 1 Å ?

85. If the uncertainty in position of an electron is 4×10^{-10} m, calculate the uncertainty in its momentum.
86. An electron has a speed of 4×10^5 m/s accurate to 0.01%. With what accuracy can we locate the position of the electron?
87. The time period of radar vibration is 0.25 μ-s. What is the order of uncertainty in the energy of the photon?
88. The life time of an excited state of a nucleus is 10^{-12} s. What is the uncertainty in energy of γ-ray photon emitted?
89. The accuracy with which the frequency of the atomic radiation can be determined by an amount $\Delta\nu \sim 1.6 \times 10^7$ CPS. What percentage of the frequency of a 5000 Å photon is this?
90. Calculate the time required for the atomic system to retain the excitation energy for a line of wavelength 6000 Å and width 10^{-4} Å.
91. Show that if a component of angular momentum of an electron in a hydrogen atom in known to be $2\hbar$ within 5% error, its angular orbital position in a plane perpendicular to that component can not be specified at all.
92. A beam of silver atoms for Stern-Gerlach experiment is produced by collimating atoms that vaporize from a silver held in a furnace at 1200° C. If the beam travels 1m, use the uncertainty relation to find the order of magnitude of the smallest spot that can be obtained at the detector. Mass of hydrogen atom = 1.66×10^{-27} kg.
93. Calculate the uncertainty in the momentum of a proton which is confined to a nucleus of radius 10^{-5} m. Estimate kinetic energy of the proton inside the nucleus and the strength of nuclear interaction. What would be the kinetic energy of an electron if it had to be confined with a smaller nucleus? Explain, why electrons are not expected to be a basic constituent of the nucleus.
94. An electron in the hydrogen atom is confined to region of radius 5×10^{-15} m. Use uncertainty principle to calculate the momentum of the electron.
95. A proton is confined to one dimensional box of length 10^{-9} m. Calculate the uncertainty in its velocity.
96. Bullets each of mass 1 g take 1 s to reach their target. Considering the bullets as mass points, find the order of magnitude of spread of successive shots at the target under the best possible conditions.
97. The energy of an electron accelerated by a potential difference of 40 volts can be determined with a probable error up to 10 eV. Calculate the least error with which its position can be determined.
98. An electron has a speed of 600 m/s with an accuracy 0.005%. Calculate the certainty with which we can locate the position of the electron.
99. An electron is confined to a box of length 1.1×10^{-8} m. Calculate the minimum uncertainty in its velocity.
100. Calculate the kinetic energy of an electron if the wavelength of electron equals the yellow line of sodium.
101. An electron has a speed of 1.05×10^4 m/s within the accuracy of 0.01%. Calculate the uncertainty in the position of the electron.
102. What is the lowest energy that a neutron can have if confined to move along the edge of an impenetrable box of length 10^{-14} m?
103. Electrons with energy of 1 eV are incident on a barrier 5 eV high and 5 Å wide. Find the probability for these electrons to penetrate the barrier. How is it affected if the barrier is doubled in width?
104. A free particle is moving along a straight line between $x = 0$ and $x = 1$. The wave function for the motion of particle in the nth state is $\psi = A \sin\left(\frac{n\pi x}{l}\right)$. Obtain the normalised wave function.

105. A particle of mass 10.0 g is moving in a one-dimensional box of side 1 cm. Find the difference in its lowest two energy states. Is it possible to measure this energy difference?

106. An electron is moving in a one-dimensional box of infinite height and width 1 Å. Find the difference in its two lowest energy states.

107. The size of a nucleus is 10^{-14} m. Find out the lowest energy of the neutron inside the nucleus.

108. An electron lies in between the two walls of infinite height. The distance between the walls is 2.5×10^{-10} m. Calculate the three possible quantum energies of the electron.

109. A proton is moving in a one-dimensional box of length 10^{-9} m. Find the first three possible quantum energies of the proton.

110. An electron is bound in a one dimensional potential box which has a width of 2.5×10^{-10} m. Assuming the height of the box to be infinite, calculate the lowest two permitted energy values of the electron.

111. The radius of the first Bohr orbit in hydrogen atoms 0.53 Å. Find the velocity of the electron in that orbit, using de-Broglie theory.

112. Find the lowest energy of neutron confined to a box 10^{-14} m across.

113. Show that the lowest ground state energy level of a particle in one dimensional potential box with rigid walls is in agreement with the uncertainty principle.

MULTIPLE CHOICE QUESTIONS

1. Which of the following is the correct statement
   (*a*) Only charged particles in motion are accompanied by matter waves.
   (*b*) Any particle in motion, whether charged or uncharged, is accompanied by matter waves.
   (*c*) No particle whether at rest or in motion is ever accompanied by matter waves.
   (*d*) Only sub-atomic particles in motion are accompanied by matter waves.
2. Matter waves
   (*a*) are electromagnetic waves (*b*) are transverse waves
   (*c*) are longitudinal waves (*d*) exhibit diffraction
3. The de-Broglie wavelength of a molecule of thermal energy kT (k-Boltzmann constant and T-absolute temperature) is
   (*a*) $\lambda = h/2$ mkT (*b*) $h/\sqrt{(2 \text{ mkT})}$ (*c*) $h\sqrt{(2 \text{ mkT})}$ (*d*) $h/4 \text{ m}^2\text{k}^2\text{T}^2$
4. If the momentum of a particle is increased to four times, then the de-Broglie wavelength will
   (*a*) become twice (*b*) become four times (*c*) become half (*d*) become one-fourth
5. According to wave mechanics, a material particle is associated with
   (*a*) a single wave (*b*) a wave packet
   (*c*) two progressive waves travelling in the same direction (*d*) a ripple
6. The wavelength of the matter waves is independent of
   (*a*) mass (*b*) velocity (*c*) momentum (*d*) charge
7. An electron and a proton are accelerated through the same potential difference. The ratio of their de-Broglie's wavelengths (λ_m/λ_p) is
   (*a*) 1 (*b*) m_e/m_p (*c*) m_p/m_e (*d*) $\sqrt{(m_p/m_e)}$

8. Of the following, moving with the same velocity, the one which has largest wavelength λ is
   (*a*) an electron (*b*) a proton
   (*c*) an α-particle (*d*) all have same de-Broglie wavelength
9. Davisson and Germer experiment relates to
   (*a*) interference (*b*) polarisation (*c*) electron diffraction (*d*) phosphorescence
10. An electron, proton, and a car, all have same wavelength. The one possessing highest velocity is
    (*a*) electron (*b*) proton
    (*c*) car (*d*) all have same velocity
11. A moving particle is associated with a wave packet or group of waves. The group velocity is equal to
    (*a*) velocity of light (*b*) velocity of sound
    (*c*) velocity of particle (*d*) 1/particle velocity
12. If the wavelength associated with a proton and a photon are same then which of the following quantities will be the same for both of them
    (*a*) mass (*b*) momentum
    (*c*) velocity (*d*) kinetic energy
13. A photon and an electron have the same wavelength then
    (*a*) photon has greater momentum (*b*) electron has greater momentum
    (*c*) both have the same momentum (*d*) none of the above
14. The function of crystal in Davisson and Germer experiment is
    (*a*) to reflect the electron beam (*b*) to refract the electron beam
    (*c*) to diffract the electron beam (*d*) to focus the electron beam
15. Uncertainty principle states that the error in measurement is due to
    (*a*) dual nature of particles (*b*) due to small size of particles
    (*c*) due to large size of particles (*d*) due to error in measuring instrument
16. The uncertainty principle is applicable to
    (*a*) large systems only (*b*) small systems only
    (*c*) sub-atomic particles only (*d*) both sub-atomic and large systems
17. When electrons are acclerated through the potential difference of V volts, the de-Broglie's wavelength (λ) associated is given by
    (*a*) $\sqrt{(150/V)}$ Å (*b*) $\sqrt{(150/V)}$ m (*c*) $150/\sqrt{V}$ Å (*d*) $\sqrt{150/V}$ Å
18. The uncertainty in the location of a particle is equal to its de-Broglie wavelength, then uncertainty in its velocity will be
    (*a*) its velocity (*b*) half of its velocity
    (*c*) twice its velocity (*d*) four times its velocity
19. The de-Broglie's wavelength of a particle is
    (*a*) proportional to its mass (*b*) proportional to its energy
    (*c*) proportional to its momentum (*d*) inversely proportional to its momentum

20. Dual nature of matter was predicated by
(*a*) Davisson (*b*) Germer (*c*) Thomson (*d*) de-Broglie
21. If the momentum of particle is increased to four times, then the de-Broglie wavelength will
(*a*) become one-fourth (*b*) become four times (*c*) become half (*d*) become twice
22. The wavelength of the matter waves is independent of
(*a*) mass (*b*) charge (*c*) momentum (*d*) velocity
23. Of the following, moving with the same velocity, the one which has largest wavelength λ is
(*a*) a proton (*b*) an electron
(*c*) an α-particle (*d*) all have same de Broglie wavelength
24. An electron and a proton are accelerated through the same potential difference. The ration of their de-Broglie's wavelengths (λ_m/λ_p) is
(*a*) 1 (*b*) m_e/m_p (*c*) $\sqrt{(m_p/m_e)}$ (*d*) m_p/m_e
25. A photon and an electron have the same wavelength then
(*a*) both have the same momentum (*b*) electron has greater momentum
(*c*) photon has greater momentum (*d*) none of the above
26. Davisson and Germer experiment relates to
(*a*) electron diffraction (*b*) polarisation
(*c*) interference (*d*) phosphorescence
27. Uncertainty principle states that the error in measurement is due to
(*a*) large size of particles (*b*) due to small size of particles
(*c*) dual nature of particles (*d*) due to error in measuring instrument
28. The rest mass of a photon is
(*a*) 1.67×10^{-31} kg (*b*) zero (*c*) one a.m.u. (*d*) infinity
29. An alpha particle and a proton are accelerated through same voltage. The ratio of their de-Broglie wavelength will be
(*a*) 1 : 2 (*b*) 2 : 1 (*c*) $1:\sqrt{2}$ (*d*) $1:2\sqrt{2}$
30. Which of the following relation is correct ?
(*a*) $\Delta J \cdot \Delta x = \hbar$ (*b*) $\Delta J \cdot \Delta \theta = \hbar$ (*c*) $\Delta J \cdot \Delta \theta = \hbar$ (*d*) $\Delta \theta \cdot \Delta x = \hbar$
31. The de-Broglie's wavelength of a particle is
(*a*) inversely proportional to its momentum (*b*) proportional to its energy
(*c*) proportional to its momentum (*d*) proportional to its mass
32. According to Heisenberg's uncertainty principle
(*a*) $\Delta x \cdot \Delta p < \hbar$ (*b*) $\Delta x \cdot \Delta p \approx \hbar$ (*c*) $\Delta x \cdot \Delta v \approx \hbar$ (*d*) $\Delta E \cdot \Delta \phi \approx \hbar$
33. The resolving of a microscope is 5×10^{-3}m. It is used to observe a moving electron. The maximum uncertainty in the position of electron will be
(*a*) 6.6×10^{-34} m (*b*) 1.0509×10^{-34} m (*c*) 5×10^{-3} m (*d*) 1.32×10^{-24} m

34. An electron is accelerated through a potential difference of 150 volt. The maximum uncertainty in the position of electron will be

(*a*) 1 m (*b*) 1.5×10^{-34} m (*c*) 6.6×10^{-34} m (*d*) 6.6×10^{-24} m

35. Heisenberg's uncertainty principle

(*a*) establishes the Bohr's orbital concept

(*b*) establishes the existence of an electron inside the nucleus

(*c*) is not observable for the macroscopic objects

(*d*) does not agree with the de-Broglie hypothesis

36. The phase velocity of a particle moving with a velocity v is

(*a*) v (*b*) c (*c*) $\frac{c}{v}$ (*d*) $\frac{c^2}{v}$

37. The correct relation is

(*a*) $v_g = v_p - \lambda \frac{dv_p}{d\lambda}$ (*b*) $v_p = v_g - \lambda \frac{dv_g}{d\lambda}$ (*c*) $v_g = v_p + \lambda \frac{dv_p}{d\lambda}$ (*d*) $v_p = v_g - \frac{1}{\lambda} \frac{dv_p}{d\lambda}$

38. The property of γ-ray due to which it is used in a γ-ray microscope is

(*a*) maximum wavelength (*b*) maximum intensity

(*c*) minimum intensity (*d*) minimum wavelength

39. If E_1, E_2 and E_3 are the three respective kinetic energies of an electron, an alpha particle and a proton each having the same de-Broglie wavelength, then

(*a*) $E_1 > E_3 > E_2$ (*b*) $E_1 > E_2 > E_3$ (*c*) $E_1 = E_2 = E_3$ (*d*) $E_2 > E_3 > E_1$

40. The de-Broglie wavelength of a photon will be equal to that of an electron provided

(*a*) the kinetic energy of proton is 43 times that of electron

(*b*) the kinetic energy of proton is 1847 times that of electron

(*c*) the kinetic energy of electron is 43 times that of the proton

(*d*) the kinetic energy of electron is 1847 times that of proton

41. The relativistic expression for the wavelength of electron moving with very high speed of the order of velocity of light is

(*a*) $\lambda = \frac{h}{mc}$ (*b*) $\lambda = \frac{h}{mv}$

(*c*) $\lambda = \frac{h}{m_0 v}\sqrt{1 - v^2/c^2}$ (*d*) $\lambda = \frac{h}{m_0 \left\{\sqrt{\left[\left(1 - v^2\right)/c^2\right] - v}\right\}}$

42. Matter waves were first experimentally observed by

(*a*) de-Broglie (*b*) Frank-Hertz

(*c*) Davisson and Germer (*d*) Stern-Gerlach

43. If a paraticle with energy E is incident on a thin energy barrier of height greater than E, there is a finite probability of the particle penetrating the barrier. This phenomenon is called

(*a*) Photoelectric effect (*b*) Compton effect (*c*) Tunnel effect (*d*) Raman effect

44. The lowest quantised energy of a particle of mass m in a one-dimensional box of length L is given by

(*a*) zero (*b*) $h^2/8mL^2$ (*c*) $2h^2/8mL^2$ (*d*) $4h^2/8mL^2$

45. The tunnelling effect is a

(*a*) classical phenomenon

(*b*) quantum mechanical phenomenon

(*c*) classical as well as quantum mechanical phenomena

(*d*) none of the above

46. Wave function associated with a material particle is

(*a*) only single valued (*b*) only finite (*c*) only continuous (*d*) all the above

47. For a free particle ($V = 0$) according to quantum mechanics

(*a*) energy levels are continuous

(*b*) energy is always zero

(*c*) energy levels are discrete and equispaced

(*d*) energy levels are discrete and not equispaced

48. A particle of mass m enclosed in a box of length l will have discrete energy levels E_n as

(*a*) $\dfrac{n^2h^2}{8ml^2}$ (*b*) $\dfrac{nh^2}{2ml^2}$ (*c*) $\dfrac{n^2h^2}{8ml}$ (*d*) $\dfrac{n^2h^2}{2ml}$

49. Quantity $|\psi|^2$ represents

(*a*) Probability density (*b*) intensity of wave (*c*) energy density (*d*) charge density

50. A particle trapped in a box of length L has the average value of momentum $\langle p \rangle$ as

(*a*) 0 (*b*) $\dfrac{h}{L}$ (*c*) $\dfrac{h}{2L}$ (*d*) 1

51. Energy of a particle in one dimensional closed box of length l is

(*a*) directly proportional to l (*b*) inversely proportional to l

(*c*) directly proportional to l (*d*) inversely proportional to l^2

52. The Schrodinger's wave equation for a free particle is

(*a*) $\nabla^2\psi + \dfrac{2mE}{\hbar^2}\psi = 0$ (*b*) $\left(\dfrac{-\hbar^2}{2m}\nabla^2 + V\right)\psi = i\hbar\dfrac{\partial\psi}{\partial t}$

(*c*) $\nabla^2\psi + \dfrac{2m}{\hbar^2}(E - V)\psi = 0$ (*d*) $\dfrac{\partial^2\psi}{\partial x^2} + \dfrac{\partial^2\psi}{\partial y^2} + \dfrac{\partial^2\psi}{\partial z^2} = \dfrac{2m}{\hbar^2}(E - V)\dfrac{\partial\psi}{\partial t}$

53. The wave function associated with a material particle is

(*a*) only finite (*b*) only continuous

(*c*) only single valued (*d*) finite, continuous and single valued

54. The solution of Schrodinger's equation $\left(\frac{-\hbar^2}{2m}\nabla^2 + V\right)\psi = i\hbar\frac{\partial\psi}{\partial t}$ can be written as

(a) $\psi(r, t) = f(x, y, z)\,\phi(t)$ (b) $\psi(r, t) = \psi_0 e^{-ikx}$

(c) $\psi(r, t) = a \sin \omega t$ (d) $\psi(r, t) = \psi_0 \sin kx$

55. The time independent Schrodinger's equation is

(a) $\left(\frac{-\hbar^2}{2m}\nabla^2 + V\right)\psi = E\psi$ (b) $\frac{-\hbar^2\nabla^2\psi}{2m} = E\psi$

(c) $\left(\frac{-\hbar^2}{2m}\nabla^2 + V\right)\psi = i\hbar\frac{\partial\psi}{\partial t}$ (d) $\left(\frac{-\hbar^2}{2m}\nabla^2 + (E - V)\psi\right) = 0$

56. According to the quantum mechanics for the particle moving in a box
(a) the energy levels are discrete and equispaced
(b) the energy levels are continuous
(c) the energy lelves are discrete but not equispaced
(d) the energy is always zero

57. If a particle of mass m is moving in a linear potential region $V = 0$ when $0 < x < l$ and $V = 0$ when $x \le 0$ and $x \ge l$, then the energy eigen values of the particle are

(a) $E_n = \frac{n^2h^2}{8ml^2}$ (b) $E_n = \left(n + \frac{1}{2}\right)h\nu$ (c) $E_n = \frac{2m}{c^2}a^2$ (d) $E_n = \frac{nh}{8ma}$

EXERCISE

1. Explain de-Broglie's concept of matter wave. Find the de Broglie wavelength of an electron of energy V electron volts.
2. Find the de-Broglie wavelength of an electron of energy 5 eV.
3. Give de-Broglie theory of matter waves. Derive an expression for the de Broglie wavelength of matter waves. Show that electrons accelerated through a potential difference of V volt have a wavelength $\frac{12.27\text{ Å}}{\sqrt{V}}$ associated with them.
4. A particle of charge e and mass m is accelerated from rest through a potential difference V. Find the, de-Broglie wavelength.
5. What are de-Broglie matter waves?
6. Discuss the dual nature of matter and waves.
7. Describe Davisson-Germer experiment to demonstrate the wave nature of particle.
8. What are de-Broglie waves? How do they help in the interpretation of Bohr's quantisation rule?
9. Deduce an expression for de-Broglie wavelength of Helium atom having energy E at temperature T° K.
10. Distinguish between the phase velocity V_p and the group velocity V_g of a wave particle. Prove that $V_pV_g = c^2$.

11. Show that the phase velocity of de-Broglie wave is greater than the velocity of light but the group velocity is equal to the velocity of particle with which the waves are associated.
12. What is de-Broglie hypothesis? Show that the wavelength λ associated with an electron of mass m and kinetic energy E is given by

$$\lambda = \frac{h}{\sqrt{2mE}}$$

13. Establish Bohr's quantisation condition on the basis of de-Broglie concept of matter waves.
14. Obtain the possible energy states of a particle in a box on the basis of de-Broglie hypothesis.
15. Show that the de-Broglie wavelength for a material particle of rest mass m_0 and charge q, accelerated from rest through a potential difference of V volts relativistically is given by

$$\lambda = \frac{h}{\sqrt{2m_0qV}} \times \left(1 + \frac{qV}{2m_0c^2}\right)^{-1/2}$$

16. Name the experiments which show wave nature of material particles.
17. Why is wave nature not observed in our daily experience?
18. What is the quantum picture of a material particle?
19. What is the difference between phase velocity and group velocity?
20. In light waves the electromagnetic field varies in space and time; in sound waves the pressure varies in space and time; whose variation constitute de-Broglie waves.
21. Give an informative account of dual nature of matter and radiation.
22. Explain the wave-particle duality of matter and obtain an expression for the de-Broglie wavelength. On the basis of de-Broglie matter wave obtain the Bohr's quantum condition $J = \frac{nh}{2\pi}$.
23. What is meant by the de-Broglie waves? Show that the wavelength associated with an electron of mass m and kinetic energy K is given as $\lambda = \frac{h}{\sqrt{2mK}}$
24. Establish the relation $\lambda = \frac{h}{p}$, starting from the principle of relativity.
25. What is meant by a wave packet? Differentiate between the phase velocity and the group velocity. Show that the phase velocity of a non-relativistic free particle is half the group velocity.
26. Define phase velocity and group velocity and derive their expressions. Show that in a non-dispersive medium, the group velocity and the phase velocity are equal.
27. 'A moving particle is equivalent to a wave packet'. Explain.
28. What is wave packet? Explain its physical meaning and state how is wave packet formed?
29. Explain the significance of wave packet in representation of a matter-wave.
30. On the basis of de-Broglie principle show that the velocity of a moving material particle is equivalent to the velocity of wave packet.
31. State and explain Heisenberg's uncertainty principle? Using this principle show that electron cannot reside in an atomic nucleus.
32. What is uncertainty principle? Apply it to prove the non-existence of the electron in the nucleus.

33. State and discuss Heisenberg's uncertainty principle. Give two examples to illustrate the principle.
34. How does the concept of Bohr orbits violate the principle of uncertainty? Why the principle does not reveal itself if we work with an object of 1 g?
35. Comment on the statement that "Heisenberg's principle of uncertainty is the outcome of the description of a particle?
36. What is complementary principle?
37. What is uncertainty principle? Give a mathematical proof of the position momentum uncertainty in quantum mechanics.
38. Derive Herisenberg's uncertainty principle $\Delta x \, \Delta p \geq \frac{\hbar}{2}$ and illustrate it by an example.
39. State the uncertainty principle. Prove this principle in the case of one dimensional linear position and momentum. Describe the experiment based on γ-ray microscope experiment which supports the principle.
40. Write an essay on Heisenberg's uncertainty principle incorporating its important applications.
41. Find the uncertainty in frequency of a photon emitted from an excited atom on the basis of Heisenberg's uncertainty principle.
42. Write short notes on
    (*i*) Wave-particle duality
    (*ii*) de-Broglie wave hypothesis
    (*iii*) Heisenberg's uncertainty principle
    (*iv*) Phase velocity and group velocity
43. Differentiate between
    (*i*) the phase velocity and group velocity
    (*ii*) the classical and quantum systems
44. What is Planck's quantum theory?
45. What is meant by the effective mass of a photon?
46. Name two experiments which establish the validy of de-Broglie relation.
47. Which velocity is equal to the particle velocity phase velocity or group velocity?
48. Write down the expression for the group velocity in terms of angular frequency and propagation constant.
49. What is the unit and dimension of propagation constant?
50. Write down the relationship between the wave number and wavelength.
51. State Heisenberg's uncertainty principle for the conjugate quantities position-momentum, energy-time, angular momentum-angular displacement.
52. Comment on the statement 'Heisenberg's uncertainty principle is valid for all kinds of particles.
53. Write down the expression for the momentum of photon.
54. Derive Schrodinger's time independent wave equation. What is the physical significance of stultification ψ used in this equation? What condition must it fulfil?
55. Derive time-dependent, time-independent Schrodinger equations for a non-relativistic free particle. Give the stationary state solution of Schrodinger wave equation.
56. Find the expression for the energy state of a particle in one dimensional box.
57. Solve the Schrodinger equation for a particle enclosed in a one-dimensional rigid box of side L. Obtain its eigen values. Draw a graph of its first three eigen functions. Discuss the probability of finding the particle at different points in the box when it is in different states.
58. Find the eigen functions of a free particle using cartesian coordinates and show that the energy values are continuous.
59. Discuss the situation which led to the development of wave mechanics?

60. Show that the minimum kinetic energy of the particle confined to one-dimensional box cannot be zero. Justify the finite minimum energy on the basis of uncertainty principle. Explain, why are we not able to see energy quantisation in daily experience?
61. Solve the Schrodinger equation for a particle enclosed in a one dimentional rigid box and obtain its eigen values.
62. Can a photon and an electron of the same momentum have the same wavelength? Compare their wavelength if the two have the same energy.
63. What is Bohr's probability interpretation of wave-function?
64. Explain the concept of probability current density.
65. Write stationary state solution of Schrodinger's equation.
66. What is an operator? Write the operators associated with energy and momentum.
67. Why do you mean by expectation values of dynamical quantities?
68. A free particle can have any energy. Does this statement hold for the free particle in a box?
69. Starting from the Maxwell's electromagnetic wave equation, derive the equation of motion of matter waves. What is physical interpretation of wave function in this concept?
70. Discuss the interpretation of matter wave function as given by Schrodinger and Max-Born. How do the two concepts differ ? Which concept is more correct?
71. Give the probabilistic interpretation of the wave function ψ.
72. Describe the general properties of wave function.
73. Solve the Schrodinger's wave equation by the method of separation of variables. Obtain eigen function and eigen value and hence explain them.
74. What is meant by normalisation of a wave function? Obtain the condition of normalisation of a wave function.
75. Write down the postulates of quantum mechanics and explain them.
76. Write short notes on
    (*i*) Probability of wave function
    (*ii*) Eigen function and eigen value
    (*iii*) Normalisation of a wave function
77. What is meant by the free particle and the free particle in a box? Write down the one-dimensional Schrodinger's equation for both and explain the difference in the eigen values of the two.
78. Write down the Schrodinger's equations.
79. If $\psi(x, y, z, t)$ is the wave function of a particle, what does the quantity $\psi\psi^*$or $|\psi|^2$ represent?
80. Write down the Schrodinger's wave equatioin for a free particle.
81. In the following wave function $\psi(x)$, if $\psi(x)$ is a real function, what is the physical meaning of the quantity p_0?
82. Write down the condition of normalisation of the wave function $\psi(r, t)$.
83. Explain the statement 'the energy levels of a free particle are continuous'.
84. 'The wave function of a free particle can not be normalised.' Is this statement true or false?
85. A particle of mass m is free to move between the limits from $x = 0$ to $x = L$. Write the normalised eigen function and the energy eigen values of the particle.
86. What is the minimum energy of the particle in a box?

2 Electrical Conductivity in Metals

2.1 INTRODUCTION

The electrical conductivity of metal is mainly depends upon the energy level of an arbitrary electron, in general the valence electron. The energy level of sodium (Na) element is simply represented by energy level diagram by single line, that the neighbouring atoms have no influence on each other.

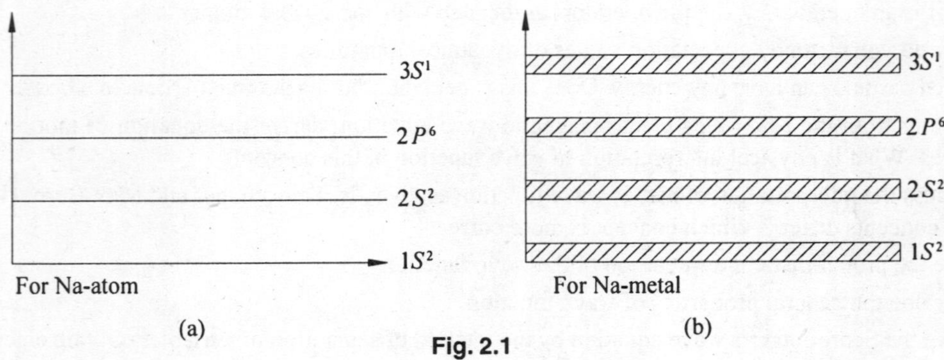

Fig. 2.1

In solid sodium Na the atoms are in close proximity, they influence on each other and raising the energy state of each electron, then each energy level will be split up into many number of lines in the solid. Each energy state will be differ from the neighbouring energy state and are slightly separated by 10^{-20} eV, the energy levels are spread between the upper limit and lower limit, such a set of energy level that are closely packed are called energy bands. The width of each energy band depends upon the proximity of atoms and the nature of an element.

There are number of atomic energy level, each atomic energy level will have corresponding energy band are called forbidden band. Each band being separated by a gap called forbidden energy gap, no electrons in the assembly can have energies laying in these gaps. The width of energy gap increases as we move to higher energy bands, there are higher energy bands above the upper most completely filled energy bands, normally these bands are occupied by the electrons. "The highest band occupied by the electrons under normal condition are called valence band". The next allowed higher energy band is called conduction band which is always above the valence band.

In sodium (Na) the electronic configuration is $1S^2\,2S^2\,2P^6\,3S^1$, if $1S^2$, $2S^2$ and $2P^6$ are filled up then $3S^1$ band is half filled, the bands above that of $3S^1$ are completely empty and are called conduction band. If the valence electron in $3S^1$ by chance gain some energy it rises to the conduction band and make the material as a conductor.

The forbidden gap represents the energy required by the valence electron get itself shifted to the conduction band, so the electrical conductivity of a solid is decided by the energy gap, valence band and a conduction band.

2.1.1 Conductors

The material in which the over laping of energy bands or the energy bands half filled is a good conductor

In the case of Sodium (Na) (monovalent) valence electron can easily occupied the place in conduction band. Because of its half filled energy band. When electric field is applied across the valence electrons aquires energy and rises to the conduction band. At this stage the valence electron becomes free electron and takes part in electrical conductivity.

In the case of Calcium (Ca) (di-valent) the valence band will be overlaping on the conduction band. Making the conduction band as half filled, then the electron behave in a similar way as the monovalent metal.

2.1.2 Insulators

In materials like diamond, glass, plastic, etc. the valence band is completely filled whereas the conduction band is 6 to 10 eV above the valence band. At room temperature the thermal energy is about 0.025 eV, so there is no possibility of an electron rises to the conduction band, *i.e.* the electrical conductivity becomes zero at room temperature. Such a solid is called as an insulator or non-conductor.

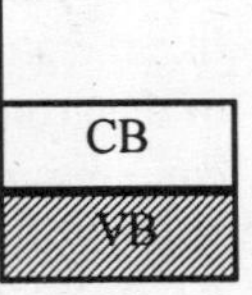

Fig. 2.2

The insulation strength depends upon the energy gap, if glass rod is heated to a moltan (liquid) state, it becomes a good conductor. Therefore, the chance of an insulator becoming a conductor depends upon the thermal energy state of material.

CB

> 0.6 eV

VB

Fig. 2.3

2.1.3 Semiconductor

In certain materials like Germanium Ge and Silicon Si the separation between valence band and conduction band is 0.7 eV and 1.1 eV respectively. As the temperature increases the electron gain energy and likely to jump across the energy gap and reach the conduction band. Even at room temperature the conduction band will have few free electron making the material a conductor the conductivity of semiconductor increases with increase of temperature. This type of conductivity is called intrensic conductivity. At 0° K the semiconductor will act as insulator.

2.2 ELECTRICAL CONDUCTIVITY IN METALS

The basic assumption of classical model is that a metal consists of positive ions, whose valence electrons are free to move between the ions, as if they constituted an electron gas.

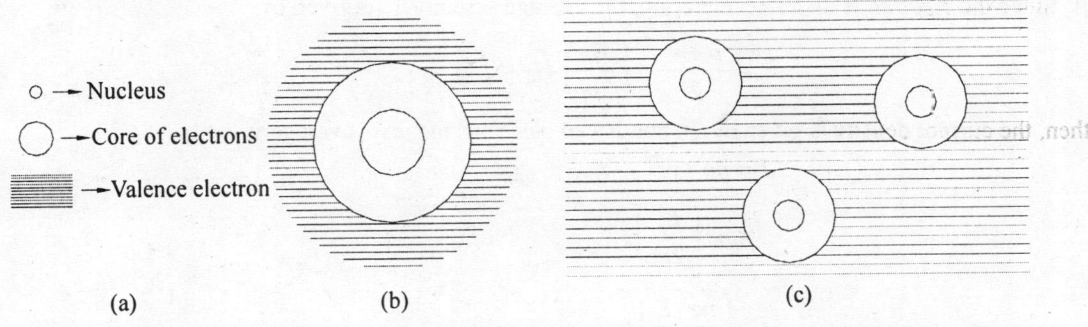

Fig. 2.4

2.2.1 Classical Free Electron Theory

The free electron model explains a number of important physical properties of metals.

All metals are good conductors of heat and electricity and obey Ohm's law because of valence electrons. These valence electrons are responsible for physical properties of metal such as electrical conductivity, thermal conductivity thermoelectricity, thermionic emission and photoelectric effects, etc.

Consider a conductor as a regular array of atoms plus a collection of free electrons. The free electrons although band to their respective atoms. When the atoms are not part of a solid, gain mobility when the free atoms condense into a solid. That is, when a large number of atoms combine to form a metal, the boundaries of the neighbouring atoms slightly overlap on each other.

Due to this overlapping, the valence electrons of all the atoms get attached from their respective atoms. Thus the valence electrons will be free to move within the metal lattice. These electrons are called free electrons and they move freely in random directions through the conductor with an average speed of the order 10^6 m/s like the atoms or the molecule in a perfect gas. So, the kinetic theory of gases can be extended to the moment of electron and it is heated as electron gas. Since the free electrons are responsible for electrical and thermal conductions in a solid free electrons are also called as conduction electrons.

The loss of the electrons make the atoms to loose their electrical neutrality and they become positive ions fixed about in their mean positions. Thus, according to Drude-Lorentz theory all metals contain free electrons which are free to move through the positive ionic core of the metals. The metal is then pictured to be held together by electrostatic forces of attraction between the positively charged ions and negatively charged electron gas.

2.3 FREE ELECTRON THEORY OF METALS

2.3.1 Drude's Classical Theory

Drude's proposed that the metal is considered as three dimensional arrangement of ions. Among the ions the free electrons move about random like the atoms or the molecule in perfect gas. So, the kinetic theory of gases can be extended to the moment of electron and it is treated as electron gas.

When an electric field is applied, the free electron in metal move along the direction of field to form electric current. If E is the strength of the electric field acting on the electric charge e.

Therefore, electrostatic force act on the electron $F = E_e$

Acceleration experiences by the electron $a = \dfrac{E_e}{m}$ $\quad\because\ F = ma$

Velocity of the electron $V = \left(\dfrac{E_e}{m}\right) t$ $\quad\because\ a = \dfrac{v}{t}$

where t = mean free time or relaxation time

Since the electron is under acceleration, the average velocity $\bar{v}$ is given by,

$$\bar{v} = \frac{o + v}{2} = \left(\frac{E_e}{2m}\right) t$$

then, the current density is given by = number of electrons moves × average velocity

$$J = ne \times \bar{v}$$

$$= ne\left(\frac{E_e}{2m}\right) t$$

$$J = \left(\frac{ne^2 E}{2m}\right) t$$

$$\text{Average velocity} = \frac{\text{mean free path}}{\text{mean free time}}$$

$$\bar{v} = \frac{\lambda}{t}$$

$$t = \frac{\lambda}{v}$$

$$\because\ J = \frac{ne^2 E}{2m}\frac{\lambda}{v}$$

Multiply and divide by $\bar{v}$ on right side, then, we get

$$J = \frac{ne^2 E\lambda}{2m} \frac{\bar{v}}{\bar{v}^2}$$

But we know that according to law of equipartitional energy,

$$\text{Kinetic energy} = \frac{3}{2} KT \qquad \text{where} \quad K = \text{Boltzman constant}$$

i.e. $\frac{1}{2} mv^2 = \frac{3}{2} KT \qquad T = \text{Absolute temperature}$

$$mv^2 = 3KT$$

On substituting this value in the above equation, we get

$$J = \frac{ne^2 E\lambda\bar{v}}{2 \times 3KT}$$

$$J = \frac{ne^2 E\lambda}{6KT}$$

The current strength i is given by

$$i = J \times A$$

$$= \frac{ne^2 E\lambda\bar{v}}{6KT} \times A$$

by diffenition, we also know that,

Electric intensity = Potential gradient

$$E = \frac{v}{x}$$

$$i = \frac{ne^2\lambda\bar{v}}{6KT} \cdot \frac{v}{x} A$$

$$= \frac{ne^2\lambda\bar{v}A}{6KTx} \cdot v$$

$$i = \sigma v$$

where $\sigma = \frac{ne^2\lambda\bar{v}A}{6KTx}$ is a constant and is called as electrical conductivity of the metal but $\sigma = \frac{1}{R}$

where R = Resistance of the metal

i.e. $i = \frac{v}{R}$ This is Ohm's law

This result is in good agreement with the experimental value. So, Ohm's law is established from free electron theory of metals.

2.4 RELAXATION TIME (T_r)

In a metal, due to the randomness in the direction of motion of the conduction electrons. The probability of finding an electron moving in any given direction is equal to that of the finding of some other electron moving in exactly the opposite direction in the absence of an electric field. As a result the average velocity of the electrons in any given direction becomes zero.

i.e. $v_{av} = 0$, in the absence of the field

However, when the metal is subjected to an external electric field, there will be a net positive value for the average velocity of the conduction electron say v_{av} in the direction of the applied field due to the drift velocity.

i.e. $V_{av} = V_{av}$ in the presence of the field.

If the field is turned off suddenly, the average velocity V_{av} reduces exponentially to zero from the value V_{av}.

The decay process is represented by the equation $V_{av} = V_{av}' \, e^{-t/T_r}$

Where t is the time counted from the instant when the field is turned off and T_r is a constant and is called the relaxation time.

if
$$t = T_r$$

Then,
$$V_{av} = V'_{av} \, e^{-1}$$

$$V_{av} = \frac{V_{av}'}{e}$$

As a consequence of the sudden disappearence of an electric field across a metal, the average velocity of conduction electrons decays exponentially to zero. The time required in this process for the average velocity to reduce to $1/e$ times of its value when the field is turned off is known as Relaxation time.

2.5 MEAN FREE PATH (λ)

According to kinetic theory of gases. It is the distance travelled by a gas molecule in between two successive collisions.

In classical theory of free electron. It is the average distance travelled by the conduction electrons between successive collisions with the lattice ions.

2.6 MEAN COLLISION TIME (*T*)

The average time that elapses between two consecutive collisions of an electron with the lattice points is called collision time.

If v is the total velocity of the electrons which is the velocity due to the combined effect of thermal and drift velocities, then the mean collision time is given by

$$T = \frac{\lambda}{v}$$

where, λ is the mean free path.

If V_d is the drift velocity, then $V_d << V_{th}$

$$\therefore \quad V = V_{th}$$

In the case of metals, it can be shown that the relaxation time T_r always refer to a single total velocity called as Fermi velocity.

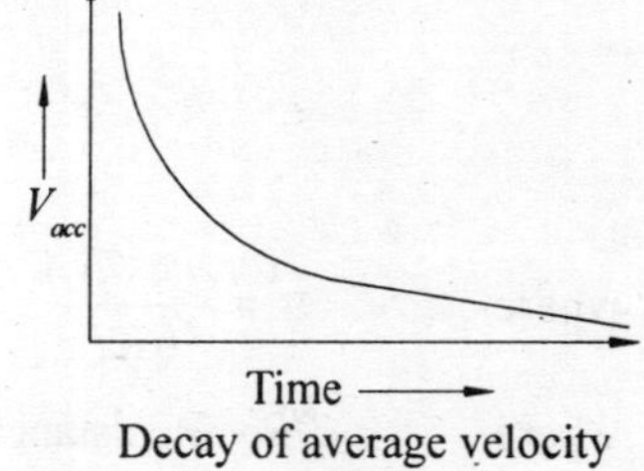

Decay of average velocity

Fig. 2.5

2.7 RELATION BETWEEN T_r AND T

Consider a conduction electron travelling inside a metal in the direction AB with a velocity V as shown in the Fig. 2.6. After scattered by the lattice point B the electron travels along BC with the same velocity making an angle θ with the original direction θ is called the scattering angle.

The value of θ is different for each case of collision between an electron and the lattice. If the average value of cos θ is taken over for a very large number of such collisions and the value is represented by $< \cos\theta >$, then the relaxation time T_r and the mean collision time T is given by,

$$T_r = \frac{T}{1 - \langle \cos\theta \rangle}$$

If the probability of scattering is same in all directions, then the scattering is called isotropic for which $\langle \cos\theta = 0$, $T_r = T$, *i.e.* the mean collision time can be treated as relaxation time itself.

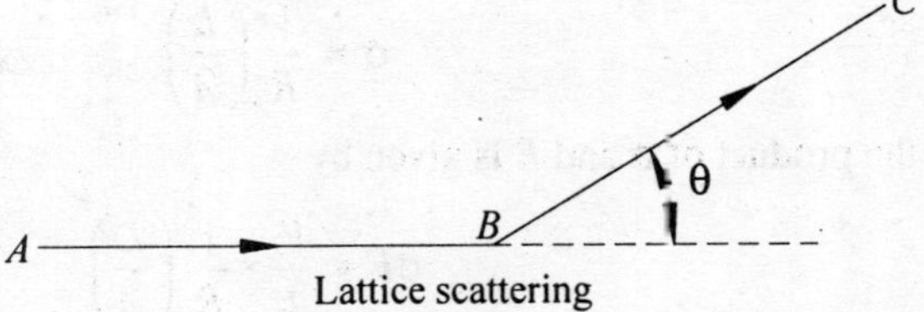

Fig. 2.6

2.8 EXPRESSION FOR DRIFT VELOCITY

The resultant velocity acquired by the electrons in the steady state of metals in the presence of an external electric field is called the drift velocity. It is denoted as V_d. Under the influence of an electric field, a conduction electron moves with a constant velocity in the steady state. If m be the mass of an electron, v is the velocity in the steady state which is the same as the drift velocity (V_d) and T is the mean collision time, then the resistance force F_r offered to its motion is given by

$$F_r = \frac{mv_d}{T}$$

If e is the charge of an electron and E is the strength of the applied field, then the force acting on the electron is $F = -eE$, in the steady state $F_r = F$

i.e. $$m\frac{v_d}{T} = -eE$$

∴ The drift velocity V_d is given by

$$V_d = -\frac{eE}{m}T$$

2.9 CURRENT DENSITY (*J*)

It is the current per unit area of cross-section of an imaginary plane held normal to the direction of current in a current carrying conductor.

If I is the current, and A is the area of cross-section then the current density J is given by

$$J = \frac{I}{A} \quad (1)$$

2.10 ELECTRIC FIELD (*E*)

In a homogeneous conductor. The electric field is defined as the potential drop/unit length of a homogeneous conductor.

If L is the length of conductor of uniform cross-section and of uniform material composition, and V is the potential difference between its two ends, then the electric field E at any point inside it is given by

$$E = \frac{V}{L} \quad (2)$$

2.11 CONDUCTIVITY (σ)

The conductivity is the physical property that characterizes the conducting ability of a material.

If R is the electrical resistance of uniform material of length L, and area of cross-section A, then the electrical conductivity is given by

$$\sigma = \frac{1}{R}\left(\frac{L}{A}\right) \tag{3}$$

the product of σ and E is given by

$$\sigma E = \frac{V}{L} \times \frac{1}{R}\left(\frac{L}{A}\right)$$

$$\sigma E = \frac{V}{RA}$$

or $$\sigma E = \frac{I}{A}\left(\text{Since } I = \frac{V}{R}\right) \tag{4}$$

on comparing Eq. (1) and (2), we have

$$J = \sigma E \tag{5}$$

This equation represents Ohm's law.

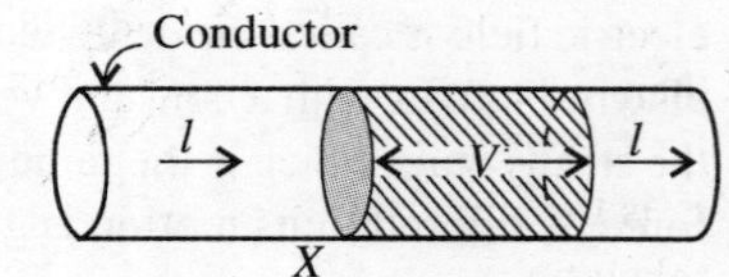

Fig. 2.7 Current carrying conductor

2.12 RESISTIVITY (ρ)

If is the property of the material and is defined as the inverse of conductivity

i.e. $$\rho = \frac{1}{\sigma}$$

2.13 EXPRESSION FOR CURRENT IN A CONDUCTOR

Let us consider a conductor of uniform area of cross-section A, carrying a current I. If v is the velocity of the electrons, then the distance transversed by the electron in unit time is v.

Let x be an imaginary plane normal to the direction of current. If we consider the motion of electrons as a group starting from X then they sweep a volume vA of the conductor in unit time.

And n be the number of electrons/unit volume, then the number of electrons in a volume $vA = n \cdot vA$, *i.e.* the number of electrons crossing any cross-section in unit time $= n \cdot vA$.

If e is the charge on each electron, then the quantity of charge crossing any section per second $= e\,(nvA)$.

The quantity of charge crossing any section per second is the salme as rate of flow of charge.

∴ Rate of flow of charge $= nevA$

Since the current is also defined as the rate of flow of charge,

then, we get

$$I = nevA \tag{1}$$

Since the velocity acquired by the electron is due to an applied electric field, it is same as drift velocity V_d,

i.e. $$V = v_d$$

∴ $$I = nev_dA \tag{2}$$

2.14 RELAXATION TIME (τ_r)

When the material is subjected to the influence of an external electric field of strength E, then the valence electrons in the metal will move with the drift velocity. As the electric field increases, the drift velocity also increases and reaches to a constant value. If the electric field is withdrawn then the drift velocity acquired by an electron assume the random motion and goes to the equilibrium state The process in which the valence electrons switching over to equilibrium state after the withdrawal of the external electric field is called as the relaxation process. The time required for this process is called the relaxation time and is denoted as τ_v.

If the field is withdrawn, E becomes zero, and the motion of the electron is given by the following equation,

$$m\frac{dV_d}{dt} = \left(\frac{1}{\tau}mv_m\right)$$

The solution of this differential equation is given by,

$$V_d = V_m e^{-t/\tau_v}$$

Where V_d is the time dependent drift velocity at time τ, V_m is the maximum constant value of the drift velocity and τ_v is the relaxation time

If $$t = \tau_v$$

then $$V_d = \frac{V_m}{e}$$

τ_v is the time at which the drift velocity V_d is $\frac{1}{e}$ times the constant value of the drift velocity V_0 is known as relaxation time τ_r.

2.15 MOBILITY OF AN ELECTRON

The mobility of electrons is defined as the magnitude of the drift velocity acquired by the electrons in a unit field.

When a material is subjected to the influence of an external electric field of strength E in which the electron acquire a drift velocity v_d, then the mobility of electron μ is given by,

$$\mu = \frac{V_d}{E}$$

But the relation between the electrical conductivity and current density is given by

$$J = \sigma E$$

or $$\sigma = \frac{J}{E} = \frac{I}{AE} \qquad \because j = \frac{I}{A}$$

$$\sigma = \frac{neV_d A}{AE} \qquad \because I = neV_d A$$

$$= ne\mu \qquad \because \mu = \frac{V_d}{E}$$

or $$\mu = \frac{\sigma}{ne}$$

$$\mu = \frac{e\tau}{m}$$

This is the expression for mobility of electrons.

2.16 EXPRESSION FOR CONDUCTIVITY IN METALS IN TERMS OF MEAN COLLISION TIME

Let us consider an electron of charge e moving in a homogeneous conductor of uniform area of cross-section A, subjected to an electric field of strength E, then the force F acting on the electron is given by

$$F = eE \tag{1}$$

The force F acting on an electron of mass m charges its velocity, then according to Newton's II law of motion, we can write.

$$F = m\frac{dv}{dt} \tag{2}$$

On comparing Eq. (1) and (2), we get

$$eE = m\frac{dv}{dt}$$

or
$$dv = \frac{eE}{m}dt \tag{3}$$

This is the change in velocity, the total change in velocity can be obtained by integrating the above expression.

$$\int_0^{\bar{v}} dv = \frac{eE}{m}\int_0^{\tau} dt$$

i.e.
$$\bar{v} = \frac{eE}{m}\tau \tag{4}$$

where τ is the mean collision time and $\bar{v}$ is the average velocity corresponds to τ. But we know that the expression for the electrical conductivity σ is given by,

$$\sigma = \frac{J}{E} \tag{6}$$

where J = Current density $= \frac{I}{A}$

I = is the current

and A = area of cross-section of a homagenous conductor

$$\therefore \quad \sigma = \frac{I}{AE} \tag{7}$$

If e is the charge of an electron and n be th number of electrons/unit volume, then the number of charges crossing/unit area/unit time is given by

i.e.
$$I = ne\bar{v}A \tag{8}$$

where $\bar{v}A$ is the volume that sweeps out by an electron in unit time.

By substituting the value of I in Eq. (7), we get

$$\sigma = \frac{ne\bar{v}A}{AE}$$

$$\therefore \quad \sigma = \frac{nev}{E}$$

Substituting the value of $\bar{v}$ from Eq. (4), we get

$$\sigma = \frac{ne}{E} \times \frac{eE}{m}\tau$$

$$\therefore \qquad \sigma = \frac{ne^2\tau}{m}$$

This is the expression for electrical conductivity in a metals.

2.17 EFFECT OF TEMPERATURE AND IMPURITY ON ELECTRICAL RESISTIVITY OF METALS

The variation of electrical conductivity with the temperature can be explained by the behaviour of the electrical resistivity ρ with *T* for sodium metal.

At very low temperatures, the resistivity of metals varies as T^5 and attains a constant value as $T \to 0$. At room temperature the conductivity of metals is independent of impurities or lattice defects. This is shown in Fig. 2.8. No crystal is completely pure or perfect, however, imperfections of any kind such as chemical impurities, lattice defects, even the crystal surface acts as scattering centres for the electrons even in the absence of lattice vibrations. Thus resistivity does not fall to zero even at *OK* and therefore resistivity curve variation is as shown in the Fig. 2.8. This indicates that, the temprature independent contribution to the resisitivity by the impurity atoms.

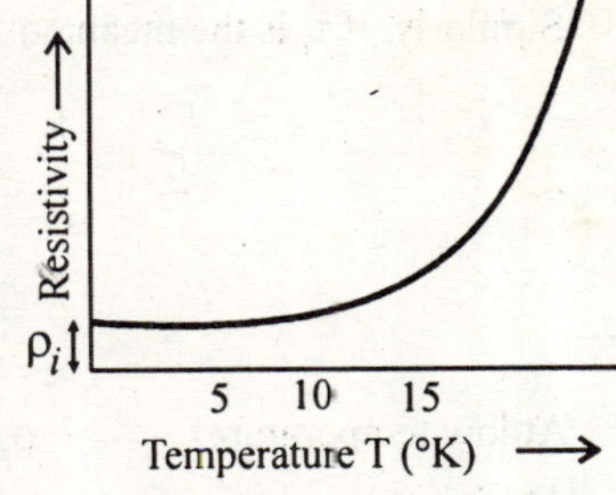

Fig. 2.8 Variation of resistivity with temperature

∴ The total resistivity ρ is given by $\rho = \rho_i + \rho_T$

At $T = OK$, electrical resistivity ρ has a small constant value. From this value ρ increases with *T*, slowly at first, later ρ increases lineraly with *T*. The linear behaviour continues essentially until the melting point is reached. At room temperature, the variation of ρ with *T* is linear.

We know that, according to quantum mechanics, electrons have wave-like properties. For an idealized conductor, no collisions with atoms would occur and the resistivity would be zero. Electron waves are scattered only if the atomic arrangement is irregular as a result of structural defects or impurities.

The resistivity is dominated by scattering of electrons while they travel in the material under the influence of the electric field. The scattering takes place in two ways which give rise to two components of resistivity.

1. At low temperatures, the residual resistivity ρ_i is caused primarily by the collision of electrons with impurities and grain boundaries and imperfections in the metal.
2. At high tempratures, the resistivity ρ_{ph} is dominated by scattering caused by collisions between electrons and atoms, *i.e.* scattering of electrons by lattice vibrations (phonons). Since it is the resistivity caused by a pure specimen which is free from all defects, it is called as the ideal resistivity.

The resistivity of the material of a conductor is given by $\rho = \dfrac{m_e}{ne^2\tau}$. In this equation the relaxation time τ depends on temprature. In metals n does not change with temperature.

If ρ is the total resistivity of the metal, then

Total resistivity, ρ = Resistivity due to phonons + Resistivity due to imperfections.

i.e.
$$\rho = \rho_{ph} + \rho_i \qquad (1)$$

This equation is known as *Matthiessen's rule*.

It states that the total resistivity of a metal is the sum of the resistivity due to phonon scattering, which is temperature dependent and the resistivity due to scattering by imperfections, which is temperature independent.

We know that, resistivity of the material is given by,

$$\rho = \frac{m_e}{ne^2\tau} \tag{2}$$

where n is the number of free electrons and $1/\tau$ is the probability of electron suffering a scattering per unit time. Thus, if $\tau = 10^{-14}$ s, then the electron undergoes 10^{14} collisions in one second.

If τ_{ph} is the mean collision time assuming there is no scattering by imperfections, then

$$\rho_{ph} = \frac{m_e}{ne^2\tau_{ph}} \tag{3}$$

Similarly, if τ_i is the mean collision time assuming there is no scattering by lattice vibrations, then

$$\rho_i = \frac{m_e}{ne^2\tau_i} \tag{4}$$

$$\therefore \qquad \rho = \frac{m_e}{ne^2\tau_{ph}} + \frac{m_e}{ne^2\tau_i} = \frac{m_e}{ne^2}\left(\frac{1}{\tau_{ph}} + \frac{1}{\tau_i}\right)$$

At low temperatures, $\rho_{ph} \to 0$, hence $\rho = \rho_i$.

At high temperatures, $\rho_i \to 0$, hence $\rho = \rho_{ph}$.

The general characteristics of electrical conduction in metals are as follows

1. The current density in the steady state is proportional to the electric field strength, *i.e.* $J \propto E$.
2. For pure metals, the electrical conductivity and thermal conductivity vary with temperature.
3. For metals that exhibit the phenomenon of superconductivity, their resistivity is zero at temperature OK and below the critical temperature for the superconducting phase transition ($T_c = 4.15\ K$) for Mercury.
4. For metals containing small amounts of impurities, the electrical resistivity can be written as $\rho = \rho_i + \rho_T$ where ρ_i = resistivity due to impurity and ρ_T that due to thermal motion and tends to zero as $T \to 0$, so $\rho\ (T = 0) = \rho_i$.
5. For most metals the electrical resistivity decreases with increasing pressure.
6. The resistivity of alloys that exhibit order-disorder transitions shows the pronounced minima corresponding to ordered phases.

2.18 FAILURES OF CLASSICAL FREE ELECTRON THEORY

The classical freeelectron theory was successful in explaining the observed mechanical properties such as elastic properties, ductivity, malliability and other related properties, however, the classical theory fails to explain the magnetic susceptibility, specific heat in metals and temperature dependance of electrical conductivity.

2.18.1 The Magnetic Susceptibility

A constant value for magnetic susceptibility is suggested by the free electron theory, but the experimental observed values are negative, small and does not depend on temperature in diamagnetic materials. For paramagnetic material the magnetic susceptibility is positive, small and inversely proportional to temperature. For ferromagnetic materials the magnetic susceptibility is very large and inversely proportional to the temperature.

2.18.2 Specific Heat

We know that the specific heat at constant volume is given by

$$C_v = \frac{3}{2}R \tag{1}$$

where R is universal gas constant.

But experimentally it was observed that the specific heat in metal is given by,

$$C_v = 10^{-4} RT \tag{2}$$

Thus, the value of specific heat as per the assumption made by classical free electron theory is far higher than that of the observed values. According to classical free electron theory [Eq. (1)] C_v is independent of temperature, but experimental facts show that C_v is directly proportional to temperature.

2.18.3 Temperature Dependance of Electrical Conductivity

According to the assumption of classical free electron theory, the kinetic energy of electron is given by

$$\frac{3}{2}kT = \frac{1}{2}mv_{th}^2$$

where k = Boltzmann constant

T = Temperature (in the absence of field)

m = mass of an electron and V_{th} = root mean square velocity

or $$V_{th}^2 = \frac{3kT}{m}$$

or $$V_{th} = \sqrt{\frac{3kT}{m}}$$

or $$V_{th} = \sqrt{T}$$

Thermal velocity or root mean square is directly proportional to square root of temperature.

But experimental observation show that the electrical conductivity σ is inversely proportional to temperature

i.e. $$\sigma_{exp} \propto \frac{1}{T} \tag{1}$$

Since, the mean collision time τ is inversely proportional to the root mean square velocity then, we can write

$$\tau \propto \frac{1}{v_{th}}$$

or $$\tau = \frac{1}{\sqrt{T}} \qquad \because v_m \propto \sqrt{T} \tag{2}$$

But, we know that the expression for electrical conductivity σ is given by,

$$\sigma = \frac{ne^2\tau}{m}$$

where n = electron concentration

e = charge of an electron

m = mass of an electron

and τ = mean of collision time

i.e. $$\sigma \propto \tau$$

or $$\sigma \propto \frac{1}{\sqrt{T}} \qquad \because \tau \propto \frac{1}{\sqrt{T}} \tag{3}$$

From Eqs (1) and (2) it is clear that the assumption made by classical theory of free electron is not agreeing with the experimental facts.

2.18.4 Heat Capacity of the Electron Gas

Classical free electron theory explains that all the valence electrons in a metal can absorb thermal energy. According to the law of equipartition of energy, every free electron has an average kinetic energy $\frac{3}{2}kT$ so that in 1 kmol of a metal, in which there are N_A atoms and, therefore, N_A free electrons, assuming that each atom contributes one valence electron to the electron gas, the total energy of the electrons is given by

$$U = \frac{3}{2} N_A k_B T \tag{1}$$

If heat is supplied to the material, the free electrons also absorb part of the heat, and the 'electronic specific heat' is given by

$$C_v = \left(\frac{dU}{dT}\right) = \frac{3}{2} k_B N_A = \frac{3}{2} R \tag{2}$$

$$= \frac{3}{2} \times 1.38 \times 10^{-23} \times 6.02 \times 10^{26}$$

That is, molar specific heat = 12.5 kJ/kmol/K.

The above value, $C_v = \frac{3}{2} R$ due to free electrons is about hundred times greater than the experimentally predicted value. Since the heat capacity of a solid due to atomic vibrations is 3 R, free electrons should make a significant contribution to the total specific heat of a metal. We know that, however, at least at high temperatures, Dulong and Petit law holds good and the total specific heat of a solid is given by 3 R; this means that the free electrons do no contribute significantly to the heat capacity of a metal. We can, therefore, conclude that the law of equipartition and hence classical Maxwell-Boltzmann statistics cannot be applied to the free electrons in a metal.

2.18.5 Calculation of Mean Free Path

The general expression for resistivity of a metal is given by the equation,

$$\rho = \frac{m}{ne^2\tau}$$

The resistivity of copper at 20° C is 1.69 × 10^{-8} ohm-m and density of free electrons, $n = 8.5 \times 10^{28}$ m^3. Thus

$$\tau = \frac{m}{ne^2\rho}$$

$$= \frac{9.11 \times 10^{-31}}{8.5 \times 10^{28} \left(1.6 \times 10^{-19}\right)^2 \times 1.69 \times 10^{-8}}$$

That is, $$\tau = 2.47 \times 10^{-14} s$$

But, $$\tau = \frac{\lambda}{c} \tag{3}$$

Referring Eq. (3)

$$\lambda = \tau\bar{c} = 2.47 \times 10^{-14} \times 1.154 \times 10^{5}$$

$$\lambda = 2.85 \text{ nm}$$

The experimentally observed value of λ is nearly ten times the above value. Classical theory could not explain the large variations in λ.

2.18.6 Relation Between Electrical Conductivity and Thermal Conductivity (Wiedemann-Franz Law)

Heat conduction in solids may take place either through the mechanism of the atoms or through that of the free electrons. In metals, electrons are the principal carriers of heat energy. Free electrons do this by being excited by energetic scattering centres and carrying the extra energy to another scattering centre in a cooler part of the metal.

Let us consider a conductor in the form of a rod of uniform cross-section. Let a thermal gradient be established. Consider lengths at *A*, *B* and *C* normal to the direction of the heat flow. Let each of them be separated by a distance λ, the mean free path of the electron gas in the metal.

$$\left[T + \left(\frac{dT}{dx}\right)\lambda\right], T \text{ and } \left[T - \left(\frac{dT}{dx}\right)\lambda\right]$$

are the steady temperatures of the layers as shown in Fig. 2.9. Let the thermal energy of an electron in these $\left[E + \left(\frac{dE}{dx}\right)\lambda\right]$, E and $\left[E - \left(\frac{dE}{dx}\right)\lambda\right]$ layers be respectively.

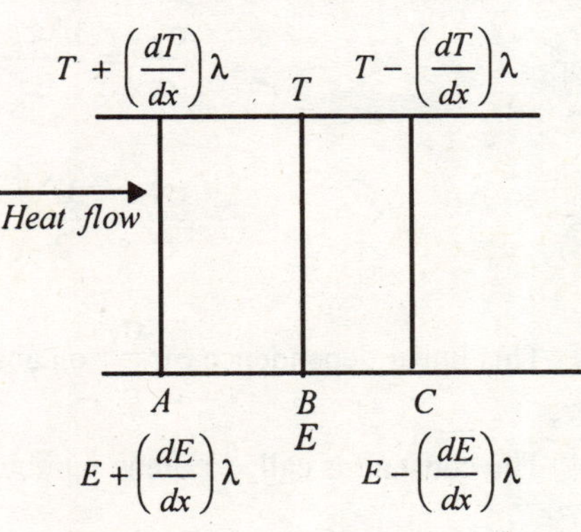

Fig. 2.9 Flow of heat through a conductor

Thus, the excess of energy carried by an electron from *A* to *B* is $\left(\frac{dE}{dx}\right)\lambda$. From kinetic theory, the number of electrons flowing in a given direction through unit area for unit times is $\frac{1}{6}n\bar{c}$, where n is density of electrons and $\bar{c}$ is the thermal velocity of the electron. Now the excess energy transported through unit area in unit time at the layer *B* is given by,

$$\frac{1}{6}n\bar{c}\lambda\left(\frac{dE}{dx}\right)$$

Similarly deficit of energy transported through *B* in the opposite direction is given by,

$$-\frac{1}{6}n\bar{c}\lambda\left(\frac{dE}{dx}\right)$$

Hence, net heat energy transported through unit area in unit time from *A* to *B*

$$\left[\frac{1}{6}n\bar{c}\lambda\left(\frac{dE}{dx}\right)\right] - \left[-\frac{1}{6}n\bar{c}\lambda\left(\frac{dE}{dx}\right)\right] = \frac{n\bar{c}\lambda}{3}\left(\frac{dE}{dx}\right) \tag{1}$$

This must be equal to the product of the thermal conductivity and the temperature gradient, *i.e.*

$$\left(\frac{n\bar{c}\lambda}{3}\right)\left(\frac{dE}{dx}\right) = \sigma_T\left(\frac{dT}{dx}\right) \tag{2}$$

Then, $$n\left(\frac{dE}{dx}\right) = n\left(\frac{dE}{dT}\right)\left(\frac{dT}{dx}\right) = n\,C_v\left(\frac{dT}{dx}\right)$$

Substituting this value of $n\left(\frac{dE}{dx}\right)$ in Eq. (2), we get

$$\frac{n\bar{c}\lambda}{3}C_v\left(\frac{dT}{dx}\right) = \sigma_T\left(\frac{dT}{dx}\right)$$

Thus, $$\sigma_T = \frac{1}{3}C_v\,\lambda n\bar{c} \tag{3}$$

Now, $$\sigma_T = \frac{1}{3}\lambda n\sqrt{\frac{3k_BT}{m}}\,(3k_B/2) = \frac{n\lambda k_B}{2}\sqrt{\frac{3k_BT}{m}}$$

$$\sigma = \frac{ne^2\lambda}{\sqrt{3mk_BT}} \tag{4}$$

or $$\frac{\sigma_T}{\sigma} = \frac{n\lambda k_B}{2}\left(\frac{3k_BT}{m}\right)^{1/2}\left(\frac{\sqrt{3mk_BT}}{ne^2\lambda}\right)$$

$$\frac{\sigma_T}{\sigma} = \frac{3T}{2}\left(\frac{k_B}{e}\right)^2 \tag{5}$$

This linear dependence of $\frac{\sigma_T}{\sigma}$ on absolute temperature is known as the law of Wiedemann and Franz law.

The constant is called Lorenz number L, that is given by

$$L = \frac{3}{2}\left(\frac{k_B}{e}\right)^2 \tag{6}$$

$$= 1.12 \times 10^{-8} \text{ watt ohm per kelvin}^2$$

The validity of Eq. (5) can be checked using experimentally determined values of σ and σ_T. Then the Eq. (5) is given by,

$$\frac{\sigma_T}{\sigma} = LT$$

or $$\frac{\sigma_T}{\sigma T} = L$$

For copper at 20 °C, the electrical resistivity and thermal conductivity are 1.72×10^{-8} Ω m and 386 $\text{Wm}^{-1}\,\text{K}^{-1}$.

Now, $$L = \frac{\sigma_T}{\sigma T} = \frac{386}{293 \times 5.81 \times 10^7}$$

i.e. $$L = 2.26 \times 10^{-8}\ \text{W}\Omega\text{K}^{-2}$$

This value of Lorentz number does not agree with the value calculated from the classical formula given in Eq. (6). Thus, the classical assumption that all the free electrons of a metal participate in thermal conduction is not correct. Quantum theory with Fermi-Dirac statistics reveals that only the electrons near the Fermi level take part in thermal and electrical conductions.

2.19 ASSUMPTION OF QUANTUM FREE ELECTRON THEORY

The assumption of quantum free electron theory are

(*i*) The interaction between the electrons and the lattice ions are neglected.

(*ii*) The electrons confined in the boundaries of a solid moves with a constant potential.

(*iii*) The possible energy levels of the electrons are governed by the Pauli's exclusion principle.

2.20 EXPRESSION FOR DENSITY OF STATES

We know that, the possible energy values of an electron in a box is given by

$$E = \frac{n^2 h^2}{8mL^2} \tag{1}$$

where n = 1, 2, 3, ⋯
h = Planck's constant
m = mass of an electron
and L = the length of the box

Let us consider an electron enclosed in a three dimensional solid of length L, then the possible energy levels of an electron in a three dimensional solid given by

$$E = \frac{h^2}{8mL^2}\left(n_x^2 + n_y^2 + n_z^2\right) \tag{2}$$

where n_x, n_y and n_z are integers (> 0)

Let $$\frac{h^2}{8mL^2} = E_0 \quad \text{and} \quad r^2 = n_x^2 + n_y^2 + n_z^2 \tag{3}$$

$$\therefore \quad E = E_0 r^2 \tag{4}$$

The possible energy states depends on the values of n_x, n_y and n_z, E can be represented as a point in the system for all the values of n_x, n_y and n_z. But for a large number of n_x, n_y and n_z we get a spherical distribution.

Since, the values of n_x, n_y and n_z are positive integers we shall consider the octant of a sphere, for which n_x, n_y and n_z are positive coordinates as shown in the Fig. 2.10.

The number of possible energy values with in the sphere of radius r = volume of one octant of the sphere of radius $r \times$ number of points /unit volume.

The number of energy states with in the sphere of radius $(r + dr)$ = volume of space between two octant of the sphere of radius r and $r + dr$ (number of points/unit volume).

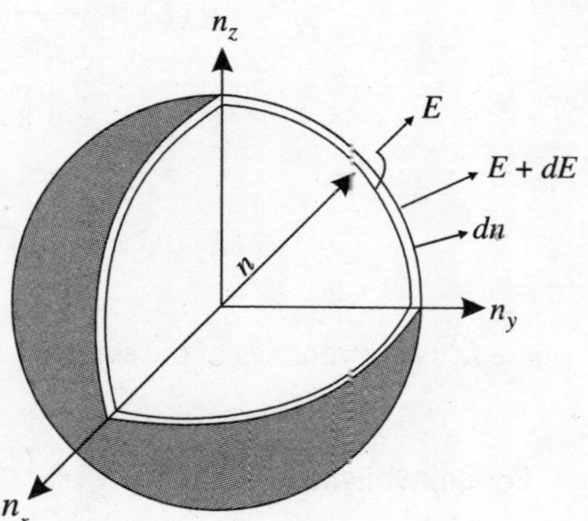

Fig. 2.10 Density of energy states

$$= \frac{1}{8}\left(\frac{4\pi}{3}\right)(r+dr)^3 - \frac{1}{8}\left(\frac{4\pi}{3}\right)r^3$$

$$= \frac{\pi}{6}\left((r+dr)^3 - r^3\right)$$

$$= \frac{\pi}{6}\left(r^3 + dr^3\, 3r^2 dr + 3dr^2 - r^3\right)$$

$$= \frac{\pi}{6}(3r^2 dr) = \frac{1}{2}\pi r^2 dr \tag{5}$$

But we know that, according to Pauli's exclusion principle each energy value has two energy state one for spin up and the other for spin down therefore, there are two energy state for each possible energy value.

$\therefore$ The number of possible energy states in the energy range E and $E + dE = 2 \times \frac{1}{2}\pi r^2 dr$

$$= \pi r^2 dr \tag{6}$$

If $g(E)$ is the number of energy state / unit energy range then, the number of energy states in the energy range

$$E \text{ and } E + dE = g(E)\, dE \tag{7}$$

which is called as density of energy states and is given by

$$g(E) = \pi r^2 dr \tag{8}$$

but $$r^2 = \frac{E}{E_0} \quad \text{or} \quad r = \sqrt{\frac{E}{E_0}}$$

$$\therefore \quad dr = \frac{1}{2}\left(\frac{1}{EE_0}\right)^{1/2} dE$$

Substituting the values in Eq. (8), we get

$$g(E)\, dE = \pi\left(\frac{E}{E_0}\right)\left[\frac{1}{2}\left(\frac{1}{EE_0}\right)^{1/2}\right] dE = \frac{\pi E^{1/2}}{2\,(E_0)^{3/2}}\, dE$$

$$\therefore \quad g(E) = \frac{\pi}{2\left(\frac{h^2}{8mL^2}\right)^{3/2}} E^{1/2} dE$$

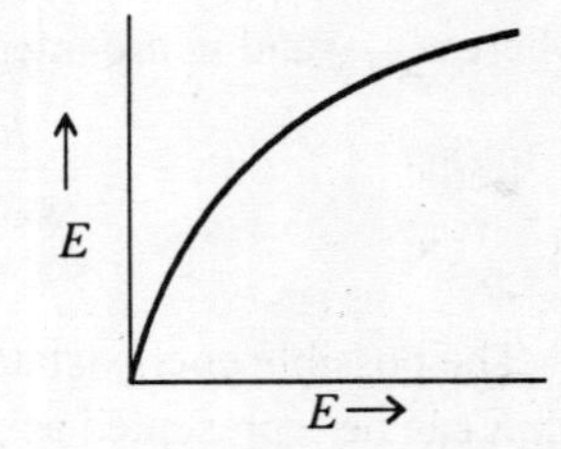

Fig. 2.11 Density of states

$$g(E)\, dE = \left(\frac{8\sqrt{2}\pi m^{3/2} L^3}{h^3}\right) E^{1/2} dE \tag{9}$$

where L^3 is the volumes of the solid.

For unit volume of the solid, $$g(E)\, dE = \left(\frac{8\sqrt{2}\pi m^{3/2}}{h^3}\right) E^{1/2} dE$$

This is the expression for density of states for the electrons in a solid.

2.21 FERMI-DIRAC STATISTICS

The Fermi-Dirac statistics is applicable to a system of identical indistinguishable particles which obey pauli's exclusion principle. These particles are called Fermions, like proton, electron, neutron etc.

According to Fermi-Dirac statistics only one particle can exist in a energy state. At 0° K all the available energy states are completely filled by the particles such that there are particle per each state, up to certain maximum energy called fermi energy. Above fermi energy no particles will present at 0° K, some of the particles will get the energy greater than fermi energy as shown in the Fig. 2.12.

The fermi energy is the maximum energy for the system of particles which obey Fermi-Dirac statistics, up to which are particle present in each state and no particles present above it at 0° K.

The number of particles present in the energy state from E to $E + dE$ is given by

$$n(E)\,dE = \frac{g(E)\,dE}{Ae^{E/KT}+1}$$

i.e. $$\frac{n(E)\,dE}{g(E)\,dE} = \frac{1}{Ae^{E/KT}+1}$$

i.e. $$f(E) = \frac{1}{Ae^{E/KT}+1}$$

Also $$f(E) = \frac{1}{e^{\frac{E-EF}{KT}}+1}$$

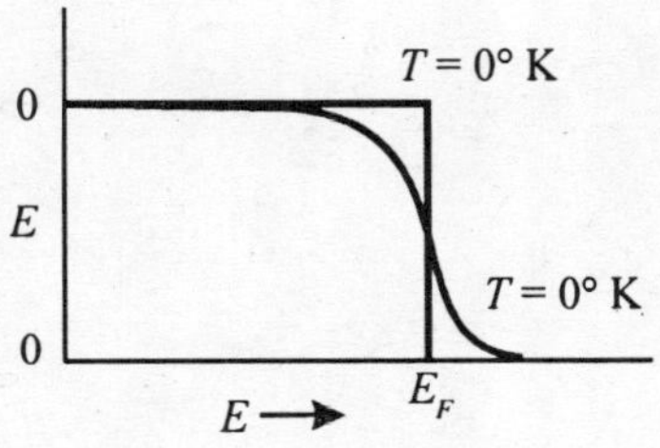

Fig. 2.12

2.22 MAXWELL-BOLTZMANN STATISTICS AS THE LIMIT OF QUANTUM STATISTICS

The distribution function for quantum statistics is given by the equation

$$f(E) = \frac{1}{Ae^{E/KT} \pm 1}$$

When $A >> 1$ $$f(E) = \frac{1}{Ae^{E/KT}}$$

$$\approx A^{-1}e^{-E/KT}$$

Which is nothing but the distribution function for Maxwell-Boltzmann statistics. Hence, the Maxwell-Boltzmann statistics is a limit of quantum statistics when $A >> 1$. When classical limit, *i.e.* when the temperature of the system is very high, mass of the system is very large and concentration of the system is very less. A becomes very much greater than 1.

We have $$n(E)\,dE = \frac{g(E)\,dE}{Ae^{F/KT} \pm 1}$$

But, $$g(E)\,d(E) = \frac{4\pi P^2 dp}{h^3}v$$

Where $$P^2 = 2mE$$

$$P = (2mE)^{1/2}$$

$$dp = (2m)^{1/2}\frac{1}{2}\cdot E^{-1/2}\,dE$$

$$= (2m)^{1/2} \frac{1}{2\sqrt{E}} dE$$

$P =$ momentum of particle

$$\therefore \; n(E)\, d(E) = \frac{4\pi P^2 dp}{h^3 \left(Ae^{E/KT} \pm 1\right)}$$

$$= \frac{4\pi \cdot 2mE \cdot (2m)^{1/2} \dfrac{1}{2\sqrt{E}} dE \cdot v}{h^3 \left(Ae^{E/KT} \pm 1\right)}$$

$$= \frac{4\pi\, (2m)^{3/2}\, E^{1/2} \dfrac{1}{2} dE \cdot v}{h^3 \left(Ae^{E/KT} \pm 1\right)}$$

$$= \frac{4\pi \left(2m^3\right)^{1/2} E^{1/2}\, dE \cdot v}{h^3 \left(Ae^{E/KT} \pm 1\right)}$$

$$n(E)\, d(E) = \frac{4\pi V}{h^3} \frac{\sqrt{2m^3} \cdot \sqrt{E} \cdot dE}{Ae^{E/KT} \pm 1}$$

$$= \frac{4\pi V}{h^3} \sqrt{2m^3} \frac{\sqrt{E}\, dE}{Ae^{E/KT}} \qquad \text{When } A >> 1$$

Total number of particles N in the system

$$N = \int_o^\infty n(E)\, dE = \int_o^\infty \frac{4\pi V}{h^3} \frac{\sqrt{2m^3}\, \sqrt{E}\, dE}{Ae^{E/KT}}$$

$$= \frac{4\pi V}{h^3} \frac{\sqrt{2m^3}}{A} \int_o^\infty \frac{\sqrt{E}\, dE}{e^{E/KT}}$$

$$= \frac{4\pi V}{h^3} \frac{\sqrt{2m^3}}{A} \int_o^\infty \sqrt{E}\, e^{-E/KT}\, dE$$

Let $E = x^2$

$dE = 2x \cdot dx$

$$N = \frac{4\pi V}{h^3} \frac{\sqrt{2m^3}}{A} \int_o^\infty x \cdot e^{-x^2/KT}\; 2x - dx$$

$$= \frac{4\pi V}{h^3} \frac{\sqrt{2m^3}}{A}\, 2 \cdot \int_o^\infty x^2 e^{-x^2/KT}\, dx$$

$$= \frac{4\pi V}{h^3} \frac{\sqrt{2m^3}}{A} 2 \cdot \frac{1}{u} \sqrt{\pi (KT)^3}$$

$$N = \frac{(2\pi \text{ mKT})^{3/2} \cdot V}{Ah^3}$$

$$A = \frac{(2\pi \text{ mKT})^{3/2}}{N/V \cdot h^3}$$

$$\therefore \; A = \frac{(2\pi \text{ mKT})^{3/2}}{ch^3} \qquad \text{where } C = \frac{N}{V} = \text{concentration of system}$$

When T is high and C is very low then $A >> 1$

$\therefore$ Maxwell-Boltzmann statistics is the limit of quantum statistics.

2.23 ELECTRON DISTRIBUTION IN A METAL AT 0° K-HENCE EXPRESSION FOR FERMI ENERGY

In a metal the valence electron are not strongly bound to the nucleus. They can move freely anywhere in the atomic structure. This is similar to the molecules of a gas which can move freely anywhere in the gas. Hence, we can apply Fermi-Dirack statistics to the system of valence electrons in a metal.

$$\therefore \qquad f(E) = \frac{n(E)\, dE}{g(E)\, dE}$$

$$n(E)\, dE = g(E)\, dE \cdot F(E)$$

$$= \frac{2 \cdot 4\pi P^2 dp \cdot v}{h^3} f(E) \text{ (For two spin states of electrons)}$$

$\therefore$ Total number of electrons in the system is

$$N = \int_o^{\infty} n(E)\, dE = \int_o^{\infty} \frac{2 \cdot 4\pi P^2 dpv}{h^3} f(E)$$

$$f(E) = 1 \quad \text{for } E < E_F$$
$$= 0 \quad \text{for } E > E_F$$

$$N = \int_o^{E_F} \frac{2 \cdot 4\pi P^2 dpv}{h^3} = \frac{8\pi V}{h^3} \frac{P^3}{3}\Bigg]_0^{E_F}$$

$$P^2 = 2mE$$
$$P = (2mE)^{1/2}$$

$$N = \frac{8\pi V}{h^3} \frac{(2mE)^{3/2}}{3}\Bigg]_0^{E_F}$$

$$N = \frac{8\pi V}{h^3} \frac{(2mE_F)^{3/2}}{3}$$

$$(2mE_f)^{3/2} = \frac{3Nh^3}{8\pi V}$$

$$2mE_F = \left(\frac{3Nh^3}{8\pi V}\right)^{2/3}$$

$$E_F = \frac{1}{2m}\left(\frac{3Nh^3}{8\pi V}\right)^{2/3}$$

$$E = \frac{h^3}{8m}\left(\frac{3N}{\pi}\right)^{2/3}$$

where $n = \frac{N}{V}$ = electron density

2.24 FERMI ENERGY

According to classical free electron theory, there is no restriction on the allowed energy levels of an electron. In quantum free electron theory, the energy of the electron in a metal is quantized. Hence due to the quantization rules the number of allowed energy levels in a metal containing N electrons is equal to N. According to Pauli's exclusion principle, an energy level can accommodate maximum two electrons, one with spin up and the other with spin down. Thus in filling the energy levels, two electron occupy the lowest level, two more the next level and so on, until all the electrons in the metal have been accommodated as shown in Fig. 2.13.

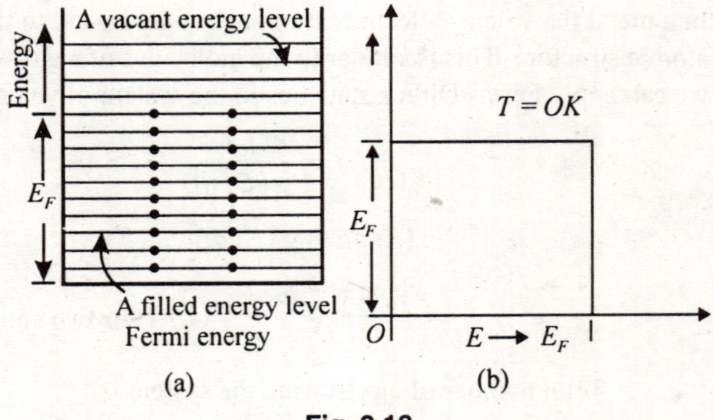

Fig. 2.13

The energy of the highest occupied level at an absolute zero temperature is called the fermi energy and the corresponding energy level is known as the fermi level. At the absolute zero temperature, the maximum kinetic energy that a free electron pocess is called the fermi energy and is denoted as E_F.

At $T = OK$, the metal does not receive any energy, all the energy levels up to fermi level are completely occupied, while all the energy levels above fermi level are vacant. If there are N electrons, the highest occupied energy level will be $\left(\frac{N}{2}\right)$ level and is called the fermi level and the corresponding energy is known as the Fermi energy.

2.25 FERMI LEVEL

The fermi level is an energy level mid way between top surface of valence band and bottom surface of conduction band. If E_v is the energy corresponding to top surface of valence band, E_c is the energy corresponding to bottom surface of conduction band and E_F is the fermi energy corresponding to fermi level, then

$$E_F = \frac{E_c + E_v}{2} = E_c + \frac{E_v - E_c}{2} = E_c + \frac{E_g}{2},$$

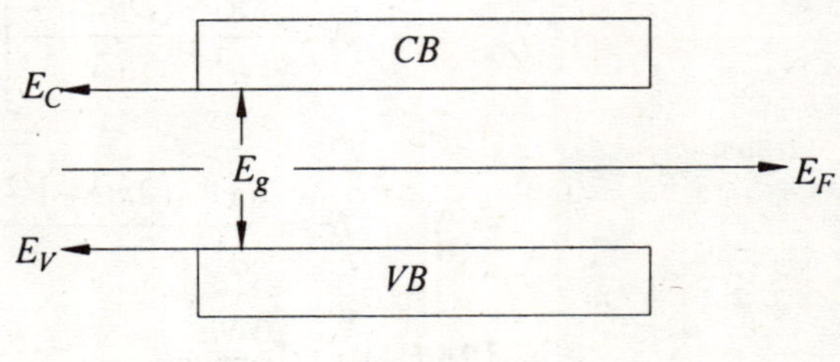

Fig. 2.14

where $E_v - E_c = E_g$ (energy gap) $\therefore\ E_F - E_C = \dfrac{E_g}{2}$

2.26 FERMI FACTOR

If the temperature is greater than OK, the the material will receive thermal energy from its surroundings. At room temperature, the available thermal energy is too small and therefore the electrons occupying energy levels below the fermi level cannot absorb the thermal energy due to non-availability of higher energy levels. But this energy can be absorbed by the electrons occupying energy levels which are just below the fermi level. Because there are a large number of unoccupied energy levels just above the fermi level to which electrons can get excited. Though these excitations seem to be random, the occupation of various energy levels obey a statistical distribution law called Fermi-Dirac distribution law.

The probability of finding an electron having any particular value of energy E at a temperature T is given by Fermi-Dirac distribution function and is as follows.

$$f(E) = \frac{1}{e^{(E-E_F)/kT} + 1}$$

Where, $f(E)$ is called the fermi factor and the expression describes how energy is distributed among the electrons.

Fermi factor is defined as the distribution function which gives the probability of occupation of a given energy state for a material in thermal equilibrium in terms of the fermi energy E_F, Boltzmann constant k and the absolute temperature T.

A graph of $f(E)$ versus energy E at $T = OK$ and $T > OK$ is shown in Fig. 2.15. It shows the Fermi-Dirac distribution function at different temperatures.

Different cases of Fermi-Dirac distribution.

At $T = OK$ two possible conditions exist:

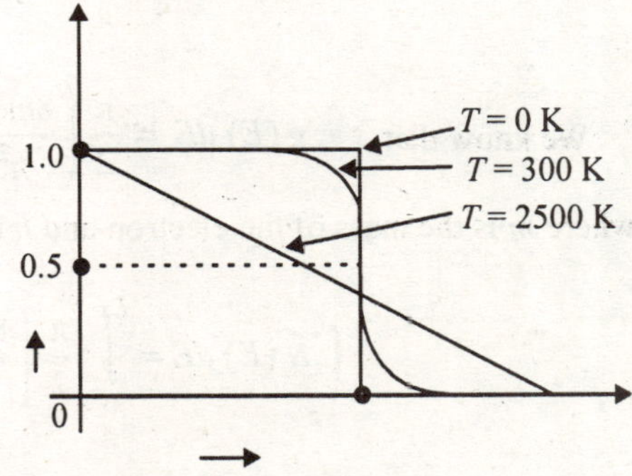

Fig . 2.15 Variation of $f(E)$ with E

2.26.1 Probability of Occupation for $E < E_F$ at $T = OK$

At $T = OK$, for $E < E_F$, $e^{(E-E_F)/KT} = 0$, so that

$$f(E) = \frac{1}{e^{-\infty} + 1} = \frac{1}{0+1} = 1 \text{ for } E < E_F$$

This shows that at $T = O\,K$, the probability of finding an electron with energy less than fermi energy is unity, *i.e.* all the energy levels between zero and fermi level E_F are occupied.

2.26.2 Probability of Occupation for $E > E_F$ at $T = OK$

At $T = OK$, for $E < E_F$, $e^{(E-E_F)/KT} = \infty$, so that

$$f(E) = \frac{1}{e^{-\infty} + 1} = \frac{1}{\infty} = 0$$

It is means that there are no electrons having energy greater than E_F, *i.e.* all the energy levels above E_F are vacant.

2.26.3 Probability of Occupation at $T > OK$

At $T > OK$ and $E < E_F$, $e^{(E-E_F)/KT} = 1$, so that

$$f(E) = \frac{1}{1+1} = \frac{1}{2}$$

At any temperature above absolute zero, the Fermi-Dirac distribution function is continuous. This implies that the probability of occupancy of fermi level at any temperature other than OK is 0.5.

Fermi level is defined as the energy level at which the probability of electron occupancy is one half.

Similarly, an operational definition can be given for fermi energy. Fermi energy is the average energy possessed by the electrons which participate in conduction process in conductors at temperatures above OK.

2.27 CALCULATION OF FERMI ENERGY AT $T = OK$

In a metal under thermal equilibrium condition the number of electrons per unit volume that have energy between E and $E + dE$ is given by

$N(E)\, dE$ = Density of available states in the energy range E and $E + dE$ × probability of occupation of these energy levels by the electron.

$$N(E)\, dE = g(E)\, dE \times f(E) \quad (1)$$

i.e.

$$N(E)\, dE = g(E) \frac{dE}{e^{(E - E_F)/kT} + 1} \quad (2)$$

A plot of $N(E)$ versus E for two temperatures is shown in Fig. 2.16.

At $T = OK$, the fermi distribution function $f(E) = 1$ for $E < E_F$ and $f(E) = 0$ for $E > E_F$. Therefore, at $T = OK$,

$$N(E)\, dE = \int_0^{E_F} g(E)\, dE \quad (3)$$

We know that,

$$g(E)\, dE = \frac{\pi}{2}\left(\frac{8ma^2}{h^2}\right)^{3/2} E^{1/2}\, dE \quad (4)$$

where m is the mass of the electron and h is the Planck's constant.

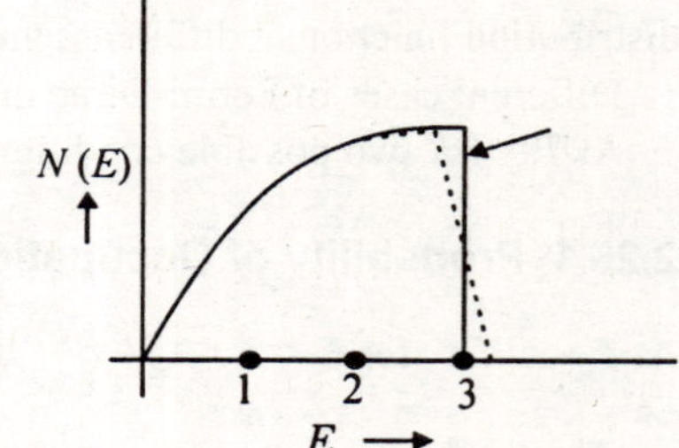

Fig. 2.16

$$\therefore \quad \int N(E)\, dE = \int_0^{E_F} \frac{\pi}{2}\left(\frac{8ma^2}{h^2}\right)^{3/2} E^{1/2}\, dE$$

or

$$N = \frac{\pi}{2}\left(\frac{8ma^2}{h^2}\right)^{3/2} \int_0^{E_F} E^{1/2}\, dE$$

$$N = \left(\frac{\pi}{2}\right)\left(\frac{8ma^2}{h^2}\right)^{3/2} \left(E^{3/2}\right)_0^{E_F} \left(\frac{2}{3}\right)$$

$$N = \left(\frac{\pi}{3}\right)\left(\frac{8m}{h^2}\right)^{3/2} V E_F^{3/2} \quad (\because \text{volume, } V = a^3) \quad (5)$$

Total number of electrons per unit volume is called as density of electrons and is given by

$$n = \frac{N}{V} = \frac{\pi}{3}\left(\frac{8m}{h^2}\right)^{3/2} E_F^{3/2} \quad \text{or} \quad E_F^{3/2} = \left(\frac{3n}{\pi}\right)\left(\frac{h^2}{8m}\right)^{3/2}$$

or $$E_F(0) = \left(\frac{h^2}{8m}\right)\left(\frac{3n}{\pi}\right)^{2/3} \tag{6}$$

According to this result, E_F shows a gradual increase with increasing electron concentration. This is expected because the electrons fill the available energy states, two electrons per state, in accordance with Pauli's exclusion principle, up to the fermi energy.

Substituting $h = 6.625 \times 10^{-34}$ JS, $m_e = 9.11 \times 10^{-31}$ kg in Eq. (6), we get

$$E_F(0) = \frac{\left(6.625 \times 10^{-34}\right)^2}{8 \times 9.11 \times 10^{-31}} \times \left(\frac{3n}{3.14}\right)^{2/3}$$

$$E_F(0) = 5.85 \times 10^{-38} \times n^{2/3}\ J = 3.65 \times 10^{-19}\ \text{eV} \tag{7}$$

The order of magnitude of the fermi energy for metal is about 5 eV. Representative values for various metals are given in Table 2.1, together with values for Fermi speed v_F of the electrons, which is the speed of the electrons when their energy is equal to the fermi energy. The fermi speed is defined by the relation

$$\frac{1}{2} m_e v_F{}^2 = E_F \tag{8}$$

or Fermi velocity $$v_F = \left(\frac{2E_F}{m_e}\right)^{1/2} \tag{9}$$

We know that, $$kT_F = E_F \tag{10}$$

Fermi temperature, $$T_F = \frac{E_F}{k} \tag{11}$$

For $E_F = 5\ \text{eV} = 5 \times 1.602 \times 10^{-19}\ \text{J} = 8.01 \times 10^{-19}$ J, we have

$$T_F = \frac{8.01 \times 10^{-19}}{1.381 \times 10^{-23}} = 58000\ K \tag{12}$$

The total energy for a free electron in a metal at OK is

$$E_0 = \int_0^{E_F} E\,N(E)\,dE$$

At $$T = O\,K,\ N(E)\,dE = \frac{3N}{2} E_F{}^{-3/2} E^{1/2}\,E\,dE = \frac{3N}{2} E_F{}^{-3/2} \int_0^{E_F} E^{3/2}\,dE$$

$$\therefore \quad E_0 = \int_0^{E_F} \frac{3N}{2} E_F{}^{-3/2}\,E\,dE = \frac{3N}{2} E_F{}^{3/2} \int_0^{E_F} E^{3/2}\,dE$$

$$= \frac{3N}{2} E_F{}^{-3/2} \times \frac{2}{5} E_F{}^{5/2}$$

$$E_0 = \frac{3}{5} N E_F \tag{13}$$

The average energy of free electron in a metal of OK is

$$E_{av} = \frac{3}{5} E_F \tag{14}$$

Table 2.1

Calculated values of various parameters for metals at 300 K based on the free electron theory.

Metal	*Electron concentration* (m^{-3})	*Fermi energy (eV)*	*Fermi speed (m/s)*	*Fermi temperature (K)*
Li	4.70×10^{28}	4.72	1.29×10^{6}	5.48×10^{4}
Na	2.65×10^{28}	3.23	1.07×10^{6}	3.75×10^{4}
K	1.40×10^{28}	2.12	0.86×10^{6}	2.46×10^{4}
Cu	8.49×10^{28}	7.05	1.57×10^{6}	8.12×10^{4}
Ag	5.85×10^{28}	5.48	1.39×10^{6}	6.36×10^{4}
Au	5.90×10^{28}	5.53	1.39×10^{6}	6.41×10^{4}

We can consider a metal to be system comprising a very large number of energy levels available to the free electrons. These electrons fill these levels in accordance with the Pauli's exclusion principle, beginning with $E = 0$ and ending with E_F. At $T = OK$, all levels below the fermi energy are filled and all levels above the fermi energy are empty. Although the levels are discrete, they are so close together that the electrons have an almost continuous distribution of energy. At 300 K, a very small fraction of the free electrons are excited above the fermi energy.

2.28 FERMI LEVEL FOR AN INTRENSIC SEMICONDUCTOR

Let us imagine that, the fermi level E_F lies midway between conduction band and valence band. The derivation is based on the following assumptions.

(*i*) The widths of conduction band and valence band are small compared to forbidden gap E_g.

(*ii*) As band widths are small, all levels in the band have the same energy. The levels in conduction band have energy E_c while levels in valence band have energy E_v as shown in Fig. 2.17.

(*iii*) At absolute zero, the solid is like an insulator, *i.e.* no conduction is possible as valence bands are completely filled and conduction bands are completely empty.

At any temperature $T°$ K, let

n_e = number of electrons in conduction band

n_v = number of electrons in valence band

$N = n_v + n_e$ = number of electrons in both bands

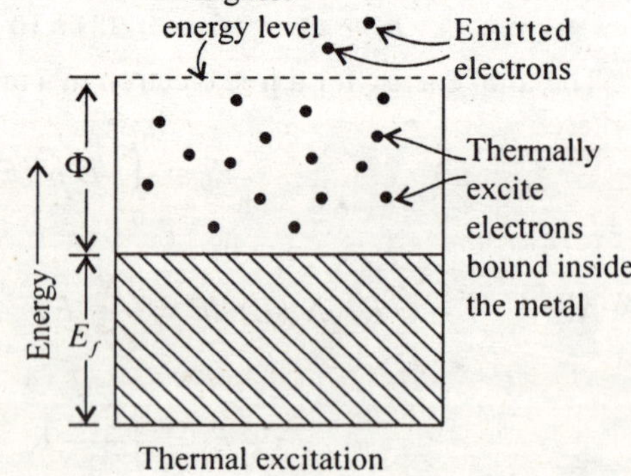

Fig. 2.17

The number of electrons n_e in conduction band is given by

$$n_e = N\,F(E_e) \qquad (1)$$

where $F(E_e)$ is the probability of an electron having energy E_e. According to Fermi-Dirac probability distribution function, we have

$$F(E_e) = \frac{1}{e^{(E_e - E_F)/kT} + 1} \qquad (2)$$

where k is Boltzmann constant.

From Eqs (1) and (2), we get

$$n_e = \frac{N}{exp.\left[(E_e - E_F)/kT\right] + 1} \qquad (3)$$

Similarly,

$$n_v = \frac{N}{exp.\left[(E_v - E_F)/kT\right] + 1} \qquad (4)$$

Now $N = n_e + n_v$

or
$$N = \frac{N}{exp.[(E_e - E_F)/kT)]+1} + \frac{N}{exp.[(E_v - E_F)/kT]+1}$$

or
$$1 = \frac{\{exp.[(E_v - E_F)/kT]+1\} + \{exp.[(E_e - E_F)/kT]+1\}}{\{exp.[(E_e - E_F)/kT]+1\} + \{exp.[(E_v - E_F)/kT]+1\}}$$

or
$$\{exp.\ [(E_v - E_F)/kT\} + 1\} + \{exp.\ [(E_e - E_F)/kT] + 1\}$$
$$= \{exp.\ [(E_e - E_F)/kT] + 1\}\ \{exp.\ [(E_v - E_F)/kT] + 1\}$$
$$= \{exp.\ [(E_e - E_F + E_v - E_F)/kT] + \{exp.\ [(E_e - E_F)/kT]\} + \{exp.\ [(E_v - E_F)/kT] + 1\}$$

or
$$1 = \{exp.\ [(E_e + E_v - 2E_F)/kT]\}$$

or
$$\log 1 = (E_e + E_v - 2E_F)/kT = 0$$

or $E_e + E_v - 2E_F = 0$

or
$$E_F = \frac{E_e + E_v}{2} \tag{5}$$

Thus fermi level is exactly in the middle of the valence band and conduction band. Expression (5) also shows that fermi level is independent of temperature.

2.29 DENSITY OF ELECTRONS IN CONDUCTION BAND

At 0° K, all the energy bands below E_F are completely filled. The conduction band at this temperature is empty. At higher temperature, some electrons at the top of valence band may be thermally excited into conduction band. According to Fermi-Dirac distribution law, the number of electrons having energies lying between E and $E + dE$ is given by

$$N(E)\,dE = \frac{g(E)\,dE}{exp.\ [(E - E_F)/kT]+1} \tag{6}$$

where $g(E)$ denotes the number of quantum states available in the given energy interval and is given by

$$g(E) = \frac{4\pi}{h^3}\,(2m)\,\sqrt{E - E_e} \tag{7}$$

when $E > E_F$ and $E_e - E_F > 4\,kT$, then

$$\frac{1}{exp.[(E - E_F)/kT]+1} \approx \frac{1}{exp.[(E - E_F)/kT]} \approx e^{-(E - E_F)/kT} \tag{8}$$

Since the single energy (E_e and E_v) for conduction and valence band is not justified, so the general form of Eq. (6) is given by

$$n_e = \int_{E_e}^{\infty} N(E)\,dE \tag{9}$$

where n_e represents the number of electrons in conduction band per unit volume of the solid. Using Eqs (6), (7) and (8) in Eq. (9), we have

$$n_e = \int_{E_e}^{\infty} \left[\frac{4\pi}{h^3}\,(2m)\,(E - E_e)^{1/2}\right] \times e^{-(E - E_F)/kT}\ dE$$

$$= C\,(kT)^{1/2} \int_{E_e}^{\infty} \left(\frac{E - E_e}{kT}\right)^{1/2} e^{-(E - E_F)/kT}\, dE$$

where $$C = \frac{4\pi}{h^3}\,(2m)$$

$$= C\,(kT)^{1/2} \int_{E_e}^{\infty} \left(\frac{E - E_e}{kT}\right)^{1/2} e^{-(E - E_e + E_e - E_F)/kT}\, dE$$

$$= C(kT)^{1/2} \int_{E_e}^{\infty} \left(\frac{E - E_e}{kT}\right)^{1/2} e^{-(E - E_e)/kT} \times e^{-(E_e - E_F)/kT}\, dE \qquad (10)$$

Now put $$x = \frac{E - E_e}{kT} \text{ or } dx = \frac{dE}{kT} \quad \therefore\; dE = (kT)\, dx$$

When $$E = E_e,\; x = 0 \text{ and when } E = \infty,\; x = \infty$$

$$\therefore\; n_e = C\,(T)^{1/2} \int_0^{\infty} (x)^{1/2}\, e^{-x} \times e^{-(E_e - E_F)/kT} \times (kT)\, dx$$

$$= C\,(kT)^{3/2} \int_0^{\infty} (x)^{1/2}\, e^{-x}\, dx \times e^{-(E_e - E_F)/kT}$$

$$= C\,(kT)^{3/2}\, e^{-(E_e - E_F)/kT} \int_0^{\infty} (x)^{1/2}\, e^{-x}\, dx$$

$$= C\,(kT)^{3/2}\, e^{-(E_e - E_F)/kT} \left(\frac{\pi^{1/2}}{2}\right)$$

$$= 2\left(\frac{2\pi\, mkT}{h^2}\right)^{3/2} e^{-(E_e - E_F)/kT} \qquad (11)$$

Equation (11) represents the density of electrons in conduction band. Since each electron which is excited upwards from the valence band leaves behind one hole, hence the number of holes in the valence band per unit volume of the solid is also the same, *i.e.* n_e.

2.30 THERMIONIC EMISSION

2.30.1 Richardson-Dushman Equation

The emission of electrons from a metal under the influence of heat is called thermionic emission. When heat energy is supplied to a metal, it will be gained by the valence electrons in the atoms. If this energy is sufficient to over come electrostatic attraction between electron and nucleus emission of electrons takes place. The minimum heat energy require for the emission of electrons is called work function of the metal (ϕ) the thermionic emission is similar to evaporation of liquid by heating. Hence, we can apply Clasius-Clapyrean equation of thermodynamics to calculate the amount of work done for the emission of electrons from the metal and it is given by

$$w = T \cdot \frac{dp}{dT}(v - u)$$

where T = temperature of the metal

$\frac{dp}{dT}$ = rate of variation of pressure from electron cloud around the metal with temperature

v = specific volume of electron cloud around the metal

u = volume occupied by the electrons inside the metal when there will be no emission

but $v >> u$

$$\therefore \quad w = T \cdot \frac{dp}{dT} v$$

The energy supplied during the work will be used in two ways.

(*i*) To remove the electrons from the atoms. If N_o is the number of electrons emitted, then the energy required for this is $N_o \phi$.

(*ii*) The remaining energy will be used to do the work to over come the pressure from electron cloud around the metal during emission, and is given by

$$dw = \int_o^v P \cdot dv = Pv$$

From the law of conservation of energy the total work done is equal to total energy spplied

$$\therefore \quad N_o\phi + pv = T \cdot \frac{dp}{dT} \cdot v$$

$$pv = KT = N_o KT$$

$$\therefore \quad v = \frac{N_o KT}{P}$$

by substituting in the above equation we get,

$$N_o\phi + N_o KT = T \cdot \frac{dp}{dT} \frac{N_o KT}{P}$$

$$N_o(\phi + KT) = N_o \frac{KT^2}{P} \cdot \frac{dp}{dT}$$

$$\frac{dp}{P} = \frac{(\phi + KT)\, dT}{KT^2}$$

integrating on both sides, we get

$$\log p = \int \frac{\phi dT}{KT^2} + \log T$$

$$\log p = I + \log T + \log A$$

$$\log p = \log T - \log A = I$$

$$\log \frac{P}{AT} = I$$

$$\frac{P}{AT} = e^I \qquad P = ATe^I$$

$$Pv = RT = N_o KT$$

If there are n number of electrons/unit volume of metal, then $v = 1$, when $N_o = n$

$$\therefore \quad P \cdot 1 = nKT \qquad \therefore \quad P = nKT$$

$$p = ATe^I \qquad nKT = ATe^I$$

$$n = \frac{A}{K} e^I \qquad n = Be^I$$

The number of electrons crossing/sec through unit area is given by

$$n_o = \frac{n\bar{c}}{\sqrt{6\pi}} = \frac{Be^I}{\sqrt{6\pi}} \cdot \sqrt{\frac{3KT}{m}} = \frac{1}{\sqrt{6\pi}} \cdot B \cdot \sqrt{\frac{3K}{m}} \cdot e^I \sqrt{T}$$

$$n_o = Ce^I \sqrt{T}$$

where
$$C = \frac{1}{\sqrt{6\pi}} B \sqrt{\frac{3K}{m}}$$

$$I = \int \frac{\phi dT}{KT^2} = \frac{-\phi}{KT}$$

$$\therefore \quad n_o = Ce^{-\phi/KT} \sqrt{T}$$

Hence, current density J is given by

$$J = en_o = e \cdot ce^{-\phi/KT} \sqrt{T}$$

$$J = D\, e^{-\phi/KT} \sqrt{T}$$

This is the Richardson-Dushman equation for thermionic emission.

3 Dielectrics

INTRODUCTION

Dielectrics are the insulators and an insulator is a material, which contains no free electrons and the number of such electrons is too low to constitute the electric current. In such substances the electrons are tightly bound to the nucleus of the atom.

When an electric field is applied between the plates of a charged capacitor in which a dielectric is placed between plates of capacitor. Then, the positive charge gets displaced in the direction of applied electric field whereas the negative charges are displaced in the opposite direction of the field. The separation of positive charge from the negative charge is due to the application of external field. This effect is known as the polarization of dielectric.

Basically dielectric materials are non conductors of electrical energy and they show special properties in the presence of an electric field. Because of these behaviour they find important applications in electronics. Glasses are widely used in optical communication. **Ex:** Glass, Ceramics and Polymers.

3.1 NON-POLAR DIELECTRICS

The dielectrics which are polarized only when they are placed in an electric field are called non-polar dielectrics. Thus, a non-polar dielectric does not possess any permanent electric dipole moment. They acquire dipole moment by induction when placed in an electric field and as shown in Fig. 3.1. The non-polar dielectrics have symmetrical structure, e.g. H_2, N_2, O_2, CO_2, CCl_4, etc.

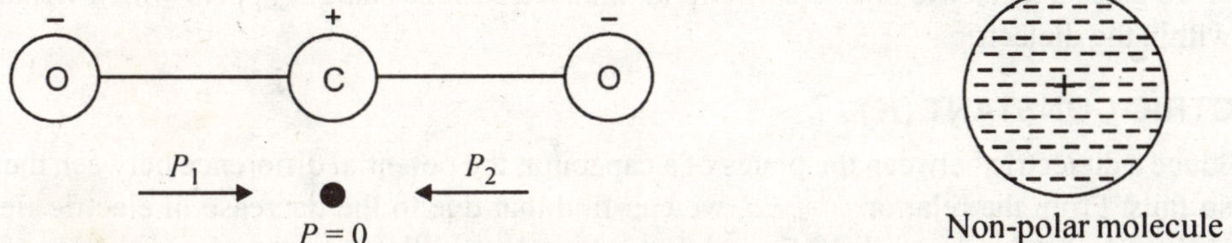

Fig. 3.1 Shows CO_2 molecule has zero electric dipole moment hence is non-polar

3.2 POLAR DIELECTRICS

A polar molecule is one in which the center of gravity of positive charge does not coincide with the center of gravity of the negative charges. Such dielectric have permanent electric dipole moments with their random orientation in the absence of external electric field as shown in Fig. 3.2 (a) and (b), e.g. HCl, H_2O, etc.

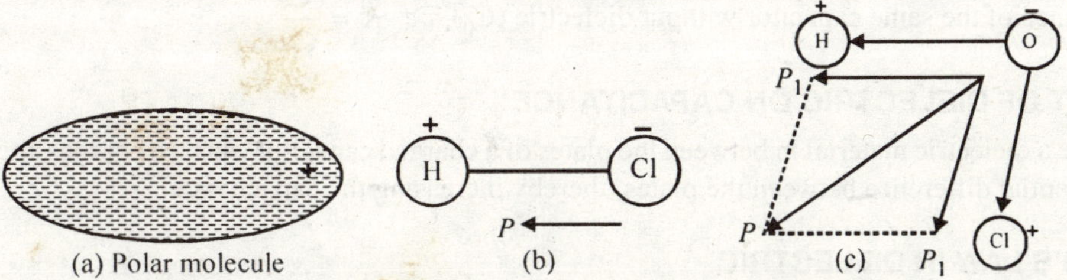

Fig. 3.2 Shows HCl molecule is polar and has net electric dipole moment (c) Shows H_2O molecule has some electric dipole moment hence H_2O is polar molecule

3.3 NON-POLAR DIELECTRIC IN AN ELECTRIC FIELD

When a non-polar dielectric is placed in an electric field E_0 between the parallel plate capacitor, then due to polarization surface charges appear, *i.e.* positive charge is induced on one surface while negative charge on other surface as shown in Fig. 3.3 (b) and (c). The induced surface charge appear in such a way that the electric field E set up by them opposes the external field E_0. The resultant field E in the dielectric is the vector sum of E and E_0, *i.e.* $E = E_0 + E$ or $E = E_0 - E$.

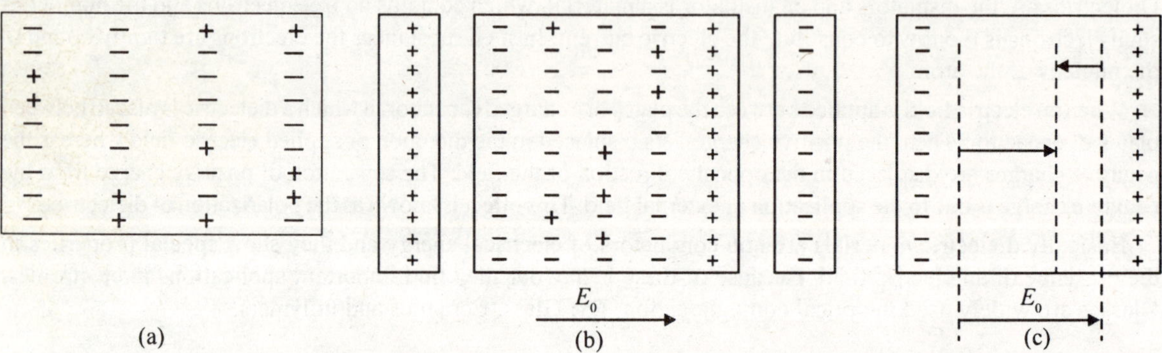

Fig. 3.3 (a) Shows the random distribution of positive and negative charges in non-polar dielectric, (b) Shows when electric field E_0 is applied then due to polarization, surface charge appears, (c) Shows the induced surface charge appear in such a way that the electric field E set up by them oppose the external field E_0

Hence, the external electric field weakens the original electric field within the dielectric. Thus, it can be conclude that if we place a dielectric in an electric field, induced surface charges appear which weakening the original field within the dielectric.

3.4 DIELECTRIC CONSTANT (*K*)

When we introduce a dielectric between the plates of a capacitor, the potential difference between the plates of a capacitor also fails. From the relation $V = Ed$, we can find that due to the decrease in electric field E, the potential difference V, should also fail. If E_0 and E are the electric fields in the absence and presence of dielectric and V_0 and V are the potential difference between the plates without the dielectric and with dielectric respectively, then

$$\frac{E_0}{E} = \frac{V_0}{V} = K \quad \text{where } K \text{ is dielectric constant}$$

As $\quad E_0 > E, K > 1.$

The dielectric constant may also be defined as the ratio of the capacitance of a capacitor with dielectric (C) to the capacitance of the same capacitor without dielectric (C_0), *i.e.* $K = \dfrac{C}{C_0}$

3.5 EFFECT OF DIELECTRIC ON CAPACITANCE

When we place a dielectric material in between the plates of a charged capacitor, surface charges appear which reduce the potential difference between the plates, thereby increasing the capacitance.

3.6 GAUSS'S LAW IN DIELECTRIC

According to Gauss's law, the electric flux through any closed surface is $1/\varepsilon_0$ times the net charge q enclosed by the surface, *i.e.*

$$\oint E \cdot \hat{n} ds = \frac{q}{\varepsilon_0} \tag{1}$$

where E = electric field vector and ds = surface area.

Figure 3.4 shows a parallel plate capacitor (a) without and (b) with dielectric. The charge q on the plates is the same. The Gaussian surfaces have also been shown in same figure.

The application of Gauss's law in the absence of dielectric gives $\oint E \cdot \hat{n} ds = \frac{q}{\varepsilon_0}$ where q = charge enclosed and $\hat{n}$ is the unit vector normal to the surface element of area ds.

or $$E_0 \oint ds = E_0 A = \frac{q}{\varepsilon_0} \quad \text{or} \quad E_0 = \frac{q}{\varepsilon_0 A} \tag{2}$$

where A = area of the capacitor plate. When the dielectric is introduced, induced surface charges appear as shown in Fig. 3.4 (b). These induced charges produce their own field which opposes the external field E_0. Let E be the resultant field within dielectric. If q is induced surface charge, then the net effective charge within Gaussian surface will be $(q - q)$. Then, the Gauss's law becomes.

$$\oint E \cdot \hat{n} ds = \frac{q - q'}{\varepsilon_0} \quad \text{or} \quad EA = \frac{q - q'}{\varepsilon_0} \quad \text{or} \quad E = \frac{q}{\varepsilon_0 A} - \frac{q'}{\varepsilon_0 A} \tag{3}$$

By definition of dielectric constant, we know that,

$$E = \frac{E_0}{K} \tag{4}$$

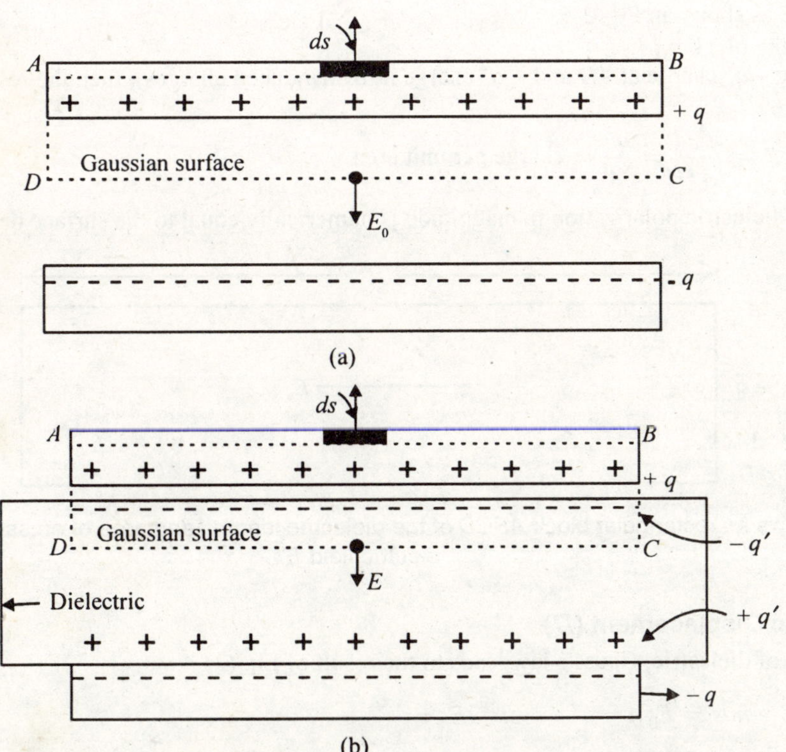

Fig. 3.4 (a) Shows parallel plate capacitor without dielectric, (b) Shows parallel plate capacitor with dielectric

From Eqs (3) and (4), we get $\frac{E_0}{K} = \frac{q}{\varepsilon_0 A} - \frac{q'}{\varepsilon_0 A}$

Substituting the value of E_0 from Eq. (2), we get

$$\frac{q}{\varepsilon_0 AK} = \frac{q}{\varepsilon_0 A} - \frac{q'}{\varepsilon_0 A} \quad \text{or} \quad q' = q\left(1 - \frac{1}{K}\right) \tag{5}$$

In the presence of dielectric, Gauss law is

$$\oint E \cdot \hat{n} ds = \frac{q - q'}{\varepsilon_0} = \frac{q}{\varepsilon_0} - q\frac{\left(1 - \frac{1}{K}\right)}{\varepsilon_0} = \frac{q}{K\varepsilon_0} \quad \text{or} \quad K \oint E \cdot \hat{n} ds = \frac{q}{\varepsilon_0} \tag{6}$$

This is Gauss's law in the dielectric.

3.7 RELATIONS BETWEEN *E*, *P* AND *D*

3.7.1 Electric Intensity (*E*)

The electric intensity at any point in the electric field is numerically equal to the force experienced by a unit positive charge placed at that point, the direction of *E* being the same as that of the force.

3.7.2 Dielectric Polarization (*P*)

The electric dipole moment per unit volume is called the dielectric polarization. It is represented by *P*. It is a vector parallel to electric field.

Consider a rectangular block *ABCD* of the dielectric of the length *l*, and area of cross-section *A* placed in an electric field *E* as shown in Fig. 3.5.

Volume of the block = *lA*.

Considering $-q'$ charge at *AD* and $+q'$ charge at *BC* (induced charges), then the resultant electric dipole moment = $q' \cdot l$.

Therefore, $P = \frac{q' \cdot l}{A \cdot l} = \frac{q'}{A}$ = charge per unit area.

Hence, the dielectric polarization in magnitude is numerically equal to the surface density of charges.

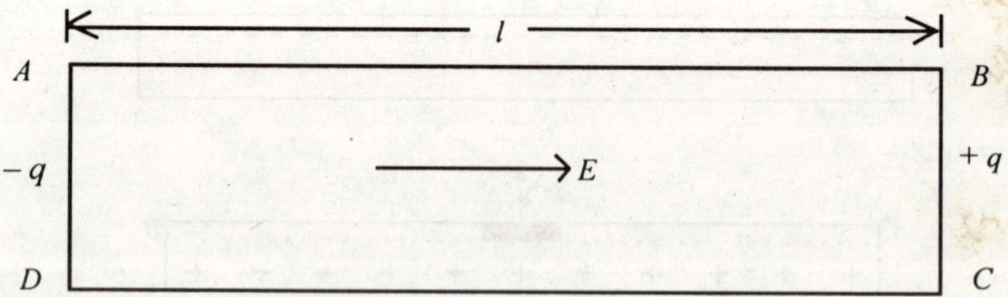

Fig. 3.5 Shows as rectangular block *ABCD* of the dielectric length *l* and area of cross-section *A* placed in electric field *E*.

3.7.3 Electric Displacement (*D*)

In the absence of dielectric, Gauss's law leads to the result of Eq. (2)

i.e. $$\oint E_0 \cdot \hat{n} ds = E_0 A = \frac{q}{\varepsilon_0} \quad \text{or} \quad E_0 = \frac{q}{\varepsilon_0 A}$$

where E_0 is the electric field and *q* is the free charge while in the presence of dielectric, induced surface charges $-q'$ appears so that Gaussian surface encloses $(q - q')$ charge. Gauss's law gives

$$\oint E \cdot \hat{n} ds = EA = \frac{q - q'}{\varepsilon_0} \quad \text{or} \quad E = \frac{q}{\varepsilon_0 A} - \frac{q'}{\varepsilon_0 A}$$

Now the field in the presence of dielectric is given by Eq. (6), *i.e.*

$$E = \frac{E_0}{K} = \frac{q}{\varepsilon_0 AK} \tag{7}$$

Putting this value of E in the preceding equation, we get,

$$\frac{q}{\varepsilon_0 AK} = \frac{q}{\varepsilon_0 A} - \frac{q'}{\varepsilon_0 A} \quad \text{or} \quad \frac{q}{A} = \frac{q}{AK} + \frac{q'}{A} \tag{8}$$

As we know, $\frac{q'}{A} = P$ and $\frac{q}{\varepsilon_0 KA} = E$

Therefore, from Eq. (8), we get, $\frac{q}{A} = \varepsilon_0 E + P$ (9)

The quantity on the left hand side of the Eq. (9) is termed as electric displacement D

$$\therefore \quad D = (\varepsilon_0 E + P)$$

where $D = \frac{q}{A}$ (10)

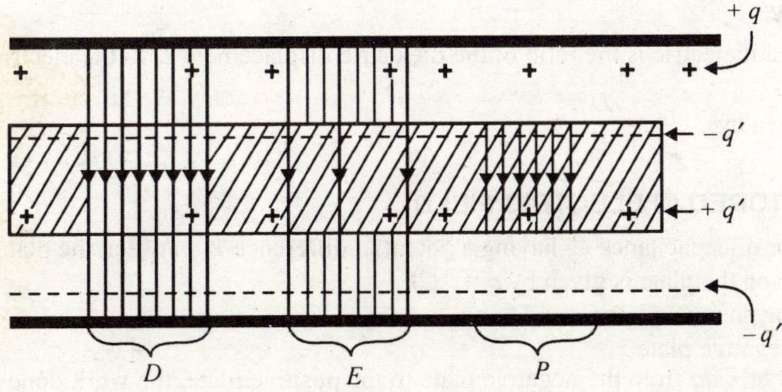

Fig. 3.6 Shows that D is connected with free charge only, P is connected with polarization charge only and E is connected with all charges

3.7.4 Relation between *D* and *E*

From Eq. (7) we know that, $E = \frac{q}{\omega_0 KA}$, which gives $\frac{q}{A} = \varepsilon_0 KE$

or $D = K\, \varepsilon_0\, E$ [obtained using Eq. (10)] (11)

3.7.5 Relation between *P* and *E*

Putting this value of electric displacement D in Eq. (10), we get,

$$\varepsilon_0 KE = \varepsilon_0 E + P \quad \text{or} \quad P = \varepsilon_0 (K - 1) E \tag{12}$$

If no dielectric is present, $K = 1$, then $P = 0$ which is correct as in the absence of any dielectric, there will be no polarization at all.

The constant $\varepsilon_0 (K - 1)$ in Eq. (12) is called as dielectric susceptibility (χ) of the dielectric.

So, $$P = \chi E \tag{13}$$

3.8 RELATION BETWEEN K AND χ

As we know $\varepsilon_0 (K - 1) = \chi$ or $\varepsilon_0 K - \varepsilon_0 = \chi$

but $K\varepsilon_0 = \varepsilon$

Therefore $\varepsilon - \varepsilon_0 = \chi$ or $\varepsilon = \varepsilon_0 + \chi$

Now $$K = \frac{\varepsilon}{\varepsilon_0} = \frac{(\varepsilon_0 + \chi)}{\varepsilon_0} = 1 + \frac{\chi}{\varepsilon_0}$$

$\therefore$ $$K = 1 + \frac{\chi}{\varepsilon_0} \tag{14}$$

3.9 ATOMIC POLARIZABILITY AND PERMITTIVITY

3.9.1 Atomic Polarizability

When a non-polar molecule is placed in electric field, the center of gravity of negative charges is displaced relative to that of positive charges. In this way the molecule acquires an induced electric dipole moment. The induced dipole moment p is directly proportional to applied electric field E.

Hence $p \propto E$ or $p = \alpha E$ where α is known as atomic polarizability.

3.9.2 Permittivity

The permittivity of a dielectric is the ratio of the dielectric displacement D to the electrical intensity E in the dielectric and is given by $\varepsilon = \dfrac{D}{E}$

3.10 ENERGY STORED IN ELECTRIC FIELD

Consider a capacitor of capacitance C having a potential difference V between the plates.

Then, the charge on the plate is given by $q = CV$

The charge on the positive plate $= +q$

and charge on the negative plate $= -q$

If we take a charge $+ dq$ from the negative plate to the positive plate, the work done against the potential difference is the given by

$$dW = Vdq = \frac{qdq}{C}$$

Therefore, work done to charge the capacitor to a total charge q from the uncharged state is obtained by integrating the above equation

$$W = \int dw = \int_o^q \frac{qdq}{c} = \frac{q^2}{2C}$$

This energy $U = W = \dfrac{q^2}{2C}$ is stored in the capacitor as potential energy of the charge. This may also be put in the form

$$U = \frac{q^2}{2C} = \frac{qV}{2} = \frac{CV^2}{2}$$

The capacitance of a parallel plate capacitor with plate area A, separation d and filled with some dielectric of dielectric constant K is given by

$$C = \frac{K\varepsilon_o A}{d}$$

∴ Energy stored $$U = \frac{CV^2}{2} = \frac{K\varepsilon_0 AV^2}{2d} = K\varepsilon_0 AD\frac{\left(\frac{V}{d}\right)^2}{2} = \frac{K\varepsilon_0 AD(E)^2}{2} \quad \left(\because \; E = \frac{V}{d}\right)$$

The energy per unit volume is given by $$u = \frac{U}{V} = \frac{U}{Ad} = \frac{K\varepsilon_0 (E)^2}{2}$$

which is required expression for the energy per unit volume in an electric field.

3.11 DIELECTRIC POLARIZATION

When a dielectric material is subjected to the influence of an external electric field leads to development of charge separation, *i.e.* the negative charges move towards the positive potential and the positive charges move towards the negative potential of the applied field. The separation of charges in the dielectric material is known as polarization. As a consequence of the separation of the charges in the material an electric dipolemoment is produced, the dipolemoment per unit volume of the material is called as dielectric polarization.

The dielectric polarization materials are classified into two groups depending on the presence of polarization.

(*i*) Polar dielectrics **Ex:** H_2O and CH_3Cl

(*ii*) Non-polar dielectrics **Ex:** H_2, O_2, CH_4 and NH_4

3.11.1 Polarization

When a dielectric material is subjected to an external field the charges are induced in the material. These induced charges give rise to an internal field E_q. If E is the external electric field and E_i is the internal field developed in the material then the resultant electric field inside the material (E_r) is given by

$$E_r = E_i + E$$

The polarization produced in the material in the presence of an external field is given by

$$p \propto E_i \quad \text{or} \quad p = \alpha E_i$$

where ∝ is the constant of proportionality and is called as polarizability. The mechanism of polarization depends on the structure of the material and are classified as

1. Electronic polarization (P_e)
2. Ionic polarization (P_i)
3. Orientational polarization (P_o)
4. Space charge polarization (P_s)

3.11.2 Electronic Polarization

In an atom and molecules with symmetric structures, the charge distribution is also symmetric, hence, the center of gravity of positive charges coincides with the centre of gravity of negative charges, therefore, the net dipole moment of the atom or the molecule is zero in the absence of external electric field. When dielectric material is subjected to an external electric field. The centre of gravity of positive charges shift towards the negative potential and the centre of gravity of the negative charges shift towards the positive potential which leads to the separation of charges and gives rise to a dipole moment and the material is said to be polarised this kind of polarization is called electronic polarization and is characterised by electronic or optical polarizability and is given by $P_e = \dfrac{\varepsilon_0 (\varepsilon_r - 1)}{n}$

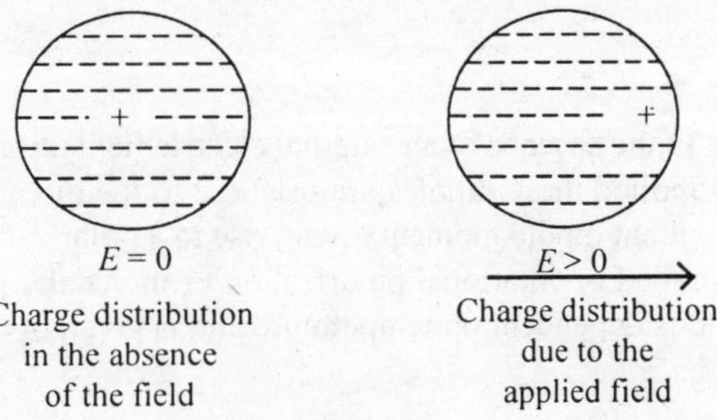

Fig. 3.7

where n = number of atoms/unit volume

The monoatomic gases exhibit the electronic polarization. The electronic polarization is independent of temperature of the material.

3.11.3 Ionic Polarization (P_i)

When an ionic solid is subjected to the influence of an electric field. The an ions and cations are separated. The extent of separation is purely depends on the strength of the applied electric field. This displacement of ions due to the electric field is called as ionic polarization. This is observed in the materials with symmetric molecules like NH_4, CH_4, CCl_4, etc. Ionic polarization is independent of temperature.

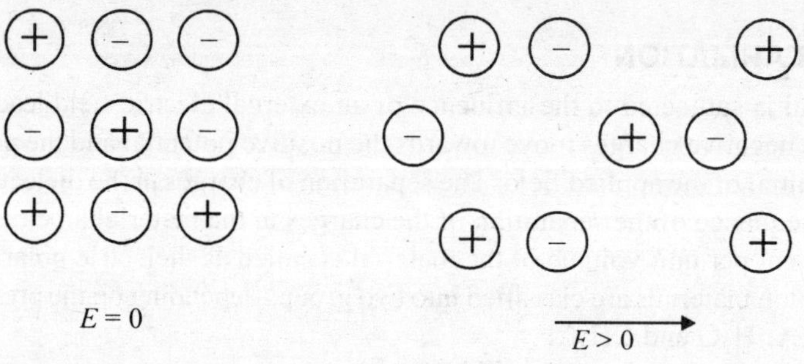

Fig. 3.8

3.11.4 Orientation Polarization (P_o)

In certain assymmetric molecules like CH_3Cl. The center of gravity of positive charges do not coinside with the center of gravity of negative charges, which leads to the development of net dipole moment even in the absence of an external electric field, the electro negativity of one of the components like Cl in CH_3Cl is more than the other. Hence, there will be an assymetric distribution of charges which gives rise to a net dipolement.

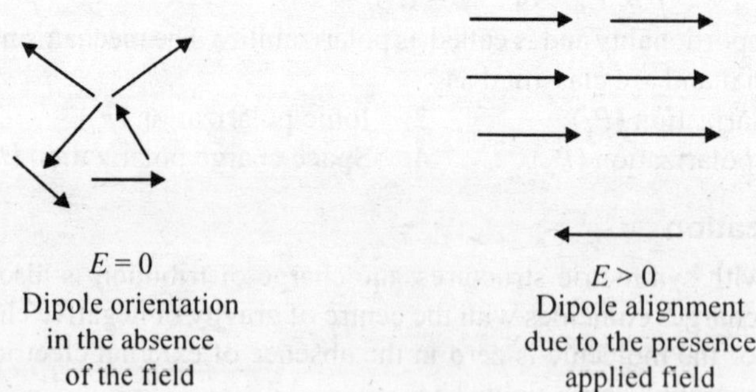

Fig. 3.9

In the absence of an external electric field the electric dipoles are randomly oriented. When an electric field is applied these dipoles are oriented to the direction of the applied electric field and the material acquires a resultant dipole moment giving rise to a polarization, which is due to the alignment of the electric dipoles and is called orientational polarization or molecular polarization. It is predominant in ionic crystalline materials and is dependent on temperature and is given by

$$P_0 = \frac{\mu^2}{3KT}$$

where μ = permanent dipole moment

3.11.5 Space Charge Polaraization (P_s)

This kind of polarization occurs due to the deposition of charges at the interface of a multiphased material. In allumina (Al_2O_3), the presence of traces of Al creates a few space charges. Similarly in (TiO_2) the presence of traces of (TiO_3) creates a few space charges. In the presence of an applied electric field, the space charges are shifted, which results in creation of dipole moment in the material. Space charge polarization is characterised by the space charge polarizibility or the infrared polarizibility.

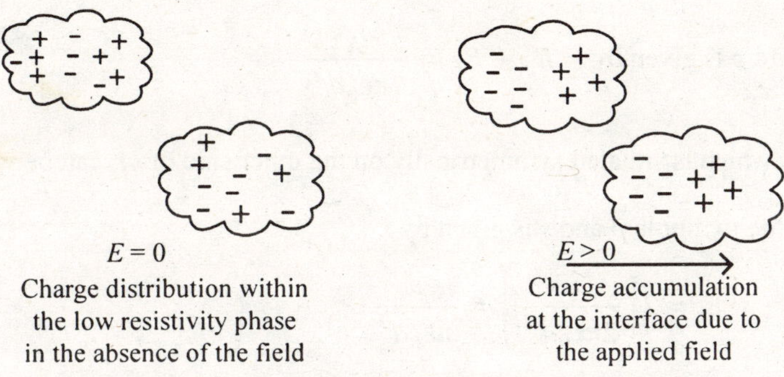

$E = 0$
Charge distribution within the low resistivity phase in the absence of the field

$E > 0$
Charge accumulation at the interface due to the applied field

Fig. 3.10

The total polarization is given by

$$P = P_e + P_i + P_o + P_s$$

and the total polarizibility is given by

$$\alpha = \alpha_e + \alpha_i + \alpha_o + \alpha_s$$

3.12 EXPRESSION FOR INTERNAL FIELD IN SOLIDS

Theory based on one dimensional atomic array. Let us consider a solid dielectric material which is subjected to an electrical field of strength *E*. Let the atomic dipoles in a solid are arranged as a linear array of equidistant atomic dipoles parallel to the direction of the applied field strength *E* as shown in Fig. 3.11.

This model is used by Epstein to evaluate the local field in one dimensional atomic array.

Consider an atom *A* in the linear array of atomic dipoles, let *d* be the interatomic distance and μ be the atomic dipole, then the total field at *A* due to all other dipole in the array is given by. (Since we know that, the polar form of components of the electric field produced at a point by an electric dipole are expressed as)

E
R Q P A S T U
d

Fig. 3.11

$$E_r = \frac{2\mu \cos\theta}{4\pi\varepsilon_0 r^3}\vec{a}_r \qquad (1)$$

and

$$E_\theta = \frac{\mu \sin\theta}{4\pi\varepsilon_0 r^3}\vec{a}_\theta \qquad (2)$$

where $\vec{a}_r$ and $\vec{a}_\theta$ are the unit vector,

μ = is the dipole moment

r = distance from the dipole to a point

and θ = is the angle between the line joining two charges.

Then, by making use of the above two expressions for the electric dipole, the electric field at A due to the dipole p which is at a distance d is given by.

Since the direction of dipole moment are co-linear

i.e. $$r = d \text{ and } \theta = 0$$

We can write E_r and $E\theta$ as

$$E_r = \frac{\mu}{2\pi\varepsilon_0 d^3} \text{ and } E_\theta = 0$$

$\therefore$ The field at A due to p is given by $= E_r + E_\theta = \dfrac{\mu}{2\pi\varepsilon_0 d^3}$

Now, for the dipole s which is situated symmetrically on the other side of A can be written as $\dfrac{\mu}{2\pi\varepsilon_0 d^3}$

$\therefore$ The field E_1 at A due to dipole p and s is given by

$$E_1 = 2\left(\frac{\mu}{2\pi\varepsilon_0 d^3}\right) = \frac{\mu}{\pi\varepsilon_0 d^3}$$

i.e. $$E_1 = \frac{\mu}{\pi\varepsilon_0 d^3}$$

If we consider the dipoles Q and T which are situated at a distance $2d$ on either side of A, then the field E_2 at A due to both atomic dipole is given by,

$$E_2 = \frac{\mu}{\pi\varepsilon_0 (2d)^3}$$

Similarly, $$E_3 = \frac{\mu}{\pi\varepsilon_0 (3d)^3}$$

$\therefore$ The total field E' induced at A due to all the dipoles in the linear array is given by

$$E' = E_1 + E_2 + E_3 + \cdots\cdots$$

$$= \frac{\mu}{\pi\varepsilon_0 (d)^3} + \frac{\mu}{\pi\varepsilon_0 (2d)^3} + \frac{\mu}{\pi\varepsilon_0 (3d)^3} + \cdots = \frac{\mu}{\pi\varepsilon_0 d^3}\left(1 + \frac{1}{2^3} + \frac{1}{3^3} + \cdots\right) = \frac{\mu}{\pi\varepsilon_0 d^3}\sum_{n=1}^{\infty}\frac{1}{n^3}$$

where n is in integer

But $$\sum_{n=1}^{\infty}\frac{1}{n^3} = 1.2$$

$\therefore$ $$E = \frac{1.2\mu}{\pi\varepsilon_0 d^3} \tag{3}$$

But, we know that, the total field at A which is the internal field E_1 and is given by the sum of applied field E and the field due to all the dipoles, *i.e.* E'

$\therefore$ $$E_i = E + E'$$

$$E_i = E + \frac{1.2\alpha_e}{\pi\varepsilon_0 d^3} \tag{4}$$

But, the electronic polarizibility and the dipolement is given by the relation

$$\mu = \alpha_e E$$

by substituting the value of μ in Eq. (3), we get

$$\therefore \quad E_i = E + \frac{1.2\alpha_e}{\pi\varepsilon_0 d^3} \quad \text{or} \quad E_i = E\left(1 + \frac{1.2\alpha_e}{\pi\varepsilon_0 d^3}\right) \tag{5}$$

This is the expression for the internal field for one dimensional dielectric solid.

Since α_e, E, ε_0 and d are positive.

$$\therefore \quad E_i > E \tag{6}$$

3.13 DEPENDENCE OF POLARIZIBILITY

The dependence of frequency on polarizability is as shown in Fig. 3.12.

When the frequency of applied field is higher than the relaxation frequency, then the dipoles cannot reorient themselves rapidly and can not remain in phase with the applied field thus they will not contribute to the total polarizibility.

The orientation polarization exist in radio and microwave region, ionic in infrared region and electronic in visible region. Both the electronic and ionic polarization function in very high frequencies hence, they contribute to polarizibility since their dipole orientations of the order of less than the atomic diameter. But the contribution of ionic polarization is up to infrared region where as the orientational polarization is limited upto microwave frequency both electronic and ionic polarization mechanism have lower relaxation time and while the orientational polarisation has high relaxation time hence, the lower relaxation frequency.

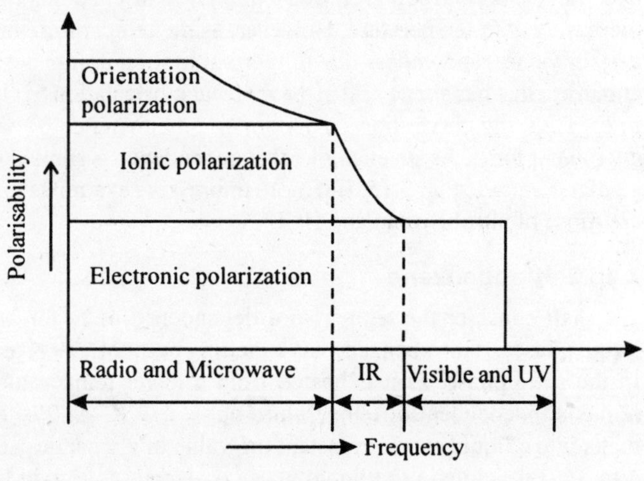

Fig. 3.12

3.14 DIELECTRIC LOSS

The loss of energy in the form of heat by a dielectric medium due to the internal friction which is developed as a consequence of shifting action of molecular dipoles under certain conditions.

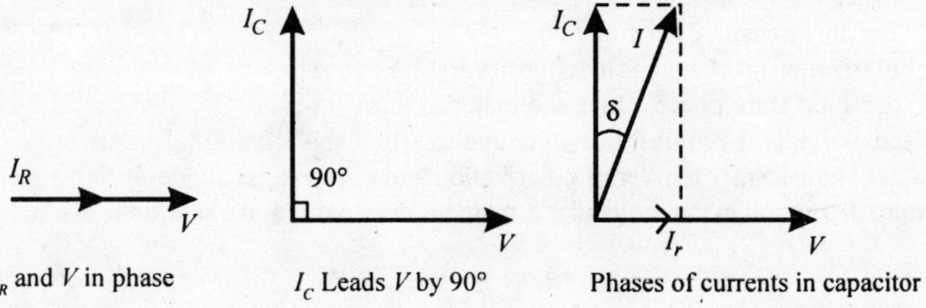

Fig. 3.13

All the different types of polarization mechanism occur in dielectric material will be effective only in static field. If τ_e, τ_i and τ_o are the relaxation time for the electronic, ionic and orientational polarization then, $\tau_e < \tau_i < \tau_o$

For every polarization there exist a minimum time for re-orientation of dipole and is called as relaxation time whose reciprocal is known as relaxation frequency.

3.15 NON-POLAR DIELECTRICS

As a property, electronic, polarization is not affected by temperature changes. However, for a non-polar dielectric material, the number of molecules which can be polarized/unit volume decreases with increase of temperature because of volumetric expansion. It leads to a decrease in ε_r. The temperature dependence of ε_r for paraffin is shown in Fig. 3.14. The slump in ε_r value that occurs around 40° C is due to the transition of paraffin from solid to liquid state at which time there will be a large increase in its value.

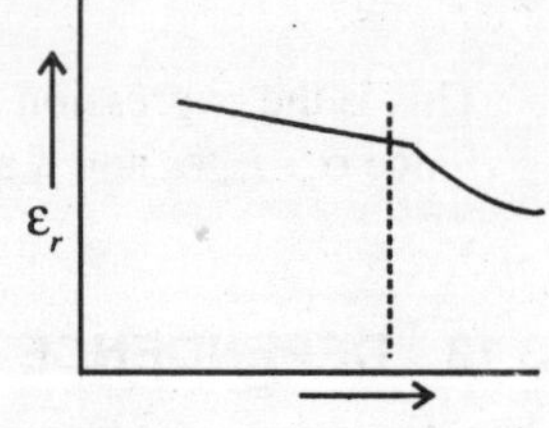

Fig. 3.14 Temperatuie dependance ε_r purafin

3.15.1 Polar Dielectrics

Polar dielectrics have the property that at low temperature the constituent molecules lay rigidly fixed in the lattice. But with increase of temperature they become relatively free and show a tendency to orient along the field. This sets ε_r increasing with temperature. However, as the temperature increases beyond a certain value, randomness due to thermal motion dominates, and disrupts the alignment that has already set in the molecular orientations. If leads to a decrease in ε_r. Thus, the curve shows a maximum on either side of which there will be decrement for ε_T. As an example, this dependence is shown for Sovol which is a polar dielectric Fig. 3.15. But more impressive examples for this dependence are those of nitrobenzene and HCl.

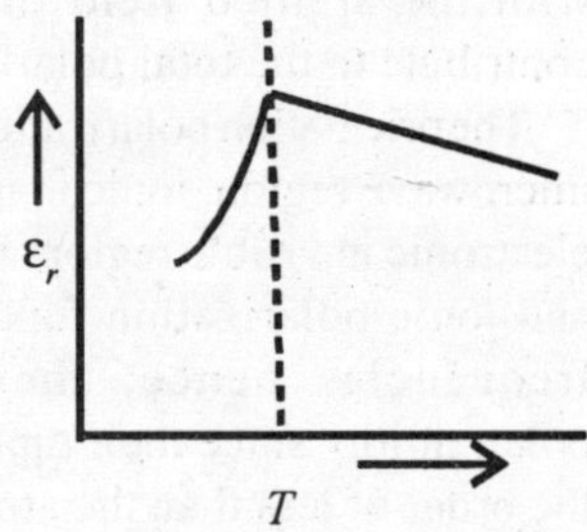

Fig. 3.15 Temperature dependance of ε_r for Sovol

3.15.2 Nitrobenzene

We shall consider the temperature dependence of ε_r for both the solid and liquid phases of nitrobenzene has a melting point of 279 K as shown Fig. 3.16. In the solid phase, as it is heated from a lower temperature, we see that ε_r remains independent of temperature up to 279 K. At 279 K the solid melts into a liquid at which instant the value of ε_r changes at once from 4 to 36. But while in the liquid phase ε_r decreases slowly but steadily with temperature. The two types of dependences in the two phases are explained as follows.

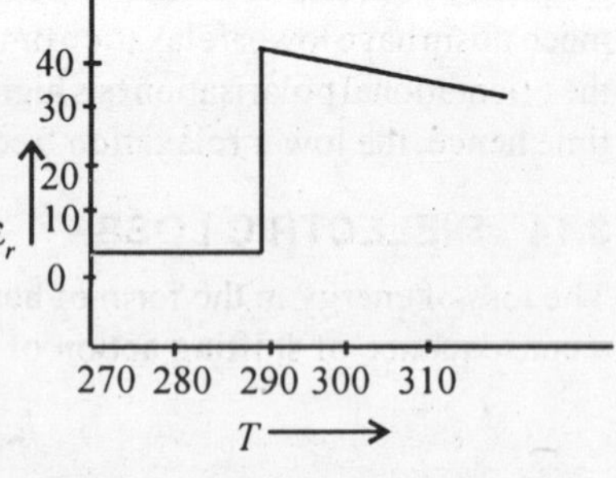

Fig. 3.16

Since nitrobenzene is a polar material, dipoles already exist in it. But in the solid phase they are not free to rotate. Hence, they lay randomly and static even when the material is subjected to a field. Thus orientational polarization is out of question. Also we know that both electronic and ionic polarization mechanisms are insensitive to temperature.

Thus with all the polarization mechanisms inactive in the solid state, ε_r remains constant in the solid state phase. Once the material turns into liquid state, the molecular dipoles become free at once and an efficient orientational polarization is brought into action. This contributes significantly to overall polarization and ε_r increases suddenly. However, with further increase in temperature disruption in the molecular alignment due to thermal agitation leads to decrease in ε_r.

3.15.3 HCl

HCl is a solid up to 159 K beyond which it is a liquid. The temperature dependence of ε_r for HCl is shown in Fig. 3.17. Comparing with the case of nitrobenzene, though it looks similar at the first instance, we must notice that the abrupt increase in ε_r value occurs here within the solid phase of HCl, whereas the same change occurs in case of nitrobenzene while there is a solid to liquid phase transition. (This strongly indicates that there must have been a change in the molecular environment in the lattice at a temperature around 100 K). In the lower temperature phase, the dipoles in the solid are virtually immobile because of which the orientational polarizational hardly changes with temperature. Thus, ε_r is low as it is not deriving contributions from any of the polarization mechanisms. In this state a slight upward trend in the direction of increasing temperature, only indicates that the dipoles are able to rotate slightly better at higher temperatures. But the new phase what HCl acquires as athe temperature crosses beyond 100 K, facilitates the dipoles for a much higher freedom of rotation in the lattice. This enables the dipoles to align along the fild easily, *i.e.* orientational polarization acts efficiently, and ε_r undergoes a large increase.

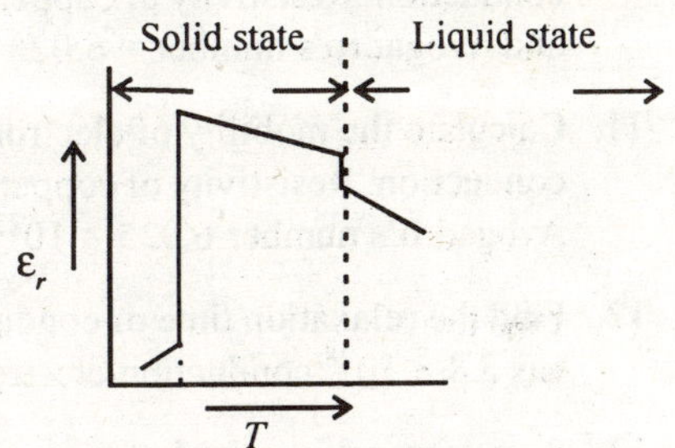

Fig. 3.17 Temperature dependance of ε_r for HCl

A further increase in temperature starts affecting the degee of order set in the preceding temperature states and ε_r begins to decrease. This trend is maintained upto 159 K at which time HCl becomes a liquid.

UNSOLVED NUMERICAL PROBLEMS

1. Calculate the Fermi energy of sodium assuming that the metal has one free electron per atom. Given $h = 6.625 \times 10^{-34}$ JS, mass of electron = 9.1×10^{-31} kg, density of sodium = 970 kg/m^3 and atomic weight of Sodium = 22.99.
2. Find the fermi energy in copper on the assumption that each copper atom contributes one free electron to the electron gas. The density of copper is 8.94×10^3 kg/m^3 and its atomic mass = 63.5 u. Given: $m = 1.66 \times 10^{-27}$ kg.
3. Determine the number density of carriers in a copper wire assuming there is one electron per copper atom. The maximum recommended current in a 14 gauge copper wire (radius = 0.81 mm, $A = 2.1 \times 10^{-6}$ m^2) used in household circuits is 15 A. Determine the drift speed of the electrons in such a case. Given: $\rho = 8.95 \times 10^3$ kg/m^3 and $M = 63.5$ g/mol.
4. Estimate the mean free time in copper at 20° C, assuming one free electron/copper atom estimate the mean free path in copper atom. Assuming that the average speed of the free electrons is about 10^6 m/s. Given: $n = 8.48 \times 10^{28}$ electrons/m^3 and $\rho = 1.673 \times 10^{-8}$ Ωm.
5. Find the drift velocity of the free electrons in a copper wire whose cross sectional area is 1.0 mm^2 when the wire carries a current of 1.0 A. Assume that each copper atom contributes one electron to the electron gas. Given: $n = 8.5 \times 10^{28}$ electrons/m^3.
6. A metallic wire has a resistivity of 1.42×10^{-8} Ωm. For an electric field of 0.14 V/m, find the average drift velocity of electrons and the mean collision time assuming that there are 6×10^{28} electrons/m^3.
7. A uniform silver has a resistivity of 1.54×10^{-8} Ωm at room temperature. For an electric field along the wire of 1 V/cm, calculate the drift velocity, the mobility and the relaxation time of electrons, assuming that there are 5.8×10^{28} electrons/m^3 of the material.
8. Calculate the drift velocity and thermal velocity of free electrons in copper at room temperature, when a copper wire of length 3 m and resistance 0.022 W carries a current of 15 A. Given: $m_{cu} = 4.3 \times 10^{-3}$ m^2/Vs.
9. Find the relaxation time of conduction electrons in a metal of resistivity 1.54×10^{-8} Ωm, if the metal has 5.8×10^{28} electrons/m^3.

10. Calculate the mobility of electrons in copper assuming that each atom contributes one free electron for conduction. Resistivity of copper = 1.7×10^{-8} Wm. Atomic weight = 63.54. Density = 8.96×10^3 kg/m^3 and Avogadro's number = 6.025×10^{23}/mole.

11. Calculate the mobility of electrons in copper assuming that each atom contributes one free electron for conduction. Resistivity of copper = 1.7×10^{-8} ohm-m. Atomic weight 63.54. Density 8.96×10^3 kg/m^3. Avogadro's number 6.025×10^{23}/mole.

12. Find the relaxation time of conduction electron in a metal of resistivity 1.54×10^{-8} ohm-m, if the metal has 5.8×10^{28} conduction electrons per m^3.

13. A uniform silver wire has a resisity of 1.54×10^{-8} ohm-m at room temperature. For an electric field of 1 volt/cm, calculate the drift velocity, the mobility, and the relaxation time of electrons assuming that there are 5.8×10^{28} conduction electrons per m^3 of the material.

14. Calculate the drift velocity and thermal velocity of conduction electrons in copper at a temperature of 300 K, when a copper wire of length 2 m and resistance 0.02 W carries a current of 15 A; given the mobility of free electrons in copper is 4.3×10^{-3} m^{2}V^{-1}s^{-1}.

15. Calculate the Fermi velocity and the mean free path for the conduction electrons in silver, given that its fermi energy is 5.5 eV, and the relaxation time for electrons is 3.97×10^{-14}s.

16. The fermi level in silver is 5.5 eV at 0° K. Calculate the number of free electrons per unit volume and the probability of occupation for electrons with energy 5.6 eV in silver at the same temperature. Given: 1 eV = 1.602×10^{-19} J.

17. Calculate the Fermi energy in eV for silver at 0° K, given that the density of silver is 10500 kg/m^3, its atomic weight is 107.9, and it has one conduction electron per atom. Given: 1 J = 6.24×10^{18} eV and $N_A = 6.025 \times 10^{28}$/k mole.

18. The Fermi level in silver is 5.5 eV. What are the energies for which the probabilities of occupancy at 300 K are 0.99, 0.01 and 0.5. Given: 1 J = 6.24×10^{18} eV.

19. Aluminimum is an FCC crystal with lattaice constant 4.05 Å, and the metal has 3 free electrons per atom. Calculate the fermi energy in eV for the metal. Also evaluate its fermi factor at an ambient temperature 300 K for an energy value 0.1 eV higher than E_F. Given: 1 Joule = 6.24×10^{18} eV.

20. Estimate the atomic spacing, *d*. Take atomic weight = 10 kg/k-mole and Avagadro number = 6.025×10^{26}/k-mole. Given the density of metallic silver is 10.5×10^3 kg/m^3.

21. Evaluate $\overline{v}$ and v_d at $T = 300$ K and $E = 1$ V/m. Take inter lattice spacing to be 3 Å. Compare their magnitudes.

22. Show that the average energy of the conduction electrons at $T = 0$ K is $\frac{3}{5} E_F$. E_F is the Fermi energy at 0 K. Given that the density of states can be written as $g(E) = A\sqrt{E}$, where A is a constant.

23. Calculate the average velocity of conduction electron on classical and quantum free electron models for silver. Comment on their variation with temperature and the type of metal.

24. Find the relaxation time of conduction electrons if the metal resistivity is 1.43×10^{-8} ohm-m. For a metal having 6.5×10^{28} conduction electrons/m^3.

25. Calculate the drift velocity of electrons in a metal of thickness 1 mm across which a potential difference of 1 volt is applied. Compare this value with the thermal velocity at temperature, 300 K. Given: The mobility of electrons = 40 cm^2/V-s.

26. A certain conductor has electron concentration $5.9 \times 10^8\ m^3$. What current density in the conductor will corresponding to a drift velocity $\frac{1}{1.6}$ m/s.

27. Metallic silver has density $10.5 \times 10^3\ kg/m^3$. Assuming one electron per atom is contributed to the conduction and, calculate the number free electron/m^3. Given that Avogadro number $N_o = 6.05 \times 10^{26}$/k-mole.

28. Show that the occupation probability at $E = E_F + DE$ is equal to the non-occupation probability at $E = E_F - \Delta E$.

29. The density of silver is $10.5 \times 10^3\ kg/m^3$. The conductivity of silver at 20° C is $6.8 \times 10^7\ ohm^{-1}\ m^{-1}$ calculate the mobility of electrons in silver.

30. Calculate the mobility and the average time of collision of the electrons in copper, obeying classical laws.

31. Sodium metal with a bcc structure has two atoms per unit cell. The radius of sodium atom is 1.85 Å. Calculate its electrical resistivity at 0° C, if the classical value of mean free time at this temperature is 3×10^{-14} s.

32. Compute the electrical resistivity of sodium at 0° C, if the mean free time at this temperature is 3.1×10^{-14} s. Furthermore, sodium builds a bcc lattice with two atoms per unit cell, and the side of the unit cell is 0.429 nm.

33. Calculate the number of states lying in an energy interval of 0.02 eV above the Fermi energy for sodium crystal of unit volume. Given: $E_F = 3.22$ eV for sodium.

34. Find the temperature at which there is 1% probability that a state with an energy 0.5 eV above fermi energy will be occupied.

35. Evaluate the Fermi function for an energy kT above the Fermi energy.

36. At what temperature we can expect a 10% probability that electrons in silver have an energy which is 1% above the Fermi energy? The fermi energy of silver is 5.5 eV.

37. A parallel plate capacitor consists of 2 plates each of area $5 \times 10^{-4}\ m^2$. They are separated by a distance 1.5×10^{-3} m and filled with a dielectric of relative permittivity 6. Calculate the charge on the capacitor if it is connected to a 100 V DC supply.

38. A parallel plate capacitor of area 650 mm^2 and a plate separation of 4mm has a charge of 2×10^{-10} C on it. What is the resultant voltage across the capacitor when a material of dielectric constant 3.5 is introduced between the plates.

39. Calculate the electronic polarizability of He atoms if the gas contains 2.7×10^{25} atoms per m^3. Given the dielectric constant of He gas at NTP is 1.0000684.

Number density of He atoms, $N = 2.7 \times 10^{25}/m^3$.

α_e = Electronic polarizability.

40. Find the polarization produced in a dielectric medium of relative permittivity 15 in presence of an electric field of 500 V/m.

41. What should be the voltage required to introduce a material of dielectric constant 4 between the plates of a parallel plate capacitor of area 1000 mm^2 having plate separation of 5 mm and a charge of 3×10^{-10} C? Also determine the applied electric field.

42. A parallel plate capacitor consists of 2 plates each of area 5×10^{-4} m^2. They are separated by a distance 1.5×10^{-3} m and filled with dielectric of relative permmittivity 6. Calculate the charge on the capacitor if it is connected to a 100V DC supply.

43. Find the polarization produced in a dielectric medium of relative permittivity 15 in the presence of an electric field of 500 V/m.

44. Find the lowest energy of an electron confined in a box of side 0.1 nm each.

45. An electron in a cubical box of metal is subjected to the influence of a magnetic field such that its length l increases by dl, while its breadth b and thickness d remain the same. Show that the energy difference between the states (311) and (111) is $\frac{h^2}{8mL^2}\left(\frac{16dL}{L}\right)$.

46. Calculate the energy of the electron in the energy level immediately after the lowest energy level, confined in a cubical box of side 0.1 mm. Also find the temperature at which the average energy of the molecules of a perfect gas would be equal to the energy of the electron in the above said level.

47. For the one-dimensional well, show that $\lim_{n \to \infty} \frac{E_{n+1} - E_n}{E_n} = 0$.

48. A parallel plate condenser has a capacitance of 2 mF. The dielectric has permmittivity $e_r = 100$. For an applied voltage of 1000 V, find the energy stored in the condenser as well as the energy stored in polarising the dielectric.

49. The following data refers to a dielectric material. $e_r = 4.94$ and $n^2 = 2.69$, where n is the index of refraction. Calculate the ratio between electronic and ionic Polarizability for this material.

50. The atoms of a paramagnetic material, each with a magnetic moment 1.2×10^{-23} amp · per m^2 are placed in an external magnetic field of induction 1.5 wb/m^2. Calculate the difference in magnetic potential energy for extreme orientations of the atomic dipoles and compare this energy with the thermal energy of translation of the atoms at a temperature of 27 °C.

51. A solid elemental dielectric, with density 3×10^{28} atoms/m^3 shows an electronic polarisability of 10^{-40} farad m^2. Assuming the internal electric field to be a Lorentz field, calculate the dielectric constant of the material.

52. Calculate the electronic polarizability of He atoms if the gas contains 2.7×10^{25} atoms per m^3. The dielectric constant of He gas at NTP is 1.0000684.

53. A parallel plate capacitor has an area of 100 cm^2, a plate separation of 1 cm and is charged to a potential of 100 V. Calculate the capacitance of the capacitor and the charge on the plates.

54. The relative dielectric constant of sulphur is 3.75 when measured at 27° C. Assuming the internal tield constant $\lambda = \frac{1}{3}$, calculate the electronic polarisability of sulphur if its density at this temperature is 2050 kg/m^3. The atomic weight of sulphur being 32.

55. Calculate the radius of atoms in α-iron belonging to BCC structure. Take the density of α-iron as 7860 kg/m^3 and atomic weight of iron as 55.85.

56. Silicon has the structure as diamond and a bond length of 2.351 Å. Estimate its density.

57. Germanium crystallizes in the diamond cubic structure with eight atoms in an unit cell. The lattice constant is 5.62 Å. Calculate the density of Germanium.

58. Zinc has a HCP structure. The height of the unit cell is 4.94 Å. The centres of the atoms are 2.7 Å. Calculate the volume of the unit cell and density of zinc.

59. For a metal having 6.5×10^{28} conduction electrons per m^3, find the relaxation time of the conduction electrons if the metal resistivity is 1.43×10^{-8} ohm-m.

60. Calculate the drift velocity of electrons in a metal of thickness 1 mm across which a potential difference of 1V is applied. Compare this value with the thermal velocity at temperature 300 K. Given: the mobility of electron = 40 cm^2/Vs.

61. The free electron density of a conductor is $10^{25}/m^3$. Find the drift velocity of the free electrons when the conductor has a current whose density is 1.12×10^6 amp/m^2.

62. Gold has one free electron/atom. Its density atomic weight and resistivity are 19300 kg/m^3, 197 and 2.21×10^{-8} Ωm. Calculate the free electron concentration, the Fermi energy in eV and the mobility of conductiion electrons. Given: $N_A = 6.025 \times 10^{26}$/k mole.

63. Calculate the free electron concentration and hence the mobility of electron in aluminium metal assuming that each of its atom contributes 3 free electrons for conduction.

Given: for aluminium, resistivity = 2.7×10^{-8} Ωm, atomic weight = 26.98, and density = 2.7×10^3 kg/m^3.

64. Calculate the drift velocity and thermal velocity of conduction electrons in aluminium at temperature 300 K, when an aluminium wire of length 5 m and resistance 0.08 Ω, carries a current of 15 A; given the mobility of free electrons in Al is 1.28×10^{-3} $m^2V^{-1}s^{-1}$.

65. The relaxation time of conduction electrons in copper is 2.6×10^{-4} s. If copper has a resistivity of 1.6×10^{-8} Ω and Fermi energy of 7 eV, find the charge carrier density and fermi velocity.

66. Calculate the mobility and the relaxation time of electrons in copper with the following data: Resistivity of copper = 1.73×10^{-8} ohm-m; atomic weight = 63.5; Density = 8.92 gm/cc; Avogadro number = 6.02×10^{23}.

67. Find the relaxation time for conduction electrons in aluminium metal, given that for aluminium, the electron concentration = $18 \times 10^{28}/m^3$ and resistivity = 2.7×10^{-8} Ωm.

68. Calculate the fermi velocity and the mean free path for conduction electron in aluminium, given that, its fermi energy is 11.63 eV, and the relaxation time for electrons is 7.3×10^{-15} s.

69. Calculate the conductivity of copper at 300 K. The collision time for electron scattering is 2×10^{-14} s at this temperature. Given: Density and atomic weight for copper are 8.96 gm/cc, and 63.5 respectively.

70. Potassium is a bcc crystal with lattice constant 5.25 Å, and the metal has one free electron per atom. Calculate the fermi energy in eV for the metal. Also evaluate its fermi factor at a temperature of 600 K foran energy value 0.075 eV higher than E_F. Given: 1 Joule = 6.24×10^{18} eV.

71. The fermi level in potassium is 2.1 eV. What are the energies for which the probabilities of occupancy at 300 K are 0.99, 0.01 and 0.5. Given: 1 J = 6.24×10^{18} eV.

72. The fermi level in potassium is 2.1 eV at a particular temperature. Calculate the number of free electrons per unit volume in potassium at the same temperature. Given: 1 eV = 1.602×10^{-19} J.
73. Cesium metal has one conduction electron/atom. Calculate the Fermi energy in eV for cesium, given that it is a bcc crystal with a lattice constant of 6.14 Å. Given: 1 J = 6.24×10^{28} eV.
74. There are 10^{20} conduction electrons per m^3 in a material having resistivity of 0.1 ohm·m. Find the charge mobility and the electric field needed to produce a drift velocity of 1 m per second.
75. What is the resulting voltage across the plates of a parallel plate capacitor with plates of area 1000 mm^2, carrying a charge of 3×10^{-10} C, and separated by a distance of 5 mm, when a material of dielectric constant 4 is introduced between them?

MULTIPLE CHOICE QUESTIONS

1. The units for permittivity of free space ε_o are
   (*a*) Hm^{-1} (*b*) Fm^{-1} (*c*) Cm^{-1} (*d*) dimensionless
2. Choose the correct relation
   (*a*) $E = \varepsilon_o (\varepsilon_r - 1) P$ (*b*) $D = \varepsilon_o (\varepsilon_r - 1) E$ (*c*) $P = \varepsilon_o (\varepsilon_r - 1) E$ (*d*) $\varepsilon_r = (\chi - 1)$
3. Insertion of a dielectric material between the plates of a capacitor
   (*a*) increases the capacitance (*b*) decreases the capacitance
   (*c*) results in no change in capacitance (*d*) none of these
4. In a capacitor of capacitance 10 μF, the spacing between the two plates is completely filled with a dielectric material of permittivity 10. The capacitance
   (*a*) decreases to 1 μF (*b*) increases to 20 μF (*c*) increases by 90 μF (*d*) does not change
5. Electronic polarization
   (*a*) decreases with increase of temperature (*b*) increases with temperature
   (*c*) may increase or decrease with temperature (*d*) is independent of temperature
6. Ionic polarization
   (*a*) decreases with increase of temperature (*b*) increases with temperature
   (*c*) may increase or decrease with temperature (*d*) is independent of temperature
7. Orientational polarization
   (*a*) decreases with increase of temperature (*b*) increases with temperature
   (*c*) may increase or decrease with temperature (*d*) is independent of temperature
8. Space charge polarization
   (*a*) decreases with increase of temperature (*b*) increases with temperature
   (*c*) may increase or decrease with temperature (*d*) is independent of temperature
9. At 10^2 Hz frequency range
   (*a*) only space charge polarization exists (*b*) only electronic polarization exists
   (*c*) both ionic and orientation polarizations exist (*d*) all the polarization process exist
10. At ~ 10^{13} Hz range
    (*a*) only electronic polarizatioin alone is present
    (*b*) ionic polarization occurs in addition to electronic polarization
    (*c*) orientational polarization occurs (*d*) space charge polarization occurs
11. Dielectric loss occurs when the dielectric is subjected to
    (*a*) DC voltage (*b*) AC voltage
    (*c*) both DC and AC voltages (*d*) none of these

12. Clausius–Mosotti equation with usual meaning of notations can be written as

(a) $\frac{\varepsilon_r + 2}{\varepsilon_r - 1} = N\alpha_e$ (b) $\frac{\varepsilon_r + 1}{\varepsilon_r - 2} = \frac{N\alpha_e}{3\varepsilon_o}$ (c) $\frac{\varepsilon_r - 1}{\varepsilon_r + 2} = \frac{N\alpha_e}{3\varepsilon_o}$ (d) $\frac{\varepsilon_r - 1}{\varepsilon_r + 2} = \frac{N\varepsilon_o}{3\alpha_e}$

13. Posistors are used to
(a) measure humidity (b) measure resistance
(c) measure and control temperature (d) measure voltage

14. Using pyroelectrometer one can measure
(a) temperature changes as small as $10^{-6\circ}$ C (b) mass variation as less as 10^{-6} gm
(c) temperature changes to the accuracy of 1° C only
(d) mass variation to the accuracy of 1 million only

15. Sommerfeld developed
(a) classical free electron theory (b) quantum free electron theory
(c) zone theory (d) corpuscular theory

16. Zone theory of metals was developed by
(a) Drude and Lorentz (b) Sommerfeld (c) Bloch (d) Newton

17. The expression for electrical conductivity in terms of mobility of electron *m* density of electrons *n* and charge of electron *e* is given by
(a) $\sigma = 1/ne\mu$ (b) $\sigma = n/e\mu$ (c) $\sigma = n\mu/e$ (d) $\sigma = ne\mu$

18. The resistivity of the material of the conductor is given by
(a) $\rho = m/ne^2\tau_c$ (b) $\rho = ne^2\tau_c/m$ (c) $\rho = n/me^2\tau_c$ (d) $\rho = me^2\tau_c/n$

19. If 2×10^{-14} s and 10^5 m/s are the collision time and the root mean square velocity of the electron respectively, at room temperature the mean free path of the electron is
(a) 2×10^{-19} m (b) 5×10^{18} m (c) 2.0 Gm (d) 2.0 nm

20. If R is the universal gas constant, according to the classical free electron theory the value of specific heat of metals is given by
(a) 3 R (b) 4.5 R (c) $\frac{3}{2}$ R (d) 0.01 R

21. If R is the universal gas constant, according to the classical free electron theory the value of electronic specific heat of metals is given by
(a) 3 R (b) 4.5 R (c) $\frac{3}{2}$ R (d) 0.01 R

22. Where K and σ are thermal and electrical conductivities in a solid, according to Widemann-Franz law

(a) $\frac{KT}{\sigma}$ = constant (b) $\frac{K\sigma}{T}$ = constant (c) $\frac{\sigma}{KT}$ = constant (d) $\frac{K}{\sigma T}$ = constant

23. If the mobility of the electrons $m = 7 \times 10^{-3}$ m²/vs, when accelerated by a field $E = 1$ v/cm, the drift velocity is given by
(a) 0.7 m/s (b) 7×10^{-3} m/s (c) 7×10^{-5} m/s (d) 0.007 m/s

24. At any temperature other than 0 K, the probability of finding an electron at fermi level is
(a) 1 (b) 0 (c) 0.5 (d) any positive value

25. The minimum possible energy of an electron in one-dimensional well of width L is given by

(a) $\frac{h}{8\text{ mL}}$ (b) $\frac{2\pi h}{\text{L}}$ (c) zero (d) $\frac{h^2}{8\text{ mL}^2}$

26. Mean energy of electrons at 0 K is given by

(a) $\bar{E}_o = \frac{E_F}{2}$ (b) $\bar{E}_o = \frac{3}{5} E_F$ (c) $\bar{E}_o = \frac{5}{3} E_F$ (d) $\bar{E}_o = \frac{3}{8} E_F$

27. The theory which proposes the free electrons inside periodic lattice field is
(*a*) Classical free electron theory (*b*) Quantum free electron theory
(*c*) Zone theory (*d*) Corpuscular theory
28. According to Matthiessen's rule, if r_0 and $r(T)$ are the temperature independent and dependent parts of the resistivity of a metal, then the total resistivity is
(*a*) $\rho = \rho_0 - \rho(T)$ (*b*) $\rho_0 = \rho + \rho(T)$ (*c*) $\rho = \rho_0 + \rho(T)$ (*d*) $\rho = \rho_0/\rho(T)$
29. Materials having different properties along different directions are called
(*a*) isotropic (*b*) anisotropic (*c*) amorphous (*d*) none of these
30. The tiny fundamental block which, when repeated in space indefinitely generates a crystal is called
(*a*) primitive cell (*b*) lattice cell (*c*) unit cell (*d*) none of these
31. There are — basic crystal systems
(*a*) four (*b*) five (*c*) six (*d*) seven
32. There are — distinguishable ways of arranging points in three dimensional space called Bravais lattices
(*a*) 7 (*b*) 14 (*c*) 8 (*d*) 5
33. Total number of symmetry elements possible for a cubic system is
(*a*) 7 (*b*) 14 (*c*) 32 (*d*) 23
34. The coordination number of simple cubic structure is
(*a*) 6 (*b*) 8 (*c*) 12 (*d*) 4
35. Effective number of atoms belonging to the unit cell of simple cubic structure is
(*a*) 8 (*b*) 1 (*c*) 4 (*d*) 6
36. Atomic packing factor of simple cubic structure is
(*a*) 0.68 (*b*) 0.74 (*c*) 1.00 (*d*) 0.52
37. The only element exhibiting simple cubic structure is
(*a*) silver (*b*) polonium (*c*) zinc (*d*) diamond
38. The relation between atomic radius r and lattice constant a in the case of simple cubic structure is
(*a*) $a = 2r$ (*b*) $a = \frac{r}{2}$ (*c*) $a = \frac{4}{\sqrt{3}}r$ (*d*) $a = 2\sqrt{2}\,r$
39. Effective number of atoms belonging to the unit cell of BCC structure is
(*a*) 8 (*b*) 1 (*c*) 2 (*d*) 9
40. Atomic packing factor of BCC structure is
(*a*) 0.68 (*b*) 0.74 (*c*) 0.52 (*d*) 1.00
41. The relation between atomic radius r and lattice constant a in the case of BCC structure is
(*a*) $a = 2r$ (*b*) $a = \frac{4}{\sqrt{3}}r$ (*c*) $a = 2\sqrt{2}\,r$ (*d*) $a = \frac{\sqrt{3}}{4}r$
42. The element crystalline in BCC structure is
(*a*) polonium (*b*) sodium (*c*) copper (*d*) zinc
43. The coordination number of BCC structure is
(*a*) 6 (*b*) 8 (*c*) 12 (*d*) 4
44. Effective number of atoms belonging to the unit cell of FCC structure is
(*a*) 14 (*b*) 8 (*c*) 4 (*d*) 2
45. The relation between atomic radius r and lattaice constant a in the case of FCC structure is
(*a*) $a = 2r$ (*b*) $a = \frac{4}{\sqrt{3}}r$ (*c*) $a = 2\sqrt{2}\,r$ (*d*) $a = \frac{r}{2\sqrt{2}}$

46. The element crystalline in FCC structure is
(*a*) copper (*b*) sodium (*c*) zinc (*d*) diamond
47. The coordination number of HCP structure is
(*a*) 6 (*b*) 8 (*c*) 12 (*d*) 9
48. Effective number of atoms belonging to the unit cell of HCP structure is
(*a*) 6 (*b*) 8 (*c*) 12 (*d*) 9
49. Atomic packing factor of FCC structure is
(*a*) 0.68 (*b*) 0.74 (*c*) 0.52 (*d*) 1.00
50. In HCP structure $\frac{c}{a}$ ratio is equal to
(*a*) $\sqrt{\frac{3}{8}}$ (*b*) $\sqrt{8 \times 3}$ (*c*) $\frac{1}{\sqrt{8 \times 3}}$ (*d*) $\sqrt{\frac{8}{3}}$
51. The number of atoms present in the unit cell of diamond cubic crystal structure is
(*a*) 2 (*b*) 4 (*c*) 8 (*d*) 16
52. Atomic packing factor of diamond crystal structure is
(*a*) 0.68 (*b*) 0.74 (*c*) 0.52 (*d*) 0.34
53. The number of sulphide atoms in the unit cell of zinc sulphide crystal is
(*a*) 2 (*b*) 4 (*c*) 3 (*d*) 6
54. The number of zinc atoms in the unit cell of zinc sulphide crystal is
(*a*) 6 (*b*) 3 (*c*) 4 (*d*) 2
55. The coordination number of NaCl structure is
(*a*) 6 (*b*) 8 (*c*) 12 (*d*) 4
56. The number of molecules present in each unit cell of NaCl is
(*a*) 8 (*b*) 4 (*c*) 27 (*d*) 12

EXERCISE

1. Explain classical free-electron theory of metals. Mention its drawbacks.
2. What are conduction electrons ? Give the assumptions on which the classical free electron theory is based.
3. Give the detail concept of free electrons as per classical free electron theory.
   Discuss briefly how valence electrons in the solid state of a metal, with the help of an example.
4. Define electrical resistance.
5. State the main assumptions of the classical free electron theory.
6. Write a note on relaxation time, mean collision time, and mean free path.
7. Describe the concept of relaxation time.
8. Explain the terms drift velocity and relaxation time for free electrons in metals.
9. Describe under what circumstances, the relaxation time is equal to the mean collision time.
10. Assuming the electron-lattice interaction to be responsible for scattering of conduction electrons in a metal, obtain an expression for conductivity in terms of relaxation time.
11. What are the effects of temperature and impurity on electrical resistivity of metals?

12. Discuss the effect of temperature and impurity on electrical resistivity of metals, and describe what are ideal resistivity and residual resistivity.
13. Discuss the effect of temperature on the electrical conductivity of metals.
14. Discuss the free electron theory of electronic conduction in metals. Comment on the effect of temperature and impurities on conductivity.
15. State Matthiessen's rule.
16. Starting from Matthiessen's rule, account for the nature of total resistivity, both at high and at low temperature.
17. Obtain an expression for electrical conductivity with the help of classical free electron theory.
18. Based on free electron theory derive an expression for electrical conductivity in a metal and discuss the failure of classical free electron theory.
19. What are relaxation time and collision time? On the basis of free electron theory of metals, obtain an expression for the electrical conductivity of a metal in terms of collisions time.
20. Explain free-electron theory of conductivity in conducting materials. Hence, derive an expression for conductivity of a conductor.
21. Explain electrical conductivity with the help of free electron theory.
22. Explain Drude-Lorentz theory of conductivity in metals. Hence derive an expression for conductivity of a conductor.
23. Using free electron theory of metals obtain an expression for electrical conductivity of a metal in terms of mean collision time.
24. Using the free electron model derive an expression for electrical conductivity in metals.
25. Explain the failure of classical free theory.
26. Describe how the classical free electron theory fails to account for the dependence of electrical conductivity on electron concentration.
27. Elucidate the difference between classical free electron theory and quantum free electron theory.
28. Discuss the quantum mechanical model of the electrons in a metal. Comment on the effect of temperature and impurities on conductivity.
29. Explain the terms Drift velocity and Relaxation time for free electrons in metals. Discuss the various drawbacks of classical free electron theory of metals and explain the assumptions made in quantum theory to overcome the draw backs.
30. State the main assumptions of quantum free electron theory along with those which are applicable from classical free electron theory also.
31. Compare classical free electron theory with quantum free electron theory. Describe how quantum free electron theory has been successful in overcoming the failures of classical free electron theory.
32. Explain density of states.
33. Derive the equation for density of states for a unit volume of solid.

34. Define fermi energy and Fermi factor. Discuss about the variation of fermi energy with temperature and energy.

35. Given an account of fermi factor.

36. What is the physical basis for concept of fermi energy? What is the assumption made for fermi energy for $T > 0°$ K? Justify the same.

37. What is fermi factor? Give the theory for calculation of fermi energy at $T = 0°$ K.

38. Discuss the dependence of fermi factor on temperature.

39. Discuss the probability of occupation of various energy states by electrons at $T = 0$, and $T > 0$, on the basis of fermi factor.

40. Define Fermi–Dirac distribution.

41. Give an account of Fermi–Dirac distribution.

42. Discuss Fermi–Dirac distribution for different temperature conditions.

43. Derive the equation for density of states for a unit volume of solid.

44. Give a qualitative account of thermionic emission.

45. Write a note on thermionic emission.

46. Write note on

(*a*) Mean free path

(*b*) Classical free electron theory of metals and its shortcomings

(*c*) Relaxation time

(*d*) Matthiessen's rule

(*e*) Quantum free electron theory

(*f*) Fermi energy

(*g*) Fermi factor

(*h*) Temperature dependence of fermi factor

(*i*) Thermionic emission

(*j*) Fermi temperature and fermi velocity

47. With usual notations show that $P = \varepsilon_o (\varepsilon_r - 1) E$

48. What is dipolar relaxation? Discuss the frequency dependence of orientational polarization?

49. Obtain an expression for the internal field seen by an atom in an infinite array of atoms subjected to an external field.

50. What is orientation polarization? Derive an expression for the mean dipole moment when a polar material is subjected to an external field.

51. Explain the electronic polarizability in atoms and obtain an expression for electronic polarizability in terms of the radius of the atom.

52. Discuss the frequency dependence of various polarization processes in dielectric materials.

53. Explain ionic polarizability of a molecule and its frequency dependence.

54. Describe the frequency dependence of various polarization process in dielectric materials.

55. Write notes on

(*i*) Relaxation time of polarization

(*ii*) Internal field of solids

56. Explain the origin of different kinds of polarization.
57. Explain ionic and orientation polarization.
58. Write a note on atomic view of dielectrics.
59. Define the terms dipole moment and polarization.
60. Obtain an expression for polarization in terms of the electric field E and the dielectric constant.
61. What are dielectric materials? Explain their properties. Give examples.
62. What are dielectrics? Discuss their important uses.
63. Discuss the effect of electric field on dielectric materials.
64. Write about important applications of dielectric materials.
65. What are dielectric materials?
66. Define dielectric constant of a material.
67. Explain the phenomenon of electric polarization in dielectric materials.
68. Explain with theory how static dielectric constant of a dielectric material is determined.
69. Explain the theory that forms the basis for determination of the static dielectric constant of dielectric materials.
70. What is electric polarization of an atom?
71. Describe in brief the various mechanisms of polarization.
72. Describe the different polarization mechanisms.
73. What are dielectric materials? Explain the dependence of polarizability and dielectric loss on frequency.
74. Explain the changes observed in a dielectric material with increase in temperature, in terms of polarization.
75. Describe in brief ionic polarization and electronic polarization.
76. Explain dielectric polarizability.
77. Discuss briefly ionic polarization and losses in dielectrics.
78. Define dielectric polarization.
79. For a dielectric material in an AC field, state the conditions under which,

    (*a*) the absorption of energy from the field becomes maximum, and

    (*b*) the polarization mechanism is halted
80. Describe the frequency dependence of various polarizabilities and absorption losses.
81. Derive an expression for internal field in a one-dimensional array of atoms in solids.
82. Explain the term internal field and derive an expression for the same in the case of liquids and solids.
83. What are dielectrics? Arrive at an expression for internal field in the case of solids and liquids.
84. Derive the expression for the internal field in case of solid and liquid of dielectric.
85. Obtain an expression for the internal field does not exist normally in gases.

86. What is internal field? Explain why internal field does not exist normally in gases.

87. Derive an expression for the internal field for a linear array of atoms in a dielectric.

88. Define internal field of dielectric and hence derive an expression for it.

89. What is Lorentz field?

90. Define dielectric loss. Obtain an expression for it and hence discuss the factors on which loss is dependent.

91. Discuss static dielectric constant.

92. Describe how static dielectric constant could be determined.

93. Discuss the temperature dependence of dielectric constant for nitrobenzene.

94. Discuss the temprature dependence of dielectric constant in the cases of nitrobenzene and HCl and compare.

95. Discuss the temperature dependence of dielectric constant in case of polar and non-polar dielectrics taking specific examples.

96. Give an account of dielectric losses.

97. Write a note on dielectric loss.

98. Define dielectric loss and derive the expression for tangent loss.

99. Define relaxation time and relaxation frequency.

100. Write a note on

(*a*) Electronic polarization

(*b*) Ionic polarization

(*c*) Orientational polarization

(*d*) Polarization mechanisms

(*e*) Lorentz field

(*f*) Temperature dependence of dielectric constant.

Magnetic Properties of Materials

3.16 MAGNETIC DIPOLE

We know that, a bar magnet or a current carrying solenoid or a coil have two poles namely the north and south which point towards north and south directions when they are suspended freely.

An arrangement consisting of two equal and opposite poles kept at a distance is called magnetic dipole. An important characteristic of magnetic dipole is that when it is kept in an external magnetic field a torque acts on it due to which the dipole is oriented in a particular direction parallel to the field. A compass needle, bar magnet, current carrying loop, coil and solenoid experience a torque in an external magnetic field and hence, all are magnetic dipoles.

Every magnetic dipole has a magnetic dipole moment denoted by $\vec{P}_m$. If a magnetic dipole of magnetic dipole moment $\vec{P}_m$ is kept in an external magnetic field $\vec{B}$, the torque acting on the dipole is given by

$$\vec{\tau} = \vec{p}_m \times \vec{B}$$

The magnitude of this torque is

$$\tau = p_m B \sin\theta$$

where θ is the angle between $\vec{P}_m$ and $\vec{B}$.

The direction $\vec{P}_m$ is taken as that from South pole to North pole.

3.17 MAGNETIC DIPOLE MOMENT OF A CURRENT LOOP

Let a coil of N turns each of area of cross-section A, carrying a current I ampere is placed in a magnetic field $\vec{B}$ in such a way that the plane of the coil makes an angle θ with the field, a torque acts on the coil whose magnitude is given by

$$\tau = NIAB \sin\theta = NIA\,(B \sin\theta)$$

Comparing this equation with the expression of torque acting on a current carrying coil placed in a magnetic field, we get

$$\tau = p_m B \sin\theta$$

Therefore the magnetic dipole moment of a current carrying coil, $p_m = NIA$

The SI unit of magnetic dipole moment is **ampere-meter² (Am^2).**

3.18 MAGNETIC SUBSTANCES AND MAGNETIC PROPERTIES

It has been observed that generally all substances exhibit magnetic properties in some or the other ways. Iron, nickel and cobalt are attracted towards the magnet in a strong way. Such substances are called as Ferro-magnetic substances. Some substances are attracted weakly by the magnets and are called as para-magnetic substances. On the other hand some substances are repelled by magnet instead of being attracted and they are known as dia-magnetic substances.

Initially the presence of magnetic property in the substances was attributed to the ultimate particles of the matter which behave like tiny magnets. In the process of magnetisation of the substances these tiny atomic or molecular magnets get arranged in such a way that their like poles lie on one side and the substances as a whole start behaving like a magnet.

After the discovery of magnetic effect of current, *i.e.* the production of magnetic field by the current flowing in a conductor, the explanation of the properties of magnetic substances was provided by considering the electrons revolving in the orbits to be equivalent to small current carrying loops. We know that the electrons in the atoms revolve in closed orbits around the nucleus and at the same time the electrons spin about their own axis. These electrons executing revolution about nucleus and spin about their own axis are treated to be equivalent to small current carrying loops and magnetic field is produced due to these current loops. In non-magnetic substances these current carrying loops equivalent to the electrons in motion are arranged at random so that the magnetic fields due to these loops inside the substances cancel each other. In the process of magnetisation of the substances these loops are so arranged that their planes become parallel to each other and currents in them are in the same direction, so that the magnetic fields of all the loops are added to produce the resultant magnetic fields.

3.19 MAGNETISATION OF SUBSTANCES

We know that, the electrons in a substance are revolving around the nucleus and spinning about their own axis. Hence, they are equivalent to current loops which give rise to the magnetic field. In an atom many current loops are set up due to the motions of all electrons in the atom which are known as atomic currents.

There are two currents in a substance

(*i*) The conduction current I_c due to the flow of free electrons.

(*ii*) Atomic currents which are completely circulating currents in which there is no displacement of charge. Due to the resultant effect of all atomic currents a current flows on the surface of the substance and is called as surface current I_s.

The magnetic field is produced due to these two currents. In ferro-magnetic and para-magnetic substances the direction of surface current I_s is in such a way that the magnetic field produced by it, is in the same direction of the magnetic field produced by the conduction current I_c. Hence, in these substances the two magnetic fields due to I_S and I_C get added to give a resultant field which is greater than that produced by I_C. On the other hand in dia-magnetic substances the magnetic field produced by surface current I_S is in opposite direction to the field produced by conduction current I_C, and hence, the resultant magnetic field is equal to the difference of two fields and consequently is smaller than the original field due to conduction current.

The behaviour of magnetic substance in a magnetic field is due to atomic or molecular dipoles in the substance. These atomic or molecular dipoles are due to the orbital motion of electrons around the nucleus and the spin motion of the electrons about their axis. The magnetic dipole moment of the nucleus due to the spin motion of nucleus about its own axis is negligible and hence not taken into consideration.

When a magnetic field is applied to a magnetic substance, the external magnetising field tries to align the atomic or molecular magnetic dipoles in its own direction which results in the magnetisation of the substance. The degree to which the magnetic dipoles get aligned and hence the degree of magnetisation of the specimen is expressed by a quantity called magnetisation of the substance or the intensity of magnetisation denoted by $\vec{M}$. The magnetisation $\vec{M}$ produced in a magnetic substance on placing it in a magnetic field is analogous to the dielectric polarisation of $\vec{p}$ of a dielectric placed in an electric field.

The intensity of magnetisation is defined as magnetic dipole moment per unit volume of the specimen

i.e.
$$\vec{M} = \frac{\vec{P}_m}{V}$$

But, the dipole moment developed in the specimen is due to the alignment of dipoles so the dipole moment is given by

$$\vec{P}_m = n\vec{P}_m$$

where $\vec{P}_m$ is the dipole moment of dipoles in the substance and n is the number of dipoles aligned by the magnetic field. Hence, $\vec{M} = \lim_{\Delta V \to 0} \frac{\Sigma \vec{P}_m}{dV}$

∴ The intensity of magnetisation in a magnetic substance is the magnetic dipole moment induced in the unit volume of the substance.

The quantity $\vec{M}$ goes on changing at every point inside the substance. In the absence of external magnetic field $\Sigma \vec{P}_m = O$ because the directions of dipole moments of the dipoles are arranged in random way. As the external magnetic field is applied the dipoles get oriented in a particular direction producing $\Sigma \vec{P}_m$ not equal to zero. The intensity of magnetisation in the substance depends on the applied magnetic field. The unit of $\vec{M}$ is A/m.

3.20 INTENSITY OF MAGNETIC FIELD

When a magnetic substance is placed in a magnetic field it gets magnetised. The quantity which measures the capacity to magnetise a substance is called intensity of magnetic field.

Let us consider a long rod PQ, as shown in Fig. 3.18 of a paramagnetic substance whose area of cross-section is A and a solenoidal coil is wrapped on it in which the conduction current I_c is flowing. Assuming a small part of length l in the middle of the rod, it is obvious that the intensity of magnetisation M in this small part will be uniform and the magnetic field induction B due to conduction current I_C will be uniform and parallel to the rod.

Fig. 3.18

Let the current at the surface of this part due to the resultant of the atomic currents of this part be I_S. Since, the rod is made up of paramagnetic substance, the directions of conduction current I_C and that of surface current I_S will be same as shown in the Fig. 3.18.

We know that, dipole moment due to current loop is equal to the product of current and area A shown in the Fig. 3.19, hence magnetic dipole moment due to this part is given by

$$\Sigma\, p_m = I_s \times A$$

where A is the area of cross-section of the rod.

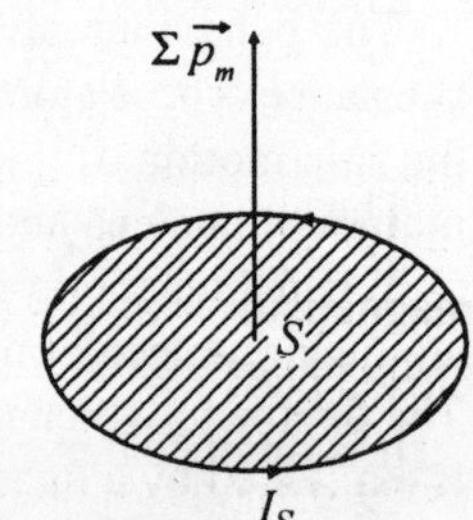

Fig. 3.19

Hence, dipole moment in unit volume of the rod or intensity of magnetisation is given by

$$M = \frac{I_S \times S}{Sl} = \frac{I_s}{l} = Js \qquad (i)$$

where Js is the surface current per unit length. Generally, Js is equal to the tangential component of M at the surface.

Let us assume a closed dotted path $abcd$ as shown in Fig. 3.18. Obviously the magnetisation inside the rod is M and outside the rod is zero. So, line integral along the path $abcd$

$$\oint \vec{M} \cdot d\vec{l} = \int_a^b \vec{M} \cdot dl + \int_b^c \vec{M} \cdot dl + \int_c^a \vec{M} \cdot dl + \int_d^a \vec{M} \cdot dl$$

$$= \int_a^b \vec{M} \cdot dl + 0 + 0 + 0 = M \int_a^b dl$$

$$= Ml = Is \qquad \left(\because\ M = \frac{Is}{l}\right)$$

Hence, $\oint \vec{M} \cdot dl = Is$ (ii)

This equation is called as Amper's law.

3.21 AMPERE'S LAW

A *line integral of the intensity of magnetisation of a closed path is equal to the surface current flowing in the area bounded by that path.*

The resultant magnetic induction B in the rod is due to both the conduction current I_C and the surface current I_S. So Ampere's law for the magnetic induction with the closed path *abcd, i.e.* line integral of B will be

$$\oint \vec{B} \bullet dl = \mu_O (I_S + I_C)$$

From the above Eq. (*ii*), we have

$$I_S = \oint \vec{M} \cdot dl$$

Hence,
$$\oint \vec{B} \bullet dl = \mu_0 \left(I_C + \oint \vec{M} \bullet dl\right)$$

or
$$\oint \left(\frac{\vec{B}}{\mu_0} - \vec{M}\right) \bullet dl = I_C \qquad (iii)$$

Where, the quantity $\left(\frac{\vec{B}}{\mu_0} - \vec{M}\right)$ is called as intensity of magnetising field denoted by $\vec{H}$. So, the intensity of magnetising field is expressed as,

$$\vec{H} = \left(\frac{\vec{B}}{\mu_0} - \vec{M}\right) \qquad (iv)$$

Therefore, from Eq. (*iii*),

$$\oint \vec{H} \bullet d\vec{l} = I_C \qquad (v)$$

The above equation is the Ampere's law for magnetising field $\vec{H}$, according to which "the line integral of magnetising field $\vec{H}$ with a closed path is equal to the conduction current through the area bounded by that closed path".

It is obvious that the surface current I_S has no contribution in the intensity of magnetising field.

3.22 MAGNETIC SUSCEPTIBILITY AND PERMEABILITY

We know that, the intensity of magnetisation M, depends on the intensity of magnetic field H and the nature of the substance. In most of the substances M is found to be directly proporional to H. In an isotropic magnetic substance.

$$M \propto H$$

or
$$M = \chi_m H \quad \text{or} \quad \chi_m = \frac{M}{H}$$

where χ_m is called magnetic suspectibility of the substance concerned. Hence,

"Magnetic suceptibility of a magnetic substance is defined as the ratio of the intensity of magnetisation to the intensity of magnetising field."

It is dimension less scalar quantity. For para-magnetic substances it is low and positive for dia-magnetic substances it is negative. And for ferro-magnetic substances its value is very high and positive. In vacuum the intensity of magnetisation is zero ($M = 0$), hence susceptibility for vacuum is zero. For most of the substances

M decreases with temperature hence, χ_m also depends on temperature. The dependence of χ_m on temperature is represented by Curie law according to which $\chi_m \propto \frac{1}{T}$ or $\chi_m = \frac{C}{T}$ where C is the Curie constant and T the absolute temperature.

3.23 RELATION BETWEEN PERMEABLITY AND SUSCEPTIBILITY

By the definition of intensity of magnetising field, we know that,

$$\frac{\vec{B}}{\mu_0} - \vec{M} = \vec{H}$$

where μ_0 is the permeability of free space

or $$\vec{B} = \mu_0 \left(\vec{H} + \vec{M}\right)$$

or $$\vec{B} = \mu_0 \left(1 + \frac{\vec{M}}{\vec{H}}\right) \vec{H}$$

but, we know that $\frac{\vec{M}}{\vec{H}} = \chi_m$ (susceptibility of the material)

Taking $1 + \chi_m = \mu_r$, we get

$$\mu_0 \left(1 + \chi_m\right) = \mu_0 \mu_r$$

Hence, $$\vec{B} = \mu_0 \mu_r \, H = \mu \vec{H}$$

Here, μ_r is the relative permeability and μ is absolute permeability.

The unit of absolute permeability of a magnetic substance is H/m.

In free space (vacuum), $\mu_r = 0$

$\therefore$ $$\mu = \mu_0$$

Hence, μ_0 is called permeability of free space whose value is $4\pi \times 10^{-7}$ H/m For dia-maganetic substances μ_r is less than 1, for paramagnetic materials greater than 1 and for ferro-magnetic materials it is very high. Thus, for para-magnetic $\mu_r > 1$ Ferro-magnetic $\mu_r >>> 1$ Dia-magnetic $\mu_r < 1$.

3.24 MAGNETISATION LOOP

When a ferro-magnetic substance is placed in a magnetising field H, it gets maganetised by the magnetic induction. When the external magnetising field H is varied the magnetisation M of the specimen also undergoes a change. The variation of magnetisation M with external magnetising field H is shown in the Fig. 3.20. This variation of M with H is for $H = 0$ to some value of H when M becomes maximum. From that value of H to zero, and from zero to particular negative value of M in positive direction. Thus, the variation of H is cyclic. The curve representing the above variation is $OPQRSTUP$. The study of the variation of M with H reveals the following facts

(*i*) The starting curve from 0 to P shows that when magnetising field H is zero to start with then M is also zero, *i.e.* substance is unmagnetised. OP part of the graph shows that M increases with increase in the value of H but this relation is not linear. At P for some positive value of H, M assumes maximum value and M does not increase further with increase in the value of H. This represents the state of magnetic saturation.

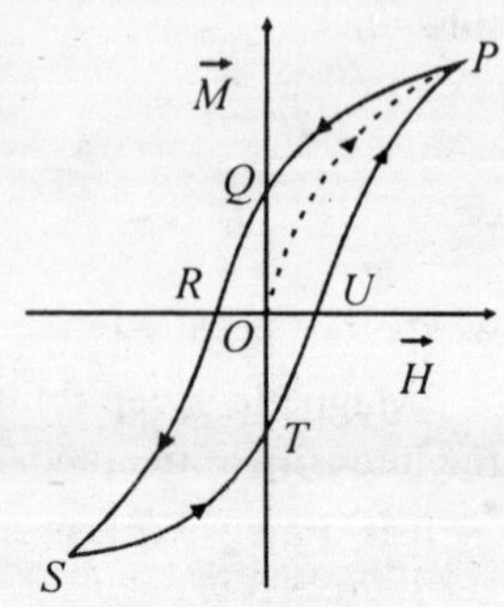

Fig. 3.20

(*ii*) The curve *PQ* shows that when the value of *H* is decreased gradually from the value at which *M* is maximum at *P*, the value of *M* decreases but the curve *PO* is not repeated and we get another curve *PQ*, *i.e.* on decreasing *H* the same values of *M* are not obtained which were obtained when it was increased.

At *Q* where magnetising field *H* becomes zero but magnetisation *M* does not reduce to zero but it is still equal to the value corresponding to *OQ*. It is obvious from this that the magnetisation *M* lags behind the magnetising field *H*.

(a) Retentivity or Residual Magnetisation

The magnetisation *M left in the specimen when magnetising field H* is decreased to zero is called residual magnetisation or retentivity of the material. The retentivity of the material is measured by the magnetisation *M* left in the specimen corresponding to the value of *M*, and corresponding to point *Q* where *H* is zero, *i.e* retentivity = value of *M* equal to *OQ*.

(b) Coercivity

(*i*) The curve *QR* shows that if the value of *H* is increased from zero at *Q* in the reverse direction, then the value of *M* decreases. At point *R* the value of *M* becomes zero when the value of *H* is equal to *OR* in reverse direction. This shows that in order to demagnetise the specimen ($M = 0$) a negative field equal 4to *OR* is required. This negative field which can destroy the residual magnetism in the specimen is called coercivity.

(*ii*) The curve *RST* represents that when the value of *H* is increased beyond *OR* in the reverse direction the magnetisation increases in reverse direction. At *S* we again get the state of saturation of magnetisation in reverse direction, *i.e.* beyond *S* the value of *M* does not increase even if reverse field is increased. If the value of *H* is decreased from negative to zero value then, the negative value of *M* also decreases and when *H* becomes zero the value of *M* is not zero but it is equal to *OT*.

(*iii*) The curve *TUP* shows that if the value of *H* is again increased in the positive direction the negative value of *M* decreases gradually and becomes zero when *H* is equal to *OU*. In increasing *H* further the value of *M* increases till the state of magnetic saturation is not attained at *P*.

(c) Magnetic Hysteresis

It is observed in the loop *PQRSTUP* that the magnetisation *M* always lags behind *H* because the value of *M* becomes zero some times later than the value of *H* becoming zero. The property or behaviour of a magnetic substance by virtue of which the magnetisation *M* lags behind the magnetising field *H* is called magnetic hysteresis.

(d) Hysteresis Loop

The curve which represents the variation of magnetisation *M* of a magnetic substance with the variation in magnetising field *H* in the cyclic magnetisation and demagnetisation of the specimen is called hysteresis loop. *PQRSTUP* is the hysteresis loop as shown in the above Fig. 3.20.

It must be understood that if a curve is drawn between *B* (magnetic induction) and *H* (magnetising field) the shape of the curve will be exactly similar but with a difference that *B* goes on increasing with *H* and attains the new saturation.

3.25 DEMAGNETISATION

It has been observed that by removing the magnetising field *H*, the intensity of magnetisation *M* does not become zero. To reduce the magnetisation to zero a negative or reverse field *H* is required and is called as coercive force. Hence, to demagnetise a magnetic specimen we magnetise and demagnetise the specimen with continuously decreasing current again and again and ultimately the current is reduced to zero. This results in the extension of hysteresis loop as shown in the Fig. 3.21.

Finally when the current becomes zero the extension of hysteresis loop becomes zero and the curve terminates at the origin O.

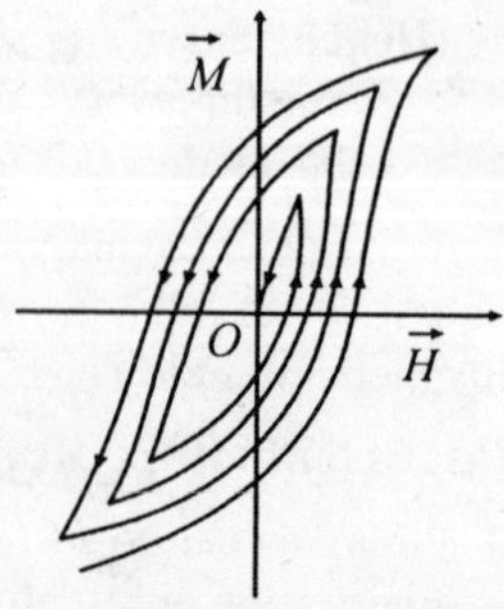

Fig. 3.21

3.26 ENERGY LOSS DUE TO HYSTERESIS

According to modern concepts a ferro-magnetic substance consists of different domains which get magnetised immediately. In an unmagnetised substance the directions of magnetisation of different domains are such that in un-magnetised state the average resultant magnetisation is zero.

When the experimental ferro-magnetic substance is placed in an external magnetising field, the domains increase in size and orient themselves in a particular favourable positions and they get rotated and orient in the direction of the field. Some external work is done in the rotation of the domains which is spent in the substance in the form of energy generally in the form of heat. On removing the external magnetising field the domains do not regain their original positions completely, which means that even on removing the external field some magnetisation remains in the substance. Thus, the energy supplied to the substance for magnetising the substance and that obtained from the substance during its demagnetisation are not equal and the difference between the supplied energy and the energy obtained from the substance remains in the substance in the form of heat and hence called energy loss. The loss or dissipation of energy is due to the characteristic of hysteresis of the substance and hence is termed as energy loss due to hysteresis. **The energy loss per unit volume in one cycle of magnetisation of the substance:** Let there be a ring of a material of permeability μ and of area of cross-section A with the length of its circumference l. A coil of N total turns is wrapped on the ring and a current of I ampere is flown in it producing a flux B. The flux passing through each turn of the coil is BA and hence, the total magnetic flux crossing the coil is given by,

$$\phi_B = NBA \text{ (weber-turn)} \qquad (i)$$

Intensity of magnetising field, $H = \dfrac{NI}{l}$ (amp.-turn / meter)

Hence, $$I = \frac{Hl}{N} \; amp. \qquad (ii)$$

The flux linked with the coil changes when the current in the coil wrapped on ring is increased. According to the law of electromagnetic induction an induced e.m.f. is produced in the coil due to change of flux through it, which is given by Faraday's law of electromagnetic induction

$$E_{in} = -\frac{d\phi_B}{dt} = -\frac{d}{dt}(NBA) = -NA\frac{dB}{dt} \text{ (volt)} \qquad (iii)$$

According to Lenz's law the induced e.m.f. opposes the growth of current as represented by a negative sign attached with the equation (iii). Hence, the work is done in growing the current in the coil against the induced e.m.f. produced in the coil. The work done by the current I in dt second against the induced e.m.f. is given by

$$dW' = E_{in} I dt$$

or $$dW' = AN\frac{dB}{dt}\left(\frac{Hl}{N}\right)dt = HAldB \text{ Joule}$$

where dB is the corresponding change of flux in dt time. This work done changes into the energy stored in the magnetic field. Where, $Al = dV$ volume of the ring.

Hence, the increase in the energy per unit volume of the substance is given by

$$dW' = \frac{dW'}{dV} = \frac{dW'}{Al} = \frac{HAldB}{Al} = HdB\left(\text{J}/\text{m}^3\right)$$

∴ The loss of energy in the substance per unit volume in increasing the flux density from B_1 to B_2 is given by

$$W = \int_{B1}^{B2} dW = \int_{B_1}^{B2} HdB\left(\text{J}/\text{m}^3\right) \qquad (iv)$$

Now, according to the B-H loop shown in the Fig. 3.22 which is represented by *PQRSTU*.

On the curve let us consider two close points *b* and *c* so that the value of *H* in this region can be taken as *ab* and that of *dB* by *ad*.

∴ *HdB* = Area of the strip *abcd*.

So, the loss of energy per unit volume in magnetising the substance from *a* state to *d* state = Area of the strip *abcd*.

Hence, the loss of energy per unit volume in magnetising the substance from *P* state to *Q* state will be equal to the sum of the areas of such strips included in the area *OPQV*.

In taking the substance from *Q* state to *R* state, *H* and *B* decrease. In this process *QVRQ* energy per unit volume is given back to the substance. Hence, the energy loss per unit volume in taking the substance from *P* to *R* states will be equal to area *OPQRO*.

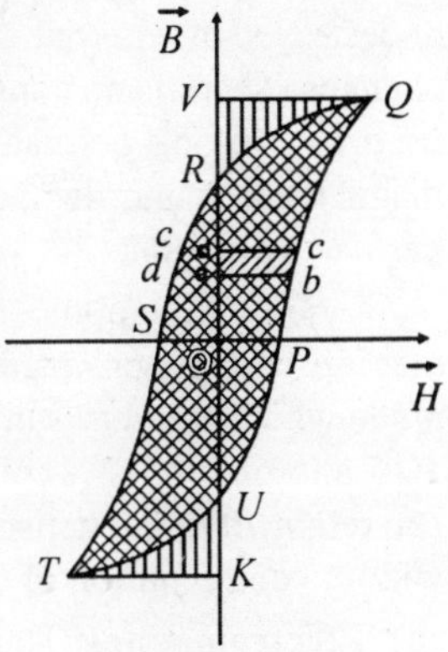

Hence, in taking the magnetic substance through one cycle the loss of energy per unit volume is equal to the area of the hysterisis cycle *PQRSTUP*, *i.e.* area of *B-H* curve, *i.e.* in the cyclic magnetisation energy loss per unit volume per cycle is given by:

Fig. 3.22

$$E_{(Cycle)} = \int HdB = Area\ of\ B\text{-}H\ loop\ \left(\text{J}/\text{m}^3\right)$$

The above equation gives, the loss of energy per unit volume per cycle and has great significance because, it helps us in the selection of materials used in the construction of cores of dynamo, motors and transformers.

3.27 DIFFERENCE BETWEEN MAGNETIC PROPERTIES OF IRON AND STEEL

The knowledge of the magnetic properties of soft iron and steel can be obtained from M-H curves of these two materials which are shown in the Fig. 3.23. The curve *P* is for soft iron and *Q* is for steel. These curves reveals the informations about the magnetic properties of soft iron and steel.

(a) Retentivity: When the magnetising field is reduced to zero soft iron retains more magnetisation than steel. Hence, the retentivity of soft iron *OA* is more than the retentivity of steel *OB*.

(b) Coercivity: Soft iron needs comparatively less coersive force for complete demagnetisation. Thus, the coercivity of soft iron *OC* is smaller than the coersivity *OD* of steel.

(c) Hysteresis Loss: The area of M-H loop of soft iron is smaller than the area of M-H loop for steel hence the energy loss per unit volume per cycle of magnetisation is smaller for soft iron than for steel.

(d) Magnetic Susceptibility: For a particular value of magnetising field *H*, the magnetisation or intensity of magnetisation *M* of soft iron is more than that for steel. This proves that the magnetic susceptibility $\chi_m = M/H$ of soft iron is more than that for steel.

(e) Permeability: If we draw B-H curves for soft iron and steel we find that for a paraticular value of H the magnetic induction B for soft iron is more than that for steel. Hence, the permeability of soft iron $\mu = B/H$ is greater than that for steel.

3.28 USES OF HYSTERESIS CURVE IN SELECTION OF MAGNETIC MATERIALS

The magnetic materials are used for making permanent magnets, electromagnets, cores of transformers, motors and dynamos, diaphragms of telephone receivers. The choice of the material generally depends on the study of hysteresis loops of these materials.

(a) Permanent Magnets: The magnetic substance used for making permanent magnet must have high retentivity so that the magnet may be powerful. It must also have high coercivity so that the magnetism of the magnet may not be decreased or destroyed by external stray magnetic fields and chemical strains and temperature changes. There is no consideration of hysteresis loss in the selection of material used for making permanent magnets because in permanent magnet the cyclic magnetisation and demagnetisation do not take place in the process of making permanent magnets.

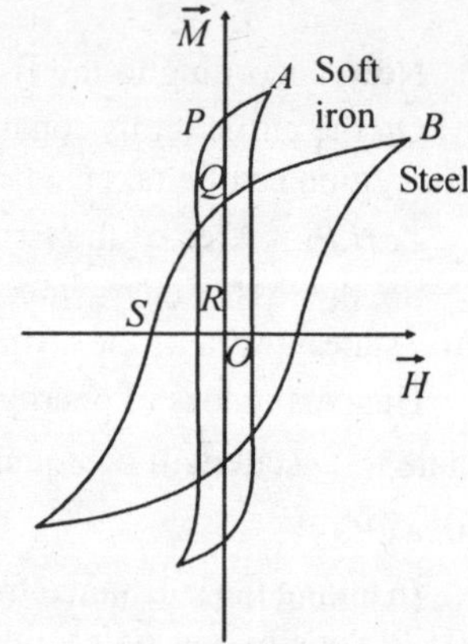

Fig. 3.23

In view of the above considerations steel is more suitable than soft iron for making permanent magnets. Recently many alloys having high coercivity are made which are very suitable to make permanent magnets. Alinco (18% nickel, 10% aluminium, 12% cobalt, 6% copper and 54% iron) has a coercivity 500 and can retain 70% magnetism upto 600° C. More recently a new alloy Vikaloy made with the combination of cobalt, vanadium has the maximum coercivity.

(b) Electromagnet: The material used to make the core of an electromagnet is very suitable which can acquire more magnetic flux density even with ordinary magnetic fields and they must have the least hysteresis loss. The M-H curve for soft iron shows that both these requirements are found in soft iron. Hence, soft iron is most suitable for the core of an electromagnet.

(c) Cores of Transformers, Motors and Dynamos: The materials used in making the cores of transformers, motors, dynamos and in the diaphragm of telephone receivers and chokes undergo cyclic magnetisation and demagnetisation, hence, these materials must possess the following characteristics:

(*a*) High initial permeability (*b*) Low hysteresis loss (*c*) High specific resistance

The slope of the hysteresis loop for soft iron, is steep and its area is small. Hence, soft iron has high permeability and low hysteresis loss. The permeability of soft iron is nearly 250. By the combination of 4% silicon in iron we get the formation of transformer steel which is an ideal material for the core of the transformers. Similarly, hypernik (50% cobalt and 49% iron) is a very good material for the core of transformer and telephone diaphragm.

3.29 TYPES OF MAGNETIC SUBSTANCES

Each substance is composed of atoms. There is nucleus at the centre of each atom in which the entire mass of the atom is concentrated at the centre. The electrons revolve round the nucleus in an elliptical orbits. This is called the *orbital motion* of electron. In addition to it, electron also rotates about its own axis which is called the *spin motion* of electron as shown in Fig. 3.23. Thus, the motion of electron is just like the orbital motion of earth round the sun and its spin motion about its own axis. Due to these motions of electron, magnetic field is produced in the atom, because electron has a negative charge and the main cause of origin of magnetism is the motion of charged particles.

A revolving or spinning electron is equivalent to a current loop with which a magnetic field is associated. Since each current loop can be treated as equivalent to a magnetic dipole, hence, due to both the orbital and spin, motions magnetic moment is produced. But major part of the total magnetic moment of electron is due to the spin motion of the electron. There is a very little contribution of the orbital motion to the total magnetic moment.

Experimentally, it is found that the atoms of every substance behave like magnetic shells. Each atom exhibits its own magnetic field. The resultant magnetic field in different materials are found to be different. In some substances it is found to be strong, in some other weak and some it is zero. According to the resultant magnetic field in the substances they are divided into the following five types:

(*i*) Para-magnetic substances
(*ii*) Ferro-magnetic substances
(*iii*) Dia-magnetic substances
(*iv*) Anti-ferromagnetic substances
(*v*) Ferri-magnetic substances or ferrites.

3.30 PARAMAGNETIC SUBSTANCES

(a) **Magnetisation:** When a paramagnetic substance is placed in an external magnetising field, each magnetic loop in the substance gets oriented in the direction of the field. The resultant magnetic dipole moment in the atoms of paramagnetic substance is produced only by the orbital electrons which do not rotate opposite. Such substance get weakly magnetised in the direction of the external field, *i.e.* the magnetic field produced in them is weak and in the direction of external field.

(b) **Magnetic Induction:** When a paramagnetic substances are kept in an external magnetic field the magnetic induction B, *i.e.* the resultant magnetic flux density in the substance is slightly greater than the magnetic flux density B_0 in vacuum, *i.e.*

$$B > B_0 \quad or \quad \frac{B}{B_0} = \mu_r > 1$$

(c) **The Relative Permeability (μ_r):** The magnetic permeability $\mu_r = \dfrac{B}{B_0}$ is slightly greater than 1. For platinum $\mu_r = \dfrac{\mu}{\mu_0} = 1.000021$.

(d) **Intensity of Magnetisation (M):** Since, these substances get weakly magnetised, when subjected to external magnetic field the magnetisation in these substances is low.

(e) **Behaviour in a Uniform Magnetic Field:** When paramagnetic substances are kept in a uniform magnetic field they experience force in the direction of the applied field. If a rod of a paramagnetic material is freelysuspended in a uniform magnetic field it is oriented parallel to the field as shown in the Fig. 3.24.

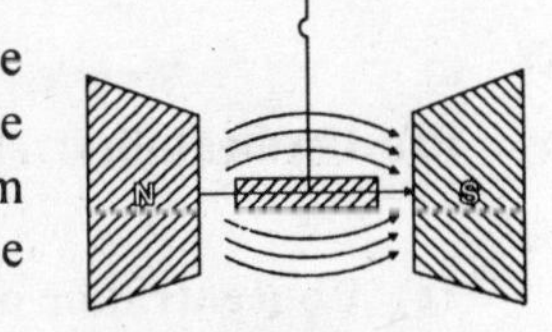

Fig. 3.24

(f) **Behaviour in a Non-uniform Field:** In a non-uniform magnetic field a para-magnetic substance experiences force in the direction from weak to strong field.

If a paramagnetic liquid in a watch glass is kept on the poles of a magnet *NS* as shown in the Fig. 3.25, the surface of the liquid is depressed in the middle as shown in the Fig. 3.26.

In the same way if a part of the U-tube filled with a paramagnetic liquid is kept vertical between the poles of magnet the level of the liquid rises in the tube as shown in Fig. 3.27.

(g) **Concentration of Magnetic Induction Lines:** The concentration of magnetic lines of force is slightly greater in the substance than outside the substance.

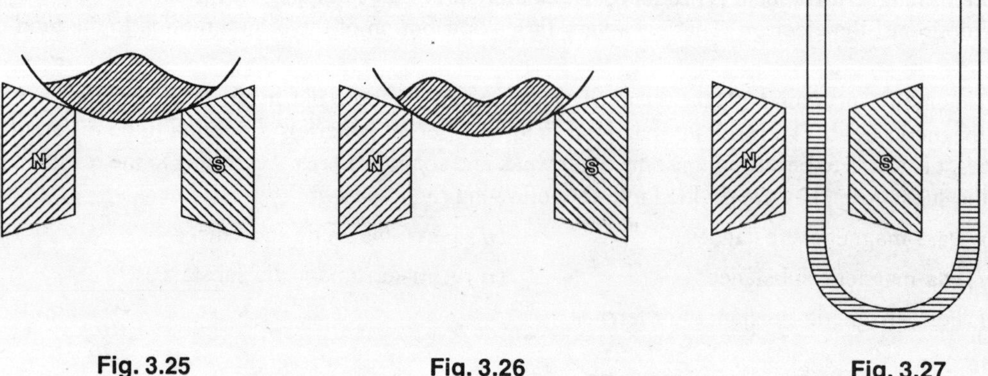

Fig. 3.25 Fig. 3.26 Fig. 3.27

(h) Magnetic Susceptibility (χ_m): The value of magnetic susceptibility χ_m is low and it varies inversely with the absolute temperature of the substance, *i.e.* $\chi_m = \frac{C}{T}$. This law is called as Curie law.

Examples: Platinum, aluminium, chromium, oxygen, mangenese and solutions of salts of iron and nickel.

3.31 DIA-MAGNETIC SUBSTANCES

(a) Atomic Shells: In dia-magnetic substance the planes of the orbits of the electrons in the atoms are so arranged that the algebraic sum of the dipole moments of the atomic shells is *zero*, *i.e.* the atoms of the dia-magnetic substance are neutral as regards to magnetic field.

(b) Magnetisation: The dia-magnetic substance get weakly magnetised in a direction opposite to that of external field, *i.e.* the internal field produced in these substances due to their own magnetisation is opposite to the external magnetising field.

(c) Magnetic Induction: In diamagnetic substances the magnetic flux density B is slightly smaller than that in vacuum B_0, *i.e.*

$$B < B_0 \quad \text{or} \quad \frac{B}{B_0} = \mu_r < 1$$

(d) Permeability: The permeability of the dia-magnetic substances $\mu_r = \frac{B}{B_0}$ is slightly less than 1.

(e) Concentration of Lines of Force: The number of lines of force in a dia-magnetic substance is slightly less than that outside the substance because μ is less than 1 for these substances.

(f) Magnetic Susceptibility (χ_m): The magnetic susceptibility for dia-magnetic substance is low and negative and it does not depend on the temperature and magnetisation.

(g) Behaviour in a Uniform Magnetic Field: When a rod of a dia-magnetic substance is freely suspended in a uniform magnetic field the rod align itself perpendicular to the field which means that these substance experience a force perpendicular to the field.

(h) Behaviour in Non-uniform Magnetic Field: In non-uniform magnetic field a diamagnetic substance experiences force from stronger to weaker part of the field.

Examples: Bismuth, antimony, copper, mercury, gold, water, alcohol, air, etc.

3.32 FERROMAGNETIC SUBSTANCES

(a) Atomic Shells: The atomic magnets of ferromagnetic substances are very strong because the dipole moment due to spin motion of electrons is very strong. According to Weiss the atoms form groups inside the substance. In the unmagnetised state these atoms form closed groups and do not show independent dipole moment. On magnetisation these closed groups break and all atomic magnets arrange themselves parallel to each other.

(b) Magnetisation: On keeping the ferromagnetic substances in an external magnetising field, they get strongly magnetised in the direction of external field. Actually these substances have the properties similar to paramagnetic materials but in magnified way.

(c) Magnetic Induction: The magnetic flux density in these substances B is much greater than that in vacuum B_0, *i.e.* $B >>> B_0$.

(d) Magnetic Permeability: The magnetic permeability of ferromagnetic substances is very large, *i.e.* $\mu_r = \dfrac{B}{B_0} >>> 1$. It is nearly 200 or more for soft iron. In these substances magnetic flux density B is not proportional to magnetising field H, *i.e.* B/H is not constant. The permeability of these substances decreases with temperature and it becomes equal to μ_0 at a particular temperature called Curie temperature.

(e) Concentration of Lines of Force: The concentration of lines of forces in these substances is very large.

(f) Magnetic Susceptibility (χ_m): It is very large and positive hence, they have large capacity of being magnetised.

(g) Variation of Susceptibility with Temperature: The susceptibility of a ferromagnetic substance varies inversely with absolute temperature of the substance, *i.e.*

$$\chi_m \propto \frac{1}{T} \quad \text{or} \quad \chi_m T = \text{constant}$$

This is called as Curie law.

In the Fig. 3.28 the variation of permeability of ferromagnetic material with absolute temperature is shown. This graph shows that at a temperature where the curve meets the X-axis the susceptibility of the material becomes zero. This temperature is called Curie temperature (T_C). Up to this temperature the substance remains ferromagnetic but it turns into paramagnetic after this temperature. This temperature is 760° C for iron. **Examples:** Iron, nickel, cobalt, steel and some alloys.

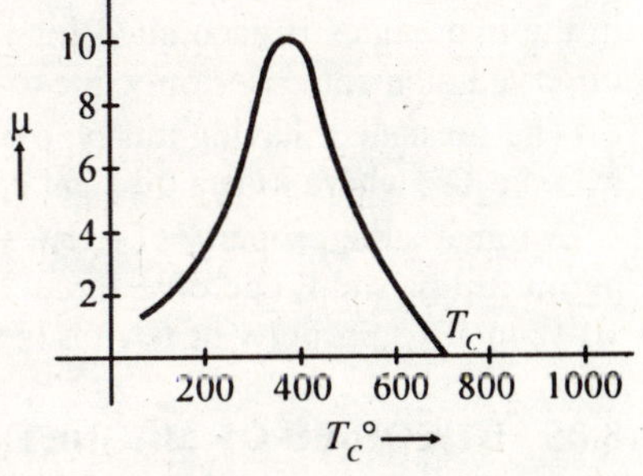

Fig. 3.28

3.33 ANTI-FERROMAGNETIC SUBSTANCES

When the neighbouring moments are aligned anti-parallel the phenomenon is known as anti-ferromagnetism. Anti-ferromagnetic materials are not so well known as the common ferromagnetic materials. Only a few, such as Mn, Cr and ferrous and nickel oxides exist. The phenomenon occurs below a certain temperature known as the neel temperature (T_N). The neel temperature is not the same as the Curie temperature (T_C). Values of T_N/T_C are usually of the order of 1.5 to 5.

The phenomenon is not so important because of the low values of susceptibility $\left(\chi_m \approx 10^{-2} \text{ to } 10^{-3}\right)$. The examples of anti-ferromagnetic materials are: MnO, $FeCl_2$, FeO, NiO, CoO, CrSb.

3.34 FERRI-MAGNETIC SUBSTANCES OR FERRITES

A group of magnetic alloys exhibit the property of magnetisation which changes with the percentage of the different constituent atoms in the alloy. A typical example of this type of alloy is $XOFe_2O_3$ with X = Mn, Co,

Ni, Cu, Mg, Zn, Cd or Fe^{++}. The alloys are called ferrites. Thus, manganese ferrite is a 1 : 1 mixture of manganese oxide and iron oxide.

Ferrites are widly used in computers and in microwave equipment. Ferrites are advantageous at high frequencies because of low eddy current losses. Ferromagnetic metals and alloy cores have to be eliminated in order to reduce the losses. At high frequencies, duct cores which consist of fine particles of ferromagnetic material insulated from each other may be used. But these have the disadvantage of diluting the ferromagnetic material and decreasing the effective relative permeability. A futher disadvantage is that the flux density varies throughout the core due to non-uniform spacing of the particles.

Ferrous ferrite, or magnetite $FeO \cdot Fe_2O_3$ was the first such material in which the phenomenon of magnetisation was observed. Other common materials of this type are nickel ferrite, $NiO \cdot Fe_2O_3$ and magnesium ferrite $MgO \cdot Fe_2O_3$. The magnetic behaviour of these materials is similar to that of iron and nickel. The magnetic susceptibility is large.

The resistivity of ferrites is very much higher than that of the ferromagnetic metals. This is because ferrites are chemical compounds.

The macroscopic behaviour of ferrites under the influence of magnetising force is similar to that of the ferromagnetic metals.

Ferrites are hard and brittle materials which cannot be sharpned by ordinary machining processes. In general diamond impregnated slitting wheel is necessary for cutting ferrite rods.

Ferrites have a lower tensile strength and a lower thermal conductivity as compared to iron. Unlike ferromagnetic materials, ferrites have a high resistivity exceeding that of iron by at least $10^6 - 10^7$ times. This property is useful in magnetic applications at high frequencies.

The external behaviour of ferrimagnetics is similar to ferromagnetic substances but the temperature dependence of their spontaneous magnetisation may be quite different. For example the magnetisation of a ferrimagnetic does not necessarily decrease monotonically with the rise in temperature but can pass through zero even before the Neel temperature. The familiar example of ferrimagnetic is magnetic Fe_3O_4 or $FeO \cdot Fe_2O_3$ ferric ions Fe^{3+} are in a state with spin $S = 5/2$ and zero orbital magnetic moment. Thus each ion should contribute 5 μ_B to the saturation moment. The ferrous ions (Fe^{2+}) have a spin 2 and should contribute 4 μ_B apart from any residual orbital magnetic contribution. Thus the effective number of Bohr magnetons should be about $2 \times 5 + 4 = 14$ per Fe_3O_4 unit if all spins were parallel. But the observed value is 4.1 μ_B. The difference in the two values is accounted for if the moments of Fe^{3+} ions are antiparallel to each other and then the observed moment arises only due to Fe^{2+} ion. Neutron diffraction verifies this model (Fig. 3.29).

The substances having this type of spin order are called Ferrites. The usual chemical formula of a ferrite is $MO \cdot Fe_2O_3$, where *M* is a divalent cateion and stands for Zn, Cd, Fe, Ni, Cu, Co or Mg.

A remarkable property of ferrites is the combination of extension of magnetic parameters (high magnetic permeability, small coercive force, high saturation magnetisation, etc.) with high resistivity of the order of $10^3 \Omega$-m. This property of ferrites is useful in the field of high and ultra high frequency electronics.

3.35 THEORIES OF MAGNETISM

According to atomic structure electrons revolve around the nucleus in closed orbits. These rotating electrons are equivalent to current loops which produce their own magnetic fields. In this way an electron revolving around the nucleus behaves like a magnetic shell or tiny magnet whose magnetic field acts in the region of revolution of the electron. Thus, every substance made up of atoms must have magnetic field in it. The planes of these magnetic shells in the atoms are generally not parallel to each other and even if they are parallel, the sense of rotation of electrons are not same. Thus, an atom shows a resultant magnetic field. The resultant magnetic field in the atom of the substances may be strong, weak or zero. According to this the substance are classified as ferro-, para- and dia-magnetic as discussed earlier.

The above discussion shows that the magnetic property in an atom is due to the rotation of electrons around the nucleus and every atom of a substance has its own dipole moment which depends on the rotation of the electron.

Stern and Gerlach established by experiment that the existance of magnetic dipole moment of an atom. According to them along with rotation around the nucleus the electron spins about its own axis called spin motion. An atom has magnetic dipole moment due to this spin motion. Thus, an atom possesses the following two dipole moments

(*i*) Due to orbital motion of electron around the nucleus.

(*ii*) Due to the spin motion of electron about its own axis.

In general the net magnetic dipole moment of an atom is the vector sum of the above two dipole moments.

When a maganetic substance is kept in an external magnetic field the magnetic dipole moments are aligned parallel to the magnetic field and the substance gets magnetised in the direction of the external field. This is the behaviour of the paramagnetic substances. When a paramagnetic substance is heated the thermal agitation of the atoms increases as a result of which the arrangement of dipoles becomes difficult. This is the reason that the magnetic property of paramagnetic materials decreases due to increase in temperature.

Dia-magnetism is found in those substances in which the number of electrons in the atoms is even. In these substances the electrons form such pairs that in one pair the orbital and spin motions are in opposite sense so that the resultant dipole moment of the atom is zero. Thus, in a dia-magnetic substance the dipoles are not set parallel to the field. The external field in this case retards the motion of those electrons which produce dipole moment in the direction of the field while it accelerates the motion of those electrons which produce dipole moments opposite to the field. That is how the net dipole moment is opposite to the field resulting in dia-magnetism.

The thermal agitation does not effect the magnetisation of dia-magnetic substance.

3.36 THEORY OF FERROMAGNETISM

Ferromagnetic substances are those which are magnetised strongly in the direction of external magnetising field. Ferromagnetic substances are all solids, and each is characterized by certain temperature known as Curie point at which its properties change abruptly. Above the Curie point the susceptibility is independent of the field strength and follows approximately the Curie-Weiss law given by the equation.

where C is Curie constant and θ is Weiss constant which is closely equal to the Curie point and the substance then becomes para-magnetic.

For iron the Curie point is 770° C, for Co, Ni, Gd and Dy it is 1127° C, 358° C, 16° C and – 168° C respectively. Below the Curie point the behaviour of the substance is quite different and very large values of magnetisation are produced by quite small fields inspite of the randomizing tendency of the thermal motions of the atoms and the magnetisation varies quite non-linearly with the field strength. This is shown in the Fig. 3.29.

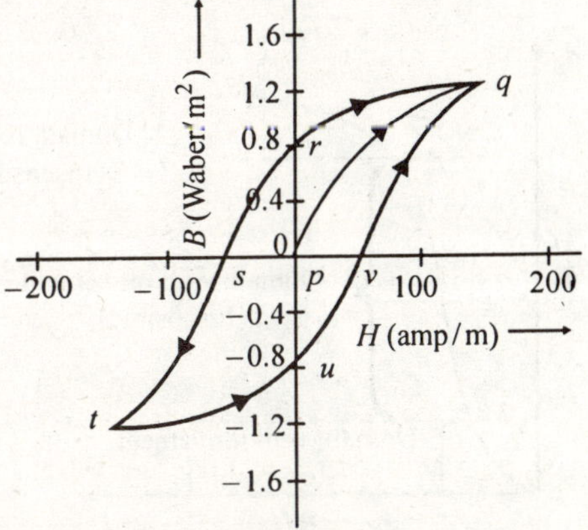

Fig. 3.29

A curve between magnetic induction B as a function of field H for a specimen of iron is drawn. If the iron is initially unmagnetised and a field of few hundred amp/m the value of B becomes practically constant. If the magnetic field H is reduced from a value corresponding to q, the induction B does not decrease along qp, but follow the path qr and even for $H = O$ at r, the induction B does not become zero but it is pr. This value of B, *i.e.* pr at $H = O$ is called residual magnetism or retentivity or remanance of the specimen. When the magnetic field H is reversed the induction falls and becomes zero for $H = ps$. This value of H, *i.e.* pr is known as Coercive force. When the magnetic field is increased still further in the reverse direction, a reverse induction is produced in the specimen which quickly attains the saturation value. If we reduce the reverse field gradually the induction follows the curve *tu* and attains a value *af* for $H = O$. When the field H is increased again in the positive direction, the induction B traces out the curve *urq*. The closed curve representing the cyclic change of B with the cyclic change in H, *qrstuq* is known as hysteresis curve. The lagging of magnetisation B behind the magnetising field H is called hysteresis.

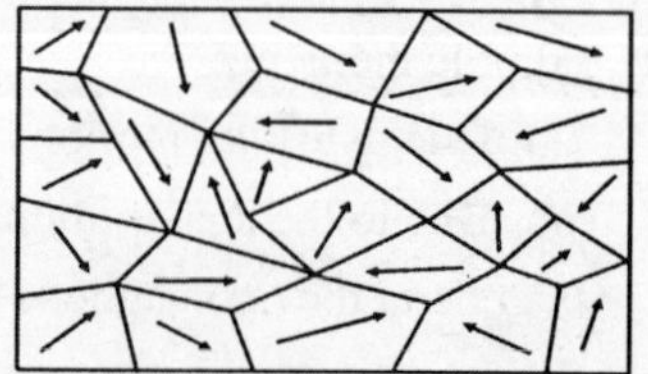

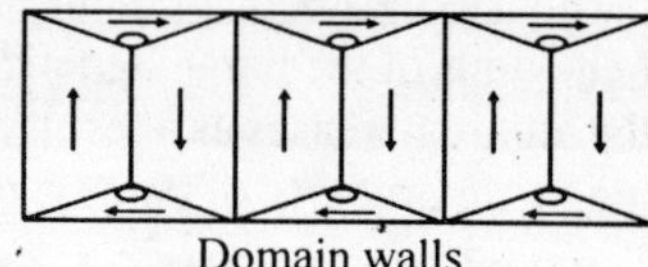

Fig. 3.30

In order to explain ferromagnetic behaviour of iron Weiss suggested that a ferromagnetic substance contains atoms with permanent magnetic dipole moment, as in the case of paramagnetic substance, but due to special form of interactions called exchange coupling occurring between adjacent atoms, there exist strong forces between neighbouring atomic dipoles which cause groups of them in the same direction. The substance would then be permanently magnetised within each group, *i.e.* the atomic dipoles in the individual groups are fully aligned. Such group of atomic dipoles are called domains whose sizes vary from 10^{-6} to10^{-2} cm^2. The mangetisation within a domain is parallel to a direction of easy magnetisation.

When the temperature of a ferromagnetic substance is increased above the Curie point, exchange coupling suddenly vanishes and the substance starts behaving like a paramagnetic. In an unmagnetised specimen the domains are randomly oriented, so that there is no resultant magnetic moment in any direction as shown in the Fig. 3.31.

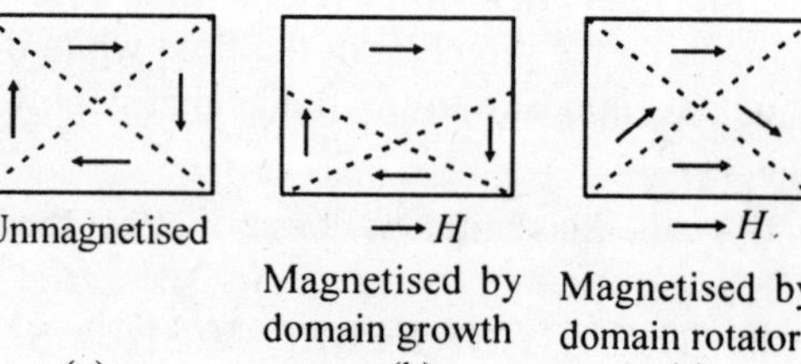

Fig. 3.31

In passing from one domain to an adjacent domain, the atomic moment vector gradually rotates from its original to its new direction over a thickness of about 1000 atomic diameters, while the domains themselves have dimensions of the order of a tenth of a millimeter. The region between the two domains is called domain wall.

When a weak magnetising field is applied, the magnetic polarisation of the specimen changes due to the motion of domain walls. In this process domain walls whose axes are parallel to the field are at small angles with the field grow at the expense of those which are oriented unfavourably to the field as shown in the Fig. 3.32. This growth of domain walls is reversible as long as the field is very small which means that on making the field zero, the magnetisation becomes *zero*, *i.e.* growth of domain walls will turn to zero. Such a magnetisation is shown in the magnetisation curve *pq* represented in the Fig. 3.32. If the magnetising field is increased then for stronger fields the magnetisation is shown by the part *qr* of the curve proceeds by irreversible wall motion.

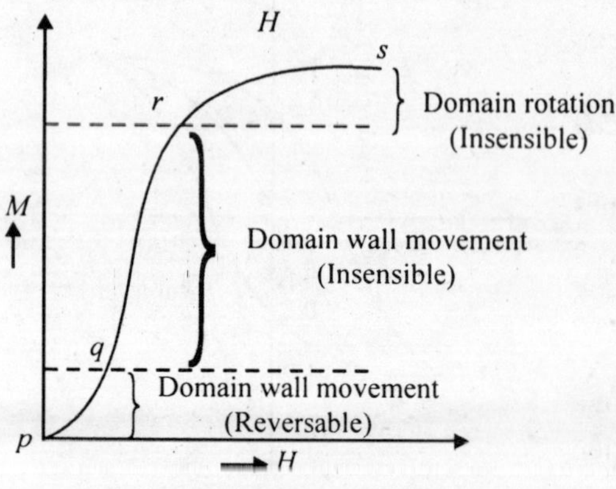

Fig. 3.32

The application of a very strong field then causes domain rotation during which the magnetic polarisation takes a direction parallel to the applied field as shown in the Fig. 3.32 and this corresponds to the part *rs* of the magnetisation curve. Thus, motion of the domain walls accounts for the portion of magnetisation curve below the knee of the curve, while domain rotation accounts for the portion above the knee. If the applied field is removed, there is a little change in the domain structure so that the magnetisation remains quite high until reverse fields are applied and this gives an account for the hysteresis.

Ferromagnetic substances are broadly divided into two classes.

(a) **Magnetically Soft Materials:** Which have high permeability and are easily magnetised and demagnetised having small coercive forces. The magnetically soft materials are used for building transformer cores,electromagnets, chokes and in the electrical machinery such as motors and dynamos. Soft magnetic materials have also small hysteresis loop area as a result of which small energy loses occur during cyclic magnetisation.

(b) **Magnetically Hard Materials:** Which have relatively low permeability and are difficult to magnetise and demagnetise having large coercivity. Hard magnetic materials have a large hysteresis loop area and hence large energy loses per cycle of magnetisation take place in these materials. They are suitable for making permanent magnets because their remanence (retentivity) is high and hence does not easily demagnetise.

APPLICATIONS OF DIFFERENT MAGNETIC MATERIALS

3.37 MAGNETIC RECORDING

Development of digital data storage devices has received enormous attention over the past 50 years. Our need for storage is explosive. With exponential growth of multimedia requirements for text, images, video and audio, an increasing amount of information is being generated or captured electronically. More than fifty percent of the generated data need storage either in analog or digital form. To accommodate ever increasing volumes of data scientists and engineers are constantly investigating new ideas and methods for data storage. As a result the dimensions of these storage devices started shrinking while performance increased.

Magnetic recording heads are the miniaturized hearts of disk drives and other magnetic storage devices. Recording heads function according to certain principles of magnetic recording, which are based directly on four magnetic phenomena.

3.38 MAGNETIC PARAMETERS FOR RECORDING APPLICATIONS

A. An electric current produces a magnetic field.

B. Since materials are easily magnetized when placed in a weak magnetic field. When the field is turned off, the material rapidly demagnetizes. These are called soft magnetic materials.

C. In some magnetically soft materials the electrical resistance changes when the material is magnetized. The resistance goes back to its original value when the magnetizing field is turned off. This is called magneto resistance or the MR effect. Giant magneto-resistance, or the GMR Effect, is much larger than the MR effect and is found in specific thin film material systems.

D. Certain other materials are magnetized with difficulty, *i.e.* they require a strong magnetic field, but once magnetized, they retain their magnetization when the field is turned off. These are called hard magnetic materials or permanent magnets.

These four phenomenon are exploited in the design and manufacture of magnetic recording heads which read and write data for storage and retrieval in computer disk drive memories, tape drives, and other magnetic storage devices. The oxides of γ-Fe_2O_3, Fe_3O_4 and CrO_2 are the most suitable materials.

3.39 MAGNETIC MEMORIES

3.39.1 Ferrite Core Memories

For random access storage in computers, Ferritecore memories are used. Ferrimagnetic material, having a nearly square-shaped hysteresis loop and a low coercivity, is taken in a shape of a toroid. A current carrying wire is sent through this. When a sufficiently high current pulse is sent through this wire, the core becomes magnetically saturated in a particular direction. Another current pulse of sufficient strength in the opposite direction magnetizes the ferrite core in the opposite direction. The two magnetization directions constitute the two possible values (0 and 1) in a binary system.

The actual configuration of a complete memory system consists of a stack of identical memory planes. A set of wires in the *x*-as well as in the *y*-directions pass through the core as shown in Fig. 3.33. In order to switch a core say x_3/y_2 from zero to one, through each of the x_3 and the y_2 wires a current of magnititude $I_s/2$ is sent where I_s is the current required to obtain saturation field. This makes only x_3/y_2 core to switch while other cores stay unaffected. This information is permanently stored. This is read by again sending a current pulse $I_s/2$ through the x_3/y_2 wires. A third wire called sensing wire passes through all the cores of a given plane. During reading process, the cores not switched are switched. This results in induced current signal through sensing wire. If the core is already recorded, there is no change and hence no induced current. Thus, the sensing wire reads the data stored. Since the reading process destroys the data stored, a special circuit is needed to rewrite the information back into the core. Since the space requirement and size are large, their usage is now limited to a few specialized aplications. Ferrites commonly used for this application are manganese magnesium ferrite, manganese copper ferrite and lithium nickel ferrite.

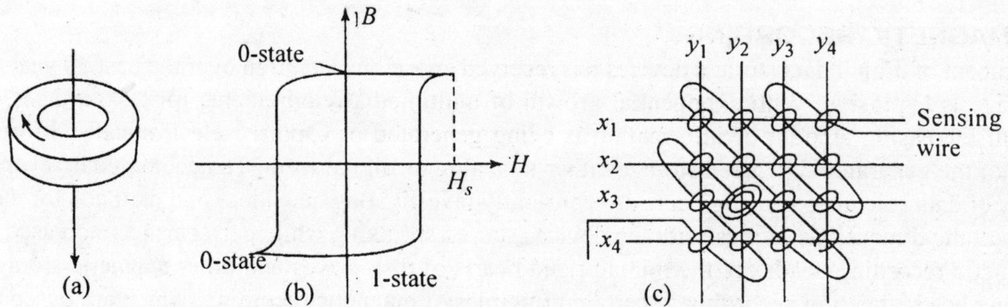

Fig.3.33 Ferrite core memories (a) toroid, (b) 0 and 1 binary states, and (c) actual configuration of memory system

3.39.2 Bubble Memories

Data can be stored in this crystals of *canted* antiferromagnetic oxides ($BaFe_{12}O_{19}$, $YFeO_3$), or in amorphous alloyed films (Gd-Co, Gd-Fe), or in ferrimagnetic materials such as Yttrium-iron-garnet ($Y_3Fe_5O_{12}$). They are tiny cylindrical regions, as small as 1 μ m in diameter having a reversed magnetization compared to the matrix. This reversal can be done by the field generated by a current carrying coil. When plane polarized light falls on these "bubbles" the plane of polarization of the reflected/transmitted light turns clockwise/anticlockwise depending on the direction of the magnetization of the bubble. Thus, the detection of direction of rotation of the plane polarized light retrieves the data stored. Since the read out is using light, during read-out process the data is not erased.

3.40 PERMANENT MAGNETS

Barium ferrite ($BaFe_{12}O_{19}$) and strontium ferrite ($SrFe_{12}O_{19}$) are the materials used for this applications. These materials are characterized by a high coercive force and high resistivity. The high coercive force allows these materials to be used where there are strong demagnetizing fields. For example, they are used as focusing magnets for television tubes. The high resistivity permits their use as permanent magnets. Even in the presence of alternating high frequency magnetic flux, without eddy current losses. Barium and strontium ferrites are used for making permanent magnets. Commercial names for barium ferrite permanent magnetic ceramic materials are ferroxdure, magnadur and feroba.

3.41 SOFT FERRITES

These ferrites are used in inductor cores, transformer cores, output transformers in television sets and rod aerials. The essential requirements are high permeability, low coercive force and low eddy current losses. Moreover the materials must be capable for operating up to 1000 MHz frequencies. Manganese zinc ferrite and nickel zinc ferrite are the materials satisfying the above conditions. A familiar commercial name for ferrites for these applications is ferrox cube.

3.42 MICROWAVE FERRITES

Microwave ferrites are those ferrites which are used at frequencies in the range 1 to 500 GHz. When a plane polarized electro magnetic wave travels through a ferrite kept in a magnetic field, the plane of polarization rotates along the applied magnetic field direction. This effect is called faraday rotation. If the wave is made to traverse back in the same ferrite, the wave suffers additional rotation. This is called non-reciprocal property. This property is made use in a variety of devices like isolators, circulators, switches, etc. In isolators, incident electro magnetic wave can propagate in forward direction while in reverse direction such propagation is not possible. Using polariser at proper angle back prapagation of the electromagnetic wave can be blocked. In circulators which are having many number of input and output teriminals, wave entering at one port can be made to emerge out through any other desired port. Circulators provide easy isolation between different channels in a microwave system such as a rader, or transmission linkage. Garnets such as YI (AI) G and a few other ferrites such as Magnesium (Mn) (Al) ferrite, $Ni_{0.3}Zn_{0.7}Fe_2O_4$ are useful in microwave applications.

3.43 MAGNETIC CERAMICS

Ferrimagnetic materials can be manufactured in the form of pressed and sintered ceramic powders. By applying magnetic field during firing and cooling processes one can obtain anisotropic properties. They have high resistivity and low eddy current losses. These materials can be used for all applications of metallic magnets. Manufacturing of such metallic magnets is relatively easy and cheap.

Many of our modern technological devices rely on the behaviour of magnetic materials. Such devices include electric motors, electrical power generators and transformers, radio, television, telephones, computers and components of audio and video recording and reproduction systems.

UNSOLVED NUMERICAL PROBLEMS

1. Find the relative permeability of a ferromagnetic material if a field of strength 220 amp/metre produces a magnetization 3300 amp/metre in it.
2. The magnetic field intensity in a piece of ferric oxide is 10^6 amp/metre. If the susceptibility of the material is 1.5×10^{-3}, calculate the magnetisation of the material and the flux density.
3. A paramagnetic material has a magnetic field intensity of 10^4 A/m. If the susceptibility of the material at room temperature is 3.7×10^{-3}, calculate the magnetization and flux density in the material.
4. The saturation magnetic induction of nickel is 0.65 wb/m^2. If the density of nickel is 8906 kg/m^3 and its atomic weight is 58.7, calculate the magnetic moment of the nickel atom in Bohr magneton.
5. A system of electron spins is placed in a magnetic field of 2 weber/m^2 at a temperature T. The number of spins parallel to the magnetic field is twice as large as the number of antiparallel spins. Determine T using classical statistics.
6. The rare earth element gadolinium is ferromagnetic below 16° C with 7.1 Bohr magneton per atom. Calculate the magnetic moment per gram. What is the value of saturation magnetization? Given that the atomic weight of gadolinium is 157.26 and its density is 7.8×10^3 kg/m^3.
7. The saturation magnetic induction of nickel is 0.65 wb/m^2. If the tensity of nickel is 8906 kg/m^3 and its atomic weight is 58.7, calculate the magnetic moment of nickel atom is Bohr magneton.

8. Calculate the magnetic susceptibility of helium gas ($Z = 2$), if the helium atom is in the ground state and the average distance of electron for it is 0.53 Å (= radius of Bohr's first orbit for the hydrogen atom). Given: Atomic concentration $n = 2.7 \times 10^{30}$ per metre^3, Permeability $\mu = 3.4 \times 10^{-7}$ N/A^2.
9. Calculate the energy loss per hour in the iron core of a transformer, if the area of B-H loop is 250 J/m^3 and the frequency of alternating current is 50 Hz. The density of iron is 7.5×10^3 kg/m^3 and mass of core is 10 kg.
10. In a B-H loop, the maximum value of $B_{max.}$ is 1.375 weber/m^2 and the area of the loop is 0.513 cm^2. If the value of 1 cm on x-axis is 10 amp/cm and the value of 1 cm on y-axis is 1 weber/m^2, calculate the hysterisis power loss when an alternating magnetic flux density of intensity 1.375 weber/m^2 and frequency 25 Hz is applied on 10^{-3} m^3 volume of the specimen.
11. An iron rod (density = 8000 kg m^{-3} and specific heat 0.11 cal g^{-1o} C^{-1}) is subjected to cycles of magnetisation at a rate of 50 Hz. If the area enclosed by the hysterisis loop corresponds to an energy loss of 100 J, find the rise in temperature per minute. Assume that there is no radiation losses.
12. A magnetic circuit of length 1.2 m and area of cross-section 100 cm^2 is made of silicon steel of permeability 7.3×10^{-3} NA^{-2} and the winding on it has 150 turns of wire. Calculate:
    (*i*) the magnetising current to obtain a magnetic field 0.30 Wb m^{-2}, and
    (*ii*) the maximum magnetic flux through the core if the magnetising current is 1.5 A.
13. A circular magnet made of silicon steel (permeability $\mu = 7.3 \times 10^{-3}$ NA^{-2} is of mean length 3.0 m and it has an air gap of width 0.10 cm. If its area of cross-section is 100 cm^2 and it has 500 turns of wire on it, assuming the area of cross-section of the air gap to the same as 100 cm^2, calculate:
    (*i*) the magnetising current for the magnetic field strength to be 0.43 Wb m^{-2}
    (*ii*) the magnetic field in the air gap if the magnetising current is 3.0 A.
14. An iron rod of 0.2 cm^2 cross-sectional area is subjected to a magnetising field of 1200 amp-m^{-1}. The susceptibility of iron is 599. Find the permeability and the magnetic flux produced.
15. A material core has 10 turns per cm of wire wound uniformly upon it, which carries a current 2.0 ampere. The flux density in the material is 1.0 weber/metre^2. Calculate the magnetising force H and the magnetisation I of the material. What would be the relative permeability of the core? ($m_0 = 4p \times 10^{-7}$ weber/amp-metre).
16. An iron rod 20 cm long, 1 cm in diameter and of a permeability 1000, is placed inside a solenoid, 1 metre long wound uniformly with 600 turns. If the current of 0.5 ampere is passed through the solenoid, find the magnetic moment of the rod.
17. An iron rod 0.2 m long, 10 mm in diameter and of permeability 100 is placed inside a long solenoid wound with 300 turns/metre. If a current of 0.5 ampere is passed through the rod, find the magnetic moment of the rod.
18. A paramagnetic salt contains 10^{28} ions per m^3 with magnetic moment of 1 Bohr magneton (9.28×10^{-24} amp-m^2). Calculate the paramagnetic susceptibility and the magnetisation produced in a uniform magnetic field of 0.1 weberm^2 when the temperature is 27° C. Given: Boltzmann constant $k = 1.38 \times 10^{-23}$ joule/K.
19. The maximum value of the permeability of m metal (77% Ni, 16% Fe, 5% Cu, 2% Cr) is 0.126 T-m/A. Find the maximum relative permeability and susceptibility.
20. A long cylindrical iron core of cross-sectional area 5.0 cm^2 is inserted into a long solenoid having 2000 turns/m and carrying a current of 2.00 amp. The magnetic field inside the core is found to be 1.57 Tesla. Neglecting the end effects, find the magnetisation I of the core and the pole strength developed.
21. For soft iron, hysteresis loss per unit volume per cycle is 10^3 joule. Take the density of iron 7.5 gm/cm^3 and specific heat 100 calories/kg° C to calculate the temperature increase per minute when an alternating magnetic field of 50 cycle per second is applied on the iron specimen.
22. A magnetic material has a magnetization of 3000 A/m and flux density of 0.044 wb/m^2. Calculate the magnetising force and the relative permeability of the material.

23. The magnetic field intensity of a ferric oxide piece is 10^6 Am^{-1}. If the susceptibility of the material at room temperature is 1.5×10^{-3}, calculate the flux density magnetization of the material.
24. A paramagnetic system of electric spin magnetic dipole moment is placed in an external field of 10^5 A/m^{-1} Calculate the average magnetic moment per dipole at 300 K.
25. If a rod of length 60 cm and area of cross-section 5 cm^2 is bent in the form of closed ring. If the permeability of the material of the rod 2000, calculate the necessary ampere-turn to produce a flux of 20,000 units.
26. For 2 kg of an iron specimen the area of hysteresis loop in equivalent to 300 J/m^3. Calculate the loss of energy per hour for the frequency of 50 Hz. The density of iron is 7.5 g/cm^3.
27. A magnetic field of 2000 A-T/m produces a flux density of $8\pi Wb/m^2$ in an iron ring. Calculate the relative permeability of iron.
28. The susceptibility of a magnetic material is 25.14×10^{-11} H/m. Calculate its relative permeability.
29. The inherent nature of all matter out of the following is
(*a*) paramagnetism (*b*) ferromagnetism (*c*) diamagnetism (*d*) none of the above
30. Magnetic susceptibility *k* is given by the relation
(*a*) $k = \frac{I}{H}$ (*b*) $k = \frac{H}{I}$ (*c*) $k = \frac{B}{H}$ (*d*) $k = \frac{H}{B}$
31. The relative permeability can be expressed as
(*a*) $\mu_r = 1 + \mu_a$ (*b*) $\mu_r = 1 + k$ (*c*) $\mu_r = 1 + \frac{k}{\mu_0}$ (*d*) $\mu_r = \mu_0 + \mu_a$
32. The reluctance (*R*) of a substance of area of cross-section a is given by
(*a*) $R = \frac{l}{a\mu_0\mu_r}$ (*b*) $R = \frac{a}{\mu_0\mu_r}$ (*c*) $R = \frac{\Phi}{\text{m.m.f.}}$ (*d*) $R = I \times N$
33. If three reluctances (R_1, R_2 and R_3) are connected in series, the equivalent reluctance *R* is given by
(*a*) $R = R_1 + R_2 + R_3$ (*b*) $\frac{1}{R} = \frac{1}{R_1} + \frac{1}{R_2} + \frac{1}{R_3}$
34. The hysteresis is
(*a*) lagging of the magnetising force *H* with respect to the flux density.
(*b*) lagging of the flux density *B* with respect to the magnetising force *H*.
(*c*) the ratio of *B* and *H*
35. The quantities *B*, *H* and *I* are related as
(*a*) $B = \mu_0 (H + Ik)$ (*b*) $B = \mu_0 (H + I)$ (*c*) $B = \mu_0 H + I$ (*d*) $B = \mu_0 I + kH$
36. The permeabilities of para and ferromagnetic materials are
(*a*) greater than unity (*b*) less than unity (*c*) equal to unity (*d*) negative
37. Soft iron is used to manufacture electromagnets because their
(*a*) magnetic saturation limit is high and retentivity and coercive force are small
(*b*) retentivity is high
(*c*) coercive force is high
(*d*) area of hysteresis curve is large

38. Define the three magnetic vectors namely, magnetic induction $\vec{B}$, magnetisation $\vec{I}$ and magnetic intensity $\vec{H}$. Establish the relation $\vec{B} = \mu_0 (\vec{H} + \vec{I})$.
39. Establish the relation (*a*) $B = \mu_0 (h + I)$, and (*b*) $\mu = \mu_0 (1 + \chi_m)$.
40. Define permeability and magnetic susceptibility and show that $\mu = \mu_0 (1 + \chi_m)$
41. Define $\vec{B}$, $\vec{M}$ and $\vec{H}$. Establish the relation $\vec{B} = \mu_0 \vec{H} + \mu_0 \vec{M}$.

MULTIPLE CHOICE QUESTIONS

1. Materials do not having permanent magnetic dipoles are
   (*a*) paramagnetic (*b*) ferromagnetic (*c*) ferrimagnetic (*d*) diamagnetic
2. When a diamagnetic material is placed inside an external magnetic field
   (*a*) induced magnetic dipoles act along the applied field direction.
   (*b*) induced magnetic dipoles act opposite to the applied field direction.
   (*c*) induced magnetic dipoles act perpendicular to the applied field direction.
   (*d*) there will be not induced magnetic dipoles.
3. Diamagnetic susceptibility is
   (*a*) large, negative (*b*) small, positive (*c*) small, negative (*d*) large, positive
4. Which of the following statement is wrong?
   (*a*) Diamagnetic property is independen to temperature.
   (*b*) Paramagnetic property is dependent of temperature.
   (*c*) Paramagnetic property is independent of temperature.
   (*d*) Ferromagnetic property is dependent of temperature.
5. Magnetic susceptibility χ is equal to
   (*a*) the ratio of the magnetic induction in the sample to the applied magnetic field intentisy.
   (*b*) magnetic moment per unit volume.
   (*c*) the ratio of the intensity of magnetisation produced in the sample to the magnetic field intensity producing it.
   (*d*) the ratio of the magnetic field intensity to the intensity of magnetisation.
6. The unit of magnetic field intensity is
   (*a*) Am^{-1} (*b*) Hm^{-1} (*c*) $Wb\ m^{-2}$ (*d*) no units
7. Magnetic induction *B* is related to magnetic field intensity *H* by the relation.
   (*a*) $B = \mu_0 (H - m)$ (*b*) $B = \mu_0 \dfrac{H}{M}$ (*c*) $B = \mu_0 \dfrac{M}{H}$ (*d*) $B = \mu_0 (H + M)$
8. One Bohr magneton equal to
   (*a*) 9.27×10^{-24} A-m^2 (*b*) 9.1×10^{-31} A-m^2 (*c*) 9.27×10^{-16} A-m^2 (*d*) 9.1×10^{-24} A-m^2
9. In SI units, tesla is expressed as
   (*a*) $NA^{-1}m^{-1}$ (*b*) $NA^{-1}m^{-3}$ (*c*) $NA^{-1}m^{-3}$ (*d*) $kg\ s^{-2}A^{-1}$

10. In a paramagnetic materials of susceptibility 10^{-3}, the induction B in Wb m^{-2} at an applied field 10 kA m^{-1} is ($\mu_0 = 1.257 \times 10^{-6}$ Hm^{-1})

(*a*) 0.01257 (*b*) 1.257 (*c*) 0.1257 (*d*) $4\pi \times 10^{-7}$

11. The temperature at which the transition of antiferro to para-magnetism takes place is called

(*a*) Curie-Weiss temperature (*b*) Curie temperature

(*c*) Debye temperature (*d*) Neel temperature

12. The transition from the ferromagnetic to the paramagnetic state is named after

(*a*) Curie (*b*) Curie-Weiss (*c*) Neel (*d*) Debye

13. Diamagnetic materials possess

(*a*) permanent magnetic dipoles (*b*) no permanent magnetic dipoles

(*c*) induced dipoles along field direction

(*d*) no induced dipoles even when external field is applied

14. The area of hysteresis loop of a ferromagnetic material gives

(*a*) the coercive force (*b*) the remanent flux density

(*c*) the intensity of magnetisation of the material

(*d*) the energy that is consumed in taking the material through are complete cycle of magnetisation.

15. For a ferromagnetic material, in the graph plotted between $1/\chi$ versus T, the extra polated line (on x-axis)

(*a*) makes positive intercept at a temperature below *OK*

(*b*) makes positive intercept at *OK*

(*c*) makes positive intercept at a temperature above *OK*

(*d*) none of these

16. For a para magnetic material, in the graph plotted between $1/\chi$ versus T, the extrapolated line makes an intercept which is

(*a*) positive at *OK* (*b*) passing through origin at *OK*

(*c*) positive at a temperature above *OK* (*d*) negative at *OK*

17. Ferrox cube is the commercial name of

(*a*) ferromagnetic materials (*b*) paramagnetic materials

(*c*) ferrimagnetic materials (*d*) diamagnetic materials

18. Magnetic ceramics are

(*a*) diamagnetic materials (*b*) paramagnetic materials

(*c*) ferromagnetic materials (*d*) ferrimagnetic materials

EXERCISE

1. Define magnetic moment. Explain the origin of magnetic moment at the atomic level. What is a Bohr magneton?
2. In hydrogen atom an electron *e* revolves around the nucleus at a distance of *r* meter with an angular velocity ω radians. Obtain expression for magnetic moment associated with it due to its orbital motion.
3. Define the terms magnetic susceptibility, magnetic permeability, magnetic induction and magnetization.
4. What are the sources of permanent dipolemoment in magnetic materials?
5. Explain the important properties of ferrittes.
6. How materials are classified as dia para ferromagnetic? Explain.
7. Explain ferri-magnetism and anti-ferromagnetism.
8. Explain hysteresis of a ferromagnetic material.
9. What is ferromagnetism? What are the distinguishing features of ferromagnetism.
10. What are ferrites? Explain the magnetic properties of ferrites and mention their industrial applications.
11. What is Bohr magneton? Give an account of domain theory of Ferromagnetism.
12. Describe hysteresis loop. How is it used to classify magnets?
13. Show the nature of magnetic dipole moments in ferro-ferri and anti-ferromagnetic materials.
14. Explain the origin of magnetic moment. Find the magnetic dipole moments due to orbital and spin motions of an electron.
15. Explain hysteresis curve and its uses.
16. Given the Weiss theory of ferromagnetism.
17. Explain the distinguishing features of ferro, antiferro and ferri-magnetic materials.
18. Give the important features of ferromagnetic materials. Explain the Weiss theory of ferromagnetism.
19. What is ferromagnetism? Explain the hysteresis curve on the basis of domains.
20. Explain ferri-magnetism.
21. Explain the hysterisis loop observed in ferromagnetic materials. What are hysterisis losses?
22. Explain in detail the classification of magnetic materials on the basis of electron spin.
23. Explain the classification of magnetic materials.
24. Explain how ferrites are superior to ferromagnetic materials. Mention the applications of ferrites.
25. Explain ferromagnetism and B-H curve.
26. Write the importance of hard magnetic materials in engineering applications.
27. Define the

    (*a*) magnetic induction (*b*) intensity of magnetisation

    (*c*) magnetic susceptibility (*d*) magnetic permeability (*e*) relative permeability

4 Superconductivity

INTRODUCTION

It is a fact that the electrical conductivity of metal increases with decrease in temperature. In case of mercury its resistance decreases to zero. When the temperature falls to 4.15 K, as shown in Fig. 4.1.

At this temperature mercury behaves as a superconductor. The temperature below which the resistance disappears is called transition temperature (T_C). The transition temperature is different for different metal.

(The DC resistivity of metals decreases with increasing temperature, then the resistivity decreases as the metals are cooled.) The DC resistivity of a few metals reduces to zero at very low temperatures.

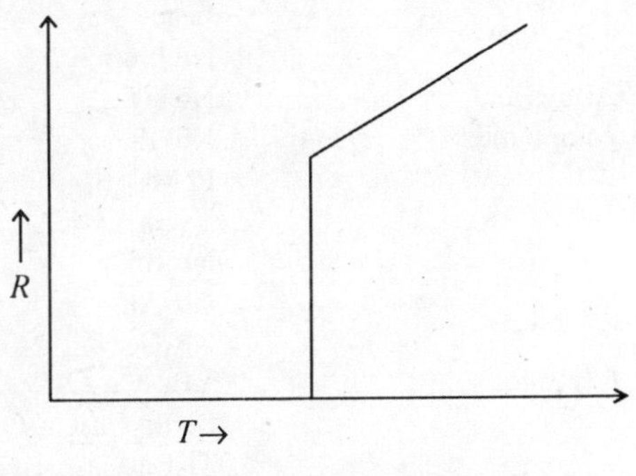

Fig. 4.1

The phenomenon of the DC resistivity of materials going to zero at lower temperatures is known as superconductivity. The materials which exhibit zero resistivity, are superconductors. superconductivity is observed in a few elements, at normal temperatures and pressure.

At high pressures a large number of elements show the superconducting behaviour. superconductivity is also observed in a few intermetallic compounds, ceramics and in semiconductors.

"There is a stage at which the material looses its resistance completely and the current set up in it continue to flow unchanged for ever" such a property pocessed by the material is called superconductivity and the substance is called superconductor.

The phenomenon of superconductivity was first observed in 1911 by Kammerlingh Onnes.

The transition temperatures of a few superconducting material and the year of observation are given in Table 4.1.

4.1 TEMPERATURE DEPENDANCE OF SUPERCONDUCTIVITY

superconductivity can be explained on the basis of quantum theory. The vibration of atom in lattice will never dies away entirely even at 0° K. Atoms in the lattice would retain irreducible motion called zero point vibration. The resistance is the consequence of interaction between the conduction electron and the vibrating atom in the lattice. In the superconducting state there will be synchronisation between the vibrating atom and conduction electron at this stage there will be no transfer of energy from the system to the lattice then the electrical resistivity becomes zero and conductivity becomes infinite.

In metals like Cu, Ag, Au, Li, Na and K a finite and constant resistivity was observed even at very low temperature. The metals which show a finite resistivity at low temperatures are normal metals.

The change in resistivity with temperature in normal metals and in superconductors are shown in Fig. 4.2.

The DC resistivity is partly due to the lattice vibrations and partly due to the impurities. If ρl and ρ_i are the resistivities due to the lattice vibrations and that due to the impurities respectively then the resultant resistivity of the material is given by

$$\rho = \rho l + \rho_i$$

But, at very low temperatures, the lattice vibrations freez and the resistivity due to the lattice component reduces to zero. Therefore, in a few pure metals, the resistivity reduces to zero at very low temperatures and the

Table 4.1

Type of material	*Name of the material*	T_c *(K)*	*Year of observation*
Elements	Tungsten	0.02	
	Titanium	0.40	
	Zinc	0.85	1911
	Aluminium	1.18	1913
	Tin	3.72	1913
	Mercury	4.15	1930
	Lead	7.19	1933
	Niobium	9.25	1937
Alloys and compounds	Nb–Ti	10.2	
	Nb–Zr	10.8	
	$PbMO_6S_8$	14.0	
	V_3Ga	16.5	
	Nb_3Sn	18.3	1954
	Nb_3Al	18.9	
	Nb_3Ge	23.0	1973
Ceramics	$YBa_2Cu_3O_7$	92.0	1987
	$Bi_2Sr_2Ca_2Cu_3O_{10}$	10.0	
	$Tl_2Ba_2Ca_2Cu_3O_{10}$	125.0	1988
	$HgBa_2Ca_2Cu_3$	53.0	

superconductivity sets in. The reason for the superconductivity is explained by Bardeen, Cooper and Schriefer in their BCS theory.

4.2 EFFECT OF MAGNETIC FIELD

In superconducting state the magnetic flux can not penetrate through the metal, so, the magnetic induction at that stage becomes zero. If sufficient strong magnetic field is applied, the superconductor restore its resistance and the corresponding field applied is called critical field (H_C).

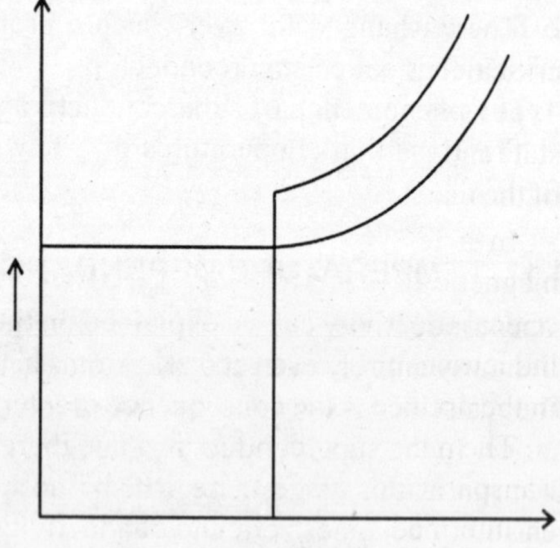

Fig. 4.2

4.3 MEISSNER EFFECT

Suppose a superconductor is kept in the magnetic field below critical field, then the magnetic flux are pushed out of the material as shown in Fig. 4.3, *i.e.* the magnetic induction inside the material is zero. This effect is called Meissener effect.

Zero resistivity and perfect diamagnetism are the two important features of superconductor.

We know that, in the superconducting state the magnetic induction inside the material is zero. If B_0 is the magnetic induction in a material due to the external field H and B_m is the magnetic induction due to the magnetisation in the material, then

$$B = B_0 + B_m$$

but, in a superconductor,

$$B = B_0 + B_m = 0$$

$$\mu_o H + \mu_o I = 0$$

$$\mu_o (H + I) = 0$$

since $\mu_o \neq 0$

$$H + I = 0$$

$$\therefore \quad I = -H$$

This implies that, the intensity of magnetisation inside the material is negative indicating that the superconducting state is diamagnetic in other words.

$$B = \mu_o H \left(1 + \frac{I}{H}\right) = 0$$

$$\mu_o H (1 + x) = 0$$

$$\therefore \quad \chi = -1$$

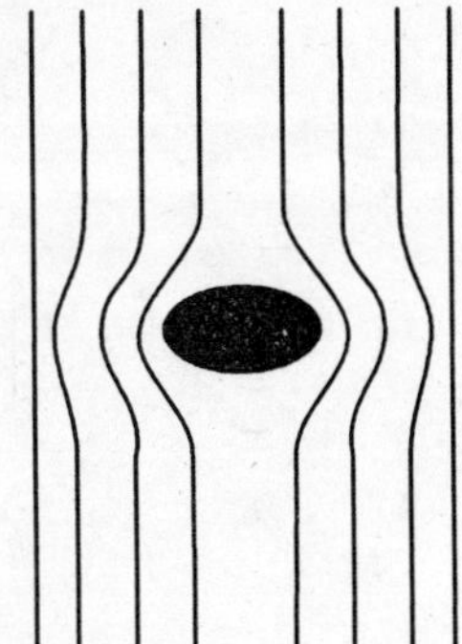

Fig. 4.3

which implies that, the magnetic susceptibility of the superconducting state is negative, indicating the superconducting state to be diamagnetic.

4.4 TYPES OF SUPERCONDUCTOR

Meissner and Ochsenfeld experimentaly showed that the extent of intensity of magnetization changes with the increasing strength of the applied magnetic field. As the intensity of the applied magnetic field increases the Meissner effect also increases. But at a particular value of the applied high magnetic field, the magnetic induction penetrates the material and the material loses the superconducting properties. The high magnetic field at which the material changes from superconducting state to normal state is called the critical field H_c.

The materials in which a sharp change of state is observed at the critical field are grouped as Type I superconductor. In other words, Type I superconductor are those in which only the superconducting state and the normal states are observed when studied as a function of the magnetic field which is as shown in Fig. 4.4.

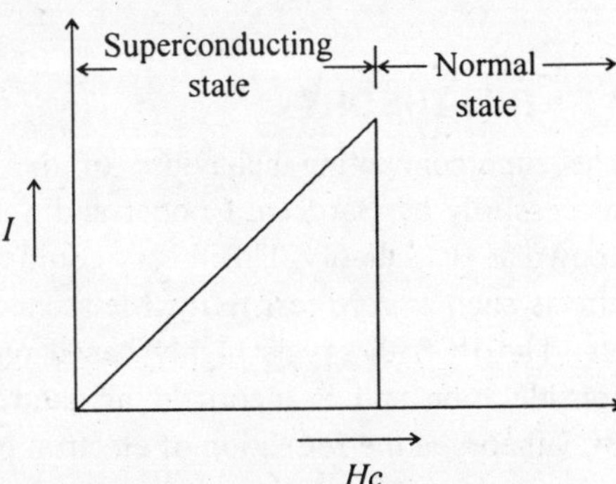

Type I Superconductor

Fig. 4.4

In a few superconductors, as the strength of the applied magnetic field increases, the Meissner effect also increases till a critical field Hc_1. Beyond this field the penetration of the magnetic induction through the materials begins, with the further increase in the applied magnetic field.

The penetration increases and at a very high field called the second critical field Hc_2, the material becomes transparent to the magnetic induction and it loses the superconducting properties. Such materials which show an intermediate mixed state between the superconducting state and the normal state are called the Type II superconductors. In other words, Type II super conductors are those which change from superconducting state to the normal state through an intermediate state. The intermediate mixed state is called the vortex state and is as shown in Fig. 4.5.

The critical field Hc is dependent of temperature of the material, the relation between the critical field and the temperature is given by

$$H_c (T) = H_c (O) \left[1 - \left(T/_{T_c}\right)^2\right]$$

Where, H_o is the value of the critical field at absolute zero temperature (T) of the material and T_c is the superconducting transition temperature.

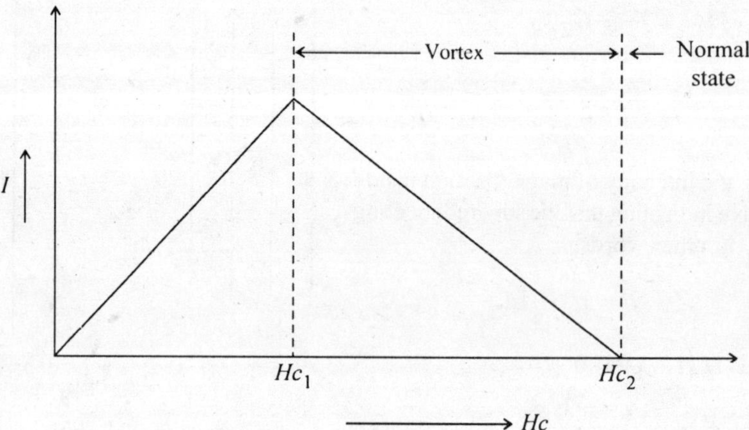

Fig. 4.5 Type II superconductor

A graph of H_c as a function of temperature $T(K)$ is as shown in Fig. 4.6.

Critical temperature of some common superconducting materials are shown in the Table 4.2.

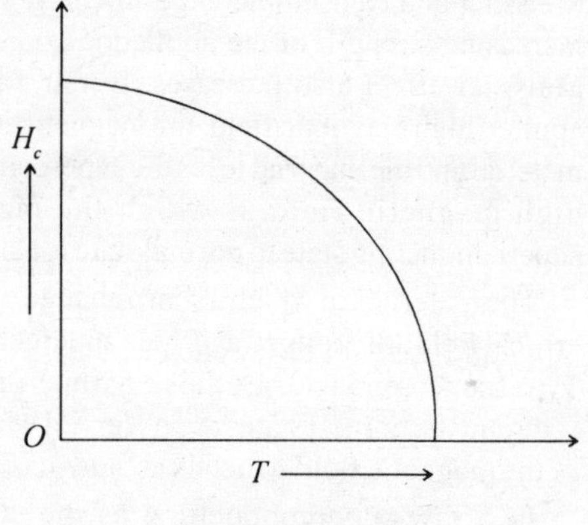

Fig. 4.6

4.5 BCS THEORY

The superconducting behaviour of the material is explained successfully by Bardeen, Cooper and Schriefer in 1957 and is known as BCS theory. This theory could explain many observed effects such as zero resistivity, Meissener effect, Isotope effect, etc. The BCS theory is highly based on quantum mechanical consideration and is theortical in nature. The essence of the BCS theory is the formation of electron pairs called the Cooper pairs in the material at very low temperature.

In the macroscopic material, the atoms or the molecules are arranged in a regular order. The positions of these atoms or molecules in the material is referred to as the lattice. Due to the binding force established between the lattice points, the atoms or the molecules in the material will be in the form of positive ions. The negatively charged electron disturbs the positions of the ions in the structure of the material and deforms the ionic potential field inside the material. At very low temperatures, where superconductivity is observed, the lattice vibrations freez and therefore the deformations in the lattice potential will be a predominent factor. Another electron passing through the same material experiences the destorted potential field, the motion of the second electron is modified by the distorted potential field. Thus, the second electron establishes the bonding with the first electron via the lattice distortion. The distortion produced by the first electron is called electron-lattice interaction, the bondage established between the electrons through the lattice is the electron-lattice-electron interaction. This interaction is responsible for the formation of cooper pair and the electron is known as Cooper electron as shown in Fig. 4.7. The Cooper pair is formed between the electrons with opposite and equal spins and the momentum. The formation of the first cooper pair follows the formation of the subsequent pairs. The paired electrons can be regarded as the Zero-Spin (Bosons) which can condense into a common ground state.

Table 4.2

Material	*Critical temperature T_c in K*	*Type*
Zinc (Zn)	0.79	I
Aluminium (Al)	1.18	I
Tin (Sn)	3.72	I
Mercury (Hg)	4.15	I
Lead (Pb)	7.19	I
Zirconium (Zr)	0.7	II
Thorium (Th)	1.4	II
Thalium (Tl)	2.38	II
Tantalum (Tq)	4.38	II
Niobium (Nb)	9.22	II
Compounds/Alloys		
Nb – Ti	10.2	II
Nb – Zr	10.8	II
NbN	14.7	II
V_3 Ga	16.5	II
Nb_3Al	18.0	II

The important steps considered in BCS theory are,

1. Cooper pairs are formed at the superconducting transition tempe-rature, the increased superconducting behaviour is due to the number of cooper pairs formed in the materials. These cooper pairs are broken by the thermal energy, high magnetic fields and the presence of the paramagnetic impurities, consequent to the breaking up of the cooper pairs, the superconductivity is destroyed.
2. The cooper pairs will establish a new set of energy states in the material in the superconducting state. They form a new ground state and the excited states and there exists an energy gap E_g between the two states, the nature of the energy gap in the superconducting state is different from that observed in the other states. The energy gap is a consequence of the lattice properties in case of insulator, but in case of superconducting state the energy gap is a consequence of the Fermi-gas. The energy gap reduces with the increasing temperature in superconducting state.

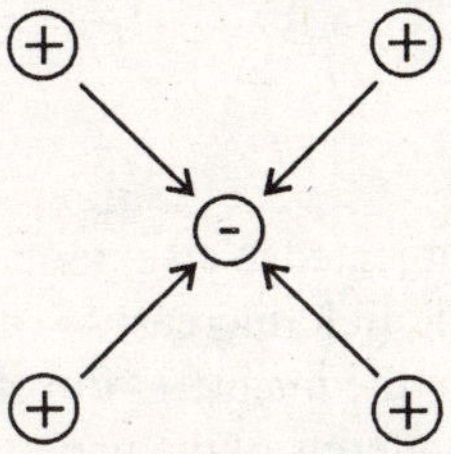

Positive ions attracted towards an electron forming a positive ion core

(a)

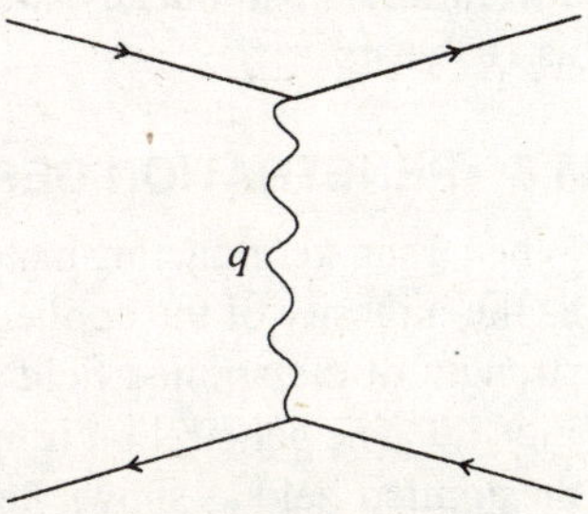

The exchange of virtual phonons between the two electrons (Cooper pair)

(b)

Fig. 4.7

3. The penetration depth and the coherance lengths are the consequence of the Cooper pairs. Penetration depth is the thickness of the material through which the magnetic induction passes in to the material. Coherence length is the distance in the material within which the energy gap parameter does not change considerably.
4. The super current through the material gives rise to a magnetic flux round the material. As the super current in the material increases. The related magnetic flux also increases, but at very high applied magnetic field superconductivity ceases. Therefore, the super current in the material has a limitation.

The magnetic flux linked with the super current is quantised. In all the calculations of the magnetic field due to the super current. The unit of electronic charge is taken as 2e, due to the formation of Cooper pairs.

5. High temperature follows the lattice vibrations. At high temperature as the lattice vibrations break the cooper pairs, the superconductivity is destroyed. And the presence of minute quantities of paramagnetic impurities do not favour the formation of the cooper pairs. Thus, the addition of minute quantity of paramagnetic impurity destroy the superconductivity. The high magnetic fields also break the electron-lattice-electron interaction subsequently the superconductivity is destroyed.

4.6 HIGH TEMPERATURE SUPERCONDUCTIVITY

The dream that superconductivity may be observed above liquid nitrogen (77° K) came true very recently, *i.e.* in 1987 by a research team lead by Paul Chu and MK Wu. The material which exhibited the high temperature superconductivity was an oxide with chemical composition of the form $Ba_2Cu_3O_7$ often referred as the 1-2-3 compound as evident from the formula it self. Now the research group is going on round in search of that magnetic material which show superconducting state at room temperature. The recent are is discovered in the year 1993 is Hg-BiBaSrC_4-O at 150 K. So far high temperature superconductivity is observed only in non-metals like ceramics. The major technical challenge is to get them in a proper shape such as wires and ribbon which facilitate the flow of electricity over a long range.

4.7 PERSISTENT CURRENT

Suppose a ring of superconductor is taken as shown in Fig. 4.8 in transition temperature and induced current is passed then the induced current continue to flow indefinitely with out any delay as long as the temperature is below T_c. It is estimated that the ring of superconductor would carry the induced current of several amperes for several years without any attenuation. Such a current is called persistant current and its delay time is estimated as 10^5 years.

4.8 PENETRATION DEPTH

When a superconducting material is subjected to an external magnetic field superconducting state also changes as the intensity of the applied magnetic field increases. The superconductors changes to normal state. As the strength of the applied field increases, the magnetic flux starts penetrating through the material as shown in Fig. 4.9. The penetration is measured interms of the penetration depth λ. The penetration depth λ is related to the applied field as shown below.

$$H = H_o e^{x/\lambda}$$

where, H_o → intensity of the applied field

x → dimension of the material

and λ → the penetration depth (300 – 500 Å)

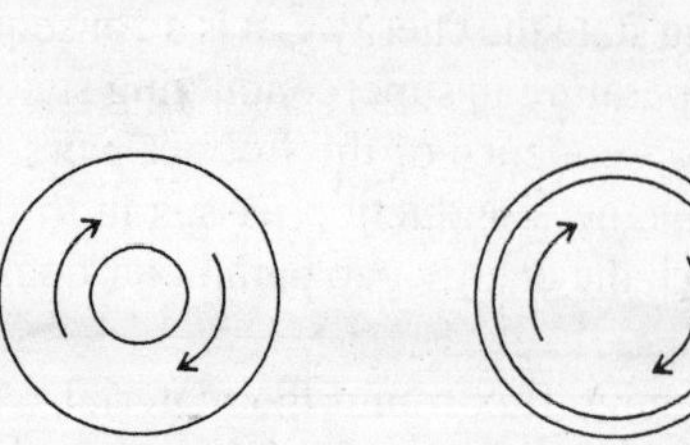

Fig. 4.8

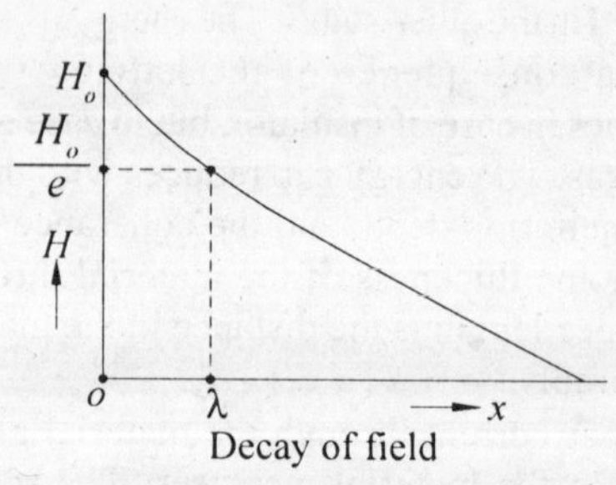

Decay of field

Fig. 4.9

The penetration depth also depends on the temperature of the material and the relation between them is given

$$\lambda = \lambda_o \left[1 - \left(\frac{T}{T_c}\right)^4\right]$$

4.9 ISOTOPE EFFECT

The superconducting transition temperature T_c is different for the different materials. T_c is one of the characteristic properties of the superconducting materials. But in case of materials which occur in isotopic forms, the superconducting transition temperature T_c depends on the atomic mass of the isotope. The experimental study of T_c in the different isotopes reveals the following relation between the T_c of isotope with its atomic mass M.

i.e. $$M^{\propto} \propto \frac{1}{T_c}$$

or $$M^{\propto} T_c = \text{constant}$$

The experimentally measured values of $\propto$ for different materials are given in the Table 4.3.

Table 4.3

S. no.	*Material*	$\propto$	*S. no.*	*Material*	$\propto$
1.	Ru	0.00 ± 0.05	7.	Cd	0.32 ± 0.07
2.	Zr	0.00 ± 0.05	8.	Zn	0.45 ± 0.05
3.	Nb_3 Sn	0.08 ± 0.02	9.	Sn	0.47 ± 0.02
4.	Os	0.15 ± 0.05	10.	Pb	0.49 ± 0.02
5.	Mo	0.33	11.	Hg	0.50 ± 0.03
6.	Mo_3 Ir	0.33 ± 0.05	12.	Tl	0.61 ± 0.10

4.10 ENERGY GAP

In the superconducting state the materials will have a special kind of energy gap. The energy gap of a superconductor is dependent on the temperature of the material and is a consequence of the Fermi energy and a consequence of the lattice parameter as in the case of other states like conductors and semiconductors.

4.11 DESTRUCTION OF SUPERCONDUCTIVITY

The thermal energy, high magnetic field and the presence of paramaganetic impurities destroy the property of superconductivity.

1. **The Thermal Energy:** As the temperature of the material increases, the material losses its superconducting property, the temperature at which the material losses superconductivity is known as transition temperature.
2. **High Magnetic Field:** As the strength of the applied magnetic field increases the penetration depth also increases. At the transition critical field, the material changes from the superconducting state to normal state.
3. **Paramagnetic Impurities:** The presence of minute quantities of paramagnetic impurities destroy the superconductivity. One part of paramagnetic impurity in million is sufficient to destroy the superconductivity.

4.12 APPLICATIONS OF SUPERCONDUCTIVITY

1. Superconductors are expected to carry the DC current without the line loss or the I^2R loss. As such they can be used in the windings of motors and generators to get uninterrupted service or power supply. But so far the superconductivity has been observed in metallic conductors at very low temperatures. Therefore, this application is possible only at lower temperatures. The high temperature superconductivity is observed in the ceramic materials. The ceramic materials are brittle in nature and hence, thin wires cannot be drawn in these materials. This nature of the high temperature superconductors limits the applications.
2. Levitation is one of the important phenomenon observed in the superconducting state. That is the superconducting material rejects the magnetic induction (Meissner effect). Due to this behaviour, the material floats in the presence of an external magnetic field. This phenomenon can be exploited for the construction of the superconducting levitation trains which float on the magnetic rails and can run at unbelieveably high speeds. This can only be realised if a suitable high temperature superconductor is found. Thus, the levitation train is a dream.
3. Using the metallic superconducting materials like zirconium, molibdinum and niobium it is possible to drive the supercurrents in the coils of wires made of these metals at very low temperatures. This technique is used to produce the electromagnets which can supply the high enough magnetic fields of the order of tens of Tesla. These superconducting magnets are used in
   (*a*) Medicine to get the three dimentional image of the human body or parts of the human body, using the technique called magnetic resonance imaging (MRI). The images obtained through this technique will have very high resolution. Even the minute disorders also can be located comfortably. Moreover the exposure of the human body to the high magnetic field will not cause any damage to the human tissues. Thus MRI is a safest, unharmful and highly useful technique for the study of disorders.
   (*b*) Another important application of the high magnetic fields produced by the superconducting magnets is in the field of research. Study of the behaviour of materials in presence of very high magnetic fields is the theme of the research work.
4. Superconducting material can be used as a memory device in a computer.
5. Superconducting coils are used in high resolution nuclear magnetic resonance to study and scanning the whole body to diagnose medical problems.
6. A superconductor wrapped with a current carrying solonoid can act like a switching element.
7. Since superconductor responds quickly for low current and signals, they are used in underground and power transmission cables.
8. The phenomenon of superconductivity is used in the measurement of extremely low voltage of the order of 10^{-15} volts and low magnetic field of 10^{-10} oyersted.
9. As the superconductivity can be destroyed by the applications of strong magnetic field, they are used as switching devices in electronic devices.
10. Superconducting materials are used in certain electrical circuit where heavy current is passed.
    The main disadvantage of superconductor is that extreemely low current is needed for its efficient working.
11. Magnetically leviated vehicles are simply called as Maglev vehicles. This phenomenon is based on the Meissener effect.
    The advantage of leviation is such that it can be used in the absence of contact between the moving and stationary body in which the friction is eliminated and great speed could be achieved with low energy consumption. The schematic diagram of Maglev vehicle is as shown in Fig. 4.10.
    The prototype of Maglev vehicle has been constructed in Japan to carry 35 passengers at a speed of about 430 km/hour for a test ride which was operated between Osaka and Tokyo.
    The system consists of a superconducting magnet mounted at the base of the vehicle. The vehicle is set to float by magnetic leviation, which is brought about by large repulsion between two highly powerful

magnetic fields. Then the forward motion of the vehicle is achieved by the synchronous lienar induction. The superconducting magnets were cooled with liquid helium. The system uses direct current from the superconducting magnets placed under the carriage, which in turn induces eddy current in aluminium guide way, thus generating repulsive force and lifts the carriage and also propell the carriage in forward direction.

12. A superconductor possess two states, the superconducting state and normal state. The application of a magnetic field greater than H_c can institute a change from superconducting to normal state and removal of the field reverses the process. This principle is applied in development of switching element **CRYOTRON**. This consists of a tantalum core around which is wounded by a niobium wire. The critical temperatures of tantalum and niobium are 4.5 K and 9.5 K respectively. Tantalum is a gate and niobium is the control.

    The schematic of the cryotron is as shown in Fig. 4.11. The current flows through the tantalum can be controlled by switching action, allowing a control current to pass through the niobium winding, the magnetic field produced is sufficient to change tantalum from its superconducting state to normal state. This closes the gate for the flow of current through tantalum. Removal of the control current reopens the gates. Switching times with a Josephson junction are of the order of 10^{-2} *ns*.

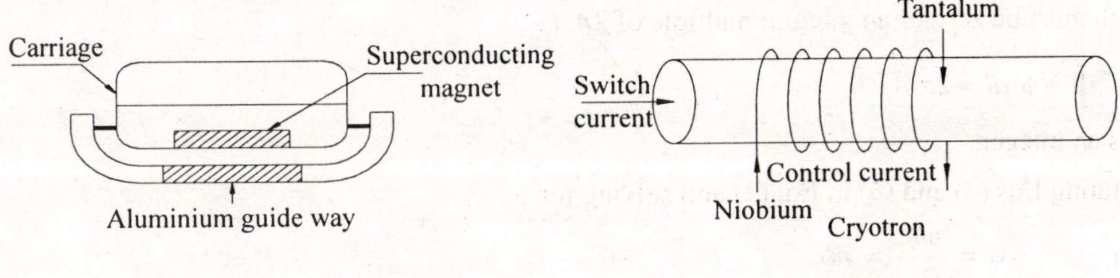

Fig. 4.10 **Fig. 4.11**

13. **SQUIDs:** SQUID is a double junction quantum interferometer. SQUID is the acronym for superconducting quantum interference device. Two Josephson junctions are mounted on a superconducting ring forms this interferometer. SQUIDs are highly sensitive devices and are widely used to detect change in magnetic flux. The SQUIDs are based on the flux quantisation in a superconducting ring. The total magnetic flux passing through the ring is quantized very minute magnetic signals are detected by these SQUID sensory. The SQUIDs are used to study tiny magnetic signals from the brain and heart. The SQUID magnetometers are used to detect the paramagnetic response in the liver. This gives the information about the amount of iron held in the lever of the body accurately. SQUIDs are used to detect the ore and oil deposit. SQUIDs are also used to detect the submerged submarines in seas.

4.13 FLUX QUANTIZATION

In the year 1950 Ginzbury and Landau introduced a complex wave function ψ to describe the superconducting state. The density of the superconducting electrons $n_s \propto |\psi|^2$, *i.e.* the complex wave function ψ is considered as an order parameter to describe the superconducting state. For a given temperature, the "order parameter" ψ is a function of position in the material. It is not a constant and vanishes above T_c.

The wavefunction ψ can be written in terms of the magnitude and phase as

$$\psi = |\psi| e^{i\phi} \tag{1}$$

∴ The current density can be written as

$$J = -\left(\frac{2e^2}{mc}A + \frac{e\hbar}{m}\nabla\phi\right)|\psi|^2 \quad (2)$$

where A is the vector potential.

Let us consider a superconducting material in the shape of a ring as shown in Fig. 4.12.

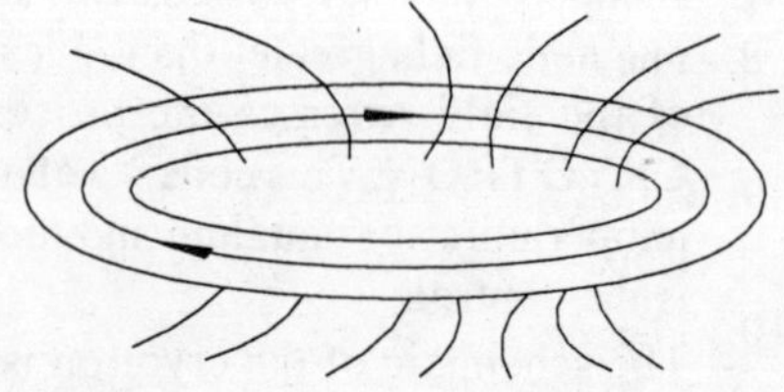

A ring shaped superconducting material

Fig. 4.12

Taking a closed path

$$\oint J\cdot dl = |\psi|^2 \oint \left(\frac{2e^2}{mc}A + \frac{e\hbar}{m}\nabla\phi\right) dl = 0 \quad (3)$$

According to Stokes theorem

$$\int_{\varepsilon} J\cdot dl = \int_{\varepsilon} \nabla\times A\cdot ds = \int_{\varepsilon} B\cdot ds = \phi \quad (4)$$

where ϕ is the flux enclosed by the ring. Since the order parameter is single valued, its phase change around the closed path must be zero or an integral multiple of 2π. *i.e.*,

$$\oint \nabla\phi\cdot dl = 2\pi n \quad (5)$$

where n is an integer.

Substituting Eqs (4) and (5) in Eq. (3) and solving for ϕ,

We get, $$\phi = \frac{nhc}{2e} = n\phi_o \quad (6)$$

Where $\phi = \dfrac{hc}{2e} = 2.07\times10^{-15}$ weber is known as fluxod or flux quantum. Thus, the magnetic flux enclosed by the ring is quantized. This equation holds good when the ring is under superconducting condition.

The flux through the ring is the sum of the flux due to the external source and the flux due to the supercurrent flowing through the ring, *i.e.*

$$\phi = \phi_{ext} + \phi_{sc} \quad (7)$$

This flux ϕ is quantized. Since ϕ_{ext} is not normally quantized, ϕ_{sc} adjusts such that ϕ is always quantized.

UNSOLVED NUMERICAL PROBLEMS

1. A superconducting tin has a critical temperature of 3.7 K at zero magnetic field and a critical field of 0.0306 Testa at OK. Calculate the critical field at 2 K.
2. Calculate the critical current for a wire of lead having a diameter of 1mm at 4.2 K. The critical temperature for lead is 7.18 K and $Ho = 6.5 \times 10^4$ A/v.
3. Calculate the critical magnetic field at 1° K for mercury if its transition temperature is 4.15 °K and critical magnetic field at 0° K is $H_c = 3.272 \times 10^6$ A/m.
4. Calculate the temperature at which the critical magnetic field is $H_c = 11.091 \times 10^4$ A/m and at 0° K is $H_c(0) = 15.6 \times 10^4$ A/m if $T_c = 9.3°$ K for niobium.
5. The transition temperature for lead is 7.19° K. Its critical magnetic field at 0° K is 6.3933×10^4 A/m. Calculate its critical magnetic field at 5°K.

6. The magnetic field B within a bar of some metal alloy is 32 tesla when a field of $H = 50 \times 10^5$ A/m is applied. Calculate the magnetic susceptibility, the relative permeability and the magnetizing field within the metal.
7. The magnetic field strength (H) in silicon is 100 A/m. if its magnetic susceptibility is 0.3×10^5, calculate the magnetization and the magnetic field B within the material.
8. The magnetic susceptibility of tungsten is 6.8×10^{-5}. Calculate its magnetization for a magnetizing field of 8000 A/m. Find its relative permeability.
9. Calculate the critical magnetic field at 0° K for mercury if its transition magnetic field strength at 3° K is 1.562×10^7 A/m, and its transition temperature is $T_c = 4.15°$ K.
10. The critical temperature of Nb_3Ge is $T_c = 23°$ and critical magnetic field strength at 0° K is 2.388 $\times 10^7$ A/m. Calculate its critical magnetic field at 15° K.
11. To what temperature must lead be cooling in a magnetic field strength of 2×10^4 A/m for it to be super conductive. Given critical temperature for lead is 7.19° K and critical magnetic field strength at 0° K is 6.3933×10^4 A/m.
12. Calculate the critical current that flow through a long thin superconducting wire of aluminium of diameter 10^{-3} metre. Given, the critical magnetic field for aluminium is 7.9×0^3 amp/metre.
13. The critical temperature, T_c for mercury is 4.185 K and isotopic mass is 199.5. If the isotopic mass changes to 203.4, calculate its critical temperature.
14. The London penetration depths for Pb at 3 K and 7.1 K are respectively 39.6 and 173 nm. Calculate its transition temperature as well as depth at OK.
15. What should be the energy of the quantum state which has 19 for the sum of the squares of the quantum numbers n_x, n_y and n_z? What is the degeneracy of this state? Assume the solid to be $10 \times 10 \times 10$ mm.
16. Show that the probability of occupancy of energy level E by an electron is (*i*) negligible for $E >> E_F$, (*ii*) 0.5 for $E = E_F$, and (*iii*) almost unity for $E << E_F$.
17. The Fermi level for potassium is 2.1 eV. Calculate the velocity of the electrons at the Fermi level.
18. The Fermi level of silver is 5.5 eV. Calculate the fraction of free electrons at room temperature located up to a width of kT on either side of E_F.
19. The resistance of a sample of copper wire of 0.1 mm diameter and 0.2 m long is 0.439 ohm. Calculate its resistivity and conductivity.
20. The resistivity of silver at room temperature is 1.6×10^{-8} ohm m. Calculate the collision time for electron scattering.

MULTIPLE CHOICE QUESTIONS

1. Below transition temperature, a superconducting material exhibits
   (*a*) only zero resistance
   (*b*) only diamagnetic property
   (*c*) zero resistance and diamagnetism
   (*d*) zero resistance and ferromagnetism
2. The dependance of the critical field upon the temperature is given by
   (*a*) $H_c(0) = H_c(T)\left[1 - \left(\frac{T}{T_c}\right)^2\right]$
   (*b*) $H_c(T) = H_c(0)\left[1 - \left(\frac{T}{T_c}\right)^2\right]$
   (*c*) $H_c(T) = H_c(0)\left[1 - \left(\frac{T_c}{T}\right)^2\right]$
   (*d*) $H_c(T) = H_c(0)\left[1 - \left(\frac{T}{T_c}\right)^2\right]^{-1}$

3. The magnetisation of a superconductor is
(*a*) O (*b*) H (*c*) 1 (*d*) – H
4. Examples of Type I superconductors are
(*a*) Al, Nb and Ta (*b*) Al, Zn and Hg (*c*) Ta, V and Nb (*d*) none of these
5. In a superconducting state
(*a*) entropy alone changes (*b*) electronic specific heat alone changes
(*c*) both entropy and electronic specific heat change (*d*) none of these
6. In Type I superconductors the coherence length is of the order of
(*a*) 10^6 m (*b*) 10^{-3} m (*c*) 10^3 m (*d*) 10^{-6} m
7. When a material becomes superconductor
(*a*) the properties of lattice structure do not change
(*b*) the properties of lattice structure do change
(*c*) it becomes ferromagnetic in nature
(*d*) magnetic property does not change
8. In superconducting state the energy gap
(*a*) is large compared to semiconductors and insulators
(*b*) is zero
(*c*) is very small as compared to semiconductors and insulators
(*d*) does not change.
9. In Type II superconductors
(*a*) the magnetic flux pass through the entire material
(*b*) the magnetic flux does not pass through the material
(*c*) the magnetic flux does not suddenly drop to zero but decreases exponentially
(*d*) none of these
10. In superconducting materials average isotopic mass M is related to the transition temprature by (where a is called the isotope effect coefficient)
(*a*) MT_c^{α} = constant (*b*) $MT_c = \infty$ (*c*) $aM = T_c$ (*d*) $M^a T_c$ = constant
11. The London penetration depth (λ) is given by the expression (with appropriate meaning)
(*a*) $\lambda = \dfrac{m}{\mu_o n_s e^2}$ (*b*) $\lambda = \left(\dfrac{m}{\mu_o n_s e^2}\right)^{1/2}$ (*c*) $\lambda = \left(\dfrac{\mu_o n_s e^2}{m}\right)$ (*d*) $\lambda = \left(\dfrac{\mu_o n_s e^2}{m}\right)^{1/2}$
12. The penetration depth is found to depend on temperature as given by the relation
(*a*) $\lambda(T) = \lambda(0)\left(1 - \dfrac{T^4}{T_c^4}\right)^{-1/2}$ (*b*) $\lambda(T) = \lambda(0)\left(1 - \dfrac{T^4}{T_c^4}\right)^{1/2}$
(*c*) $\lambda(T) = \lambda(0)\left(1 - \dfrac{T_c^4}{T^4}\right)^{-1/2}$ (*d*) $\lambda(T) = \lambda(0)\left(1 - \dfrac{T_c^4}{T^4}\right)^{1/2}$
13. In flux quantizaiton the sum of the flux due to the external source and the flux due to the supercurrent flowing through the ring is $\phi = \phi_{ext} + \phi_{sc}$.

(*a*) f, f_{ext} and f_{sc} are all individually quantized (b) f, f_{ext} and f_{sc} are not quantized

(*c*) f alone is quantized while f_{sc} adjusts according to f_{ext}

(*d*) f_{ext} and f_{sc} are quantized while f is not quantized.

14. Switching times with a Josephson junction are in order of magnitude

    (*a*) 10^{-2} ns (*b*) 1 ms (*c*) 10^{-15} s (*d*) 1 ms

EXCERCISE

1. Explain the dependance of superconducting state of a metal on temprature and strength of the magnetic field in which the metal is placed.
2. Write a note on Meissener effect.
3. Compare and contrast Type I and Type II superconductors.
4. Write a note on penetration depth of superconductivity.
5. Write a note on flux quantization.
6. Describe Josephson effects and their applications.
7. What are DC and AC Josephson effects?
8. What are Cooper pairs?
9. Write a note on BCS theory of superconductivity.
10. Write any four applications of superconductors.
11. Perfect diamagnetism is more fundamental property than perfect conductivity to assert that a material is in superconducting state. Explain the statement.
12. Explain Meissener effect. How is it used to classify the superconductors?
13. Define the terms in superconductors.

    (*i*) critical temperature (*ii*) critical magnetic field (*iii*) critical current
14. Explain the Meissener effect. Give some important applications of superconductors.
15. Explain the critical parameters and their significance in superconductors.
16. Discuss the parameters that destruct the superconductivity.
17. Give the important features and uses of superconductors.
18. Give the BCS theory of superconductivity.
19. Discuss the important experimental facts of superconductivity.
20. What is Meissener effect? Show that superconductors exhibit perfect diamagnetism.
21. What is superconductivity? Describe the effect of

    (*a*) Magnetic field (*b*) Frequency (*c*) Isotopes on superconductors

    Mention a few industrial applications of superconductors.
22. What is superconductivity? Explain Meissener effect. What are the possible applications of superconductors?
23. Explain Type I and Type II superconductors. What are Josephson's effect?
24. Distinguish between soft and hard superconductors. Explain how the penetration depth varies with (*i*) Magnetic field strength, (*ii*) Temperature

5 Applied Optics

INTRODUCTION

In interference of light, we use a term coherence between two sources of light. The two sources are said to be coherent, when they vibrate in the same phase or if they have constant phase difference. We know that, light from a source emits as the radiations by atoms or molecules of the source. The phase is different at different times. Now the question is that to what extent may the radiations from different atoms of a given source be related in phase, in the direction of emission and in polarization, *i.e.* the coherence of a given source. In recent years, some sources are developed which are highly coherent, *i.e.* the radiation given out by all the emitters in the source is in mutual agreement not only in phase but also in the direction of emission and polarization. These coherent sources are called lasers. The light beams from lasers, with coherence lengths of upto 10^{14} Hz will make possible many experiments and practical applications which would not be possible with the previously available light sources of coherence lengths extending up to 10^7 Hz.

5.1 SPONTANEOUS EMISSION

We know that an atom has discrete set of energy states. The lowest energy state of the atom is called the ground state energy. The atoms of each substances are generally in the ground state energy. When an external energy is imparted to the substance, its atoms absorb the energy and move to the higher energy state, *i.e.* the atoms get excited as shown in Fig. 5.1 (a). An atom remains in the excited state only for a period of 10^{-8} s, and then it come back to its ground state. When there is transition of atom from the excited state to the ground state, the difference in energy corresponding to these states is emitted in form of a photon as shown in Fig. 5.1 (b). If the energy difference in these states is ΔE, the frequency v of the photon emitted is given by $\Delta E = hv$ where h is Planck's constant. This emission of photon is called as *spontaneous emission.* This emission is irregular and is at different instants from different atoms, hence, the light obtained from different atoms in spontaneous emission is not coherent in both phase and direction.

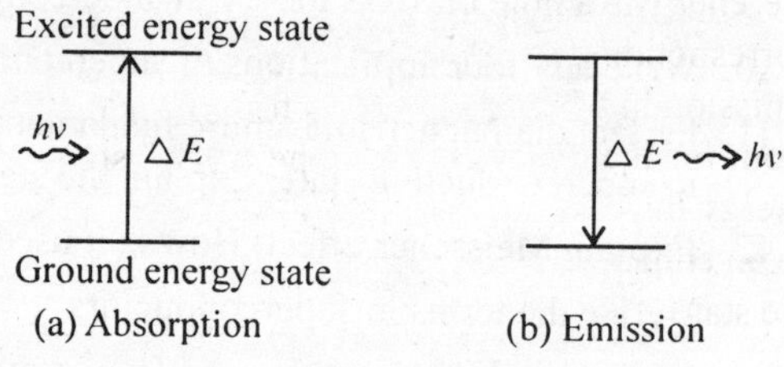

Fig. 5.1

5.1.1 Stimulated Emission

If the atom is in the excited state a photon of same frequency as that of the frequency of photon in which the atom passing from the excited state to the ground state will emit radiation then that atom comes to the ground state very soon and stimulates the light incident on it, by emitting the photon of exactly the same frequency. This is called is *stimulated emission.*

Thus, for the stimulated emission, light is made to incident on the atoms, of frequency in which the atoms emit by spontaneous emission. By doing so, the number of photons corresponding to that energy increases in the environment of atoms. In the stimulated emission, the light emitted from each atom is coherent in direction and phase with the incident light. As the incident light reaches the subsequent excited atoms, the stimulated emission becomes stronger and stronger.

Let the ground state energy of an atom of a substance is E_0 and the energy of the excited state is E. If an atom absorbs the external energy and reaches to the excited energy state. Then the atom returns back to its ground state within 10^{-8} s and it emits a photon of frequency v corresponding to the energy difference $\Delta E = E - E_0$, where $\Delta E = hv$ as shown in Fig. 5.2. Hence, stimulated emission will be possible only if some photons of frequency v or energy $E - E_0$ are made incident by making the external light incident on it, so that the number of photons corresponding to the energy $E - E_0$ increases as shown in Fig. 5.2.

As a consequence, more and more photons are emitted due to the stimulated emission. Thus, a chain process starts and almost all the atoms emit photon by the stimulated emission. This process is called the *laser action,* because in this process an intense light beam is obtained by amplifying the light by the stimulated emission of radiation. Generally, there are very few atoms in the excited state, therefore, stimulated emission is not strong enough to produce lasing action. For the strong stimulated emission, the necessary condition is that the number of excited atoms should be comparatively very large. The increase of the number of excited atoms, is called *the population inversion.* Obviously, the light beam obtained by the stimulated emission is very intense, monochromatic and coherent.

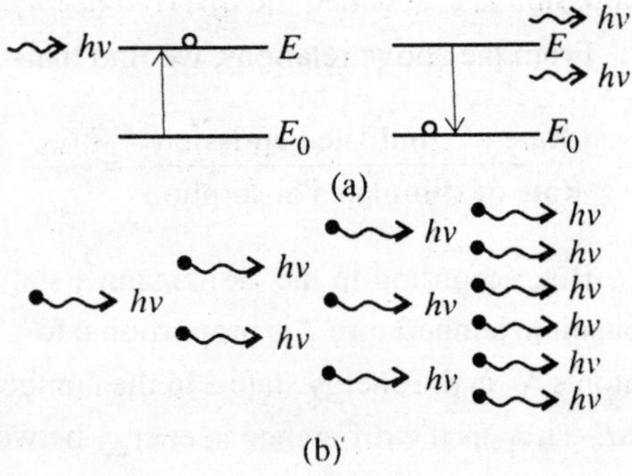

Fig. 5.2

5.2 EINSTEIN'S *A* AND *B* COEFFICIENTS

Figure 5.3 shows the two energy states 1 and 2 of an atomic system. In the diagram dots (°) represent the atoms in eDCh energy state. Three kinds of transition involving the electromagnetic radiation are possible between the two energy states. Figure 5.3 shows the self emission in which when an atom undergoes self transition from the energy state 2 to the energy state 1, a photon of energy hv_{12} is emitted. If E_0 and E_1 be the energies corresponding to these energy states 1 and 2 respectively, then $E_1 - E_0 = hv_{12}$. Figure 5.3 shows the stimulated emission in which a photon of frequency v_{12} (or energy hv_{12}) from the surrounding environment stimulates an atom for the transition from the energy state 2 to the energy state 1. As a consequence, two photons each of energy hv_{12} are emitted. In this emission, the radiated light waves are exactly in phase with the incident light, so an enhanced beam of coherent light is obtained. The absorption in which an atom move from the state 1 to the state 2 by absorbing a photon of energy hv_{12} from the surrounding environment is shown in Fig. 5.3.

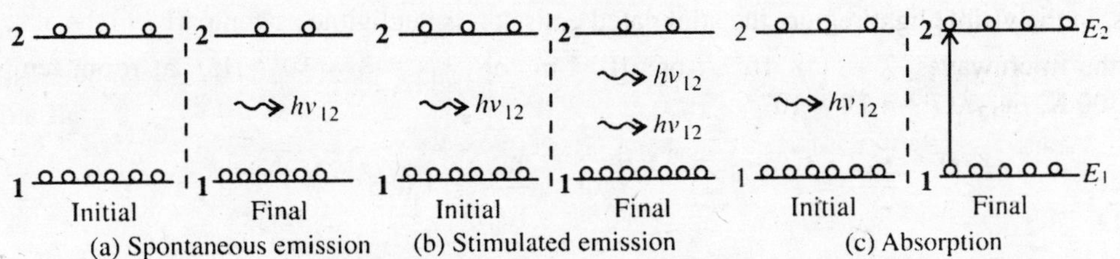

Fig. 5.3

Einstein suggested that the stimulated emission in the energy state 1 is possible just like transition from the energy state 1 to the energy state 2 by absorption. So, all the three phenomenon are statistical in nature and the relative probability of these phenomenon depends on the following *three* factors

(*i*) The number of atoms N_1 present in the energy state 1.

(*ii*) The number of atoms N_2 present in the energy state 2. and

(*iii*) The radiation density ρv_{12} corresponding to the frequency v_{12} in the surroundings.

The rate of spontaneous-emission $\quad r_{sp} \propto N_2 \quad$ or $\quad r_{sp} = A_{12} N_2$

The rate of stimulated emission $\quad r_{st_e} \propto N_2 \quad$ and $\quad r_{st_e} = \rho v_{12}$

or $\quad r_{st_e} = B_{12} N_2 \rho v_{12}$

and the rate of absorption $r_{st_a} \propto N_1$ and $r_{st_a} \propto v_{12}$ or $r_{st_a} = B_{12} N_1 \rho v_{12}$

Here the coefficients A_{12} and B_{12} are called as *Einstein's A and B coefficients*. It is clear that the rate of stimulated emission and the rate of absorption is determined by the same coefficient B. That is why the stimulated emission is also called the *inverted absorption.*

From the above relations, we find that

$$\frac{\text{Rate of stimulated emission}}{\text{Rate of stimulated absorption}} = \frac{r_{ste}}{r_{sta}} = \frac{N_2}{N_1}$$

But, according to the Boltzmann's statistics law, the probability of finding a system with energy E at an absolute temperature T is proportional to $e^{-E/kT}$. Hence, at an absolute temperature T, the ratio of number of atoms N_2 in the energy state 2 to the number of atoms N_1 in the energy state 1 will be equal to $e^{-\Delta E/kT}$, where $\Delta E = h\nu_{12}$ is the difference in energy between these states, *i.e.*

$$\frac{N_2}{N_1} = e^{-h\nu_{12}/kT}$$

∴ From the above equation, we get

$$\frac{r_{ste}}{r_{sta}} = e^{-h\nu_{12}/kT}$$

From the above expression, it is clear that

(*i*) For the visible light $\lambda = 5000$ Å or $\nu_{12} = 6 \times 10^{14}$ Hz, at room temperature $T = 300$ K, $h\nu_{12}/kT = 956.5$

$$\therefore \quad \frac{r_{ste}}{r_{sta}} = e^{-956.5} \quad \text{or} \quad \frac{r_{ste}}{r_{sta}} \approx 10^{-44}$$

Thus, in the visible light region, the stimulated emission is negligible as compared to the absorption.

(*ii*) For the microwaves, $\lambda = 1 \times 10^8$ Å or 10^{-2} m or $\lambda_{12} = 3 \times 10^{10}$ Hz, at room temperature $T = 300$ K, $h\nu_{12}/kT = 4.78 \times 10^{-3}$.

$$\text{Hence,} \quad \frac{r_{ste}}{r_{sta}} = e^{-4.78 \times 10^{-3}} \quad \text{or} \quad \frac{r_{ste}}{r_{sta}} = 1.005$$

Thus, in the microwave region, the stimulated emission is not negligible as compared to the absorption. It is clear from the above discussion that the probability of stimulated emission is more in the microwave region.

5.3 RELATION BETWEEN EINSTEIN'S *A* AND *B* COEFFICIENTS

There are three processes namely absorption, spontaneous emission and stimulated emission, due to which there is a change in number of atoms in the excited state. If N_1 is the number of atoms present in the ground energy state, N_2 is the number of atoms present in the excited energy state and $\rho\nu_{12}$ is the radiation density corresponding to the frequency ν in the surroundings (where ν_{12} = the frequency of the photon which is emitted by the self emission due to transition from the energy state 2 to the energy state 1), then

the rate of spontaneous emission $\quad r_{sp} = A_{12} N_2$

the rate of emission $\quad r_{ste} = B_{12} N_2 \rho\nu_{12}$

and the rate of stimulated absorption $\quad r_{sta} = B_{12} N_1 \rho\nu_{12}$

It is clear that, the total rate of change in the number of atoms in the excited energy state must be equal to the sum of the rate of spontaneous-emission, the rate of stimulated emission and the rate of absorption, *i.e.* the total rate of change in number of atoms in the excited energy state is

$$r = r_{sp} + r_{ste} + r_{sta}$$

or
$$r = -A_{12} N_2 - B_{12} N_2 \rho\nu_{12} + B_{12} N_1 \rho\nu_{12}$$

Here, the negative sign indicates the decrease in number of atoms in the excited energy state and the positive sign indicates the increase in number of atoms in the excited energy state.

But in equilibrium, the rate of change in the number of atoms in the excited energy state must be zero, *i.e.* $r = 0$

or
$$-A_{12} N_2 - B_{12} N_2 \rho\nu_{12} + B_{12} N_1 \rho\nu_{12} = 0$$

i.e.
$$B_{12} N_1 \rho\nu_{12} = (B_{12} \rho\nu_{12} + A_{12}) N_2$$

But according to the Boltzmann's statistics, at an absolute temperature T,

$$N_2 = N_1 e^{-h\nu_{12}/kT}$$

$\therefore$
$$B_{12} N_1 \rho\nu_{12} = \left(B_{12} \rho\nu_{12} + A_{12}\right) N_1 e^{-h\nu_{12}/kT}$$

or
$$B_{12} \rho\nu_{12} e^{h\nu_{12}/kT} = B_{12} \rho\nu_{12} + A_{12}$$

or
$$B_{12} \rho\nu_{12} \left(e^{h\nu_{12}/kT} - 1\right) = A_{12}$$

or
$$\frac{A_{12}}{B_{12}} = \rho\nu_{12} \left(e^{h\nu_{12}/kT} - 1\right)$$

But according to Planck's quantum theory, the energy density corresponding to the frequency ν_{12} is given by,

$$\rho\nu_{12} = \frac{8\pi h \nu_{12}^3}{c^3} \left(\frac{1}{e^{h\nu_{12}/kT} - 1}\right)$$

$\therefore$ we get,

$$\frac{A_{12}}{B_{12}} = \frac{8\pi h \nu_{12}^3}{c^3}$$

This is the relation between Einstein's A and B coefficients.

5.4 PRINCIPLE OF LASER, OPTICAL PUMPING AND POPULATION INVERSION

Laser is a device in which intense, monochromatic, coherent and unidirectional light beam is obtained. It was discovered in 1950, by Town and Shallow. The word laser is the acronym of *light amplification by stimulated emission of radiation.*

Thus, a laser produces a beam of light whose waves have the same frequency (monochromatic), are in phase with one another (coherent) and well collimated Fig. 5.4 shows a laser beam.

(a) Ordinary light (b) Monochromatic incoherent light (c) Laser beam (monochromatic coherent light)

Fig. 5.4

The change in number of atoms in the excited state is possible by the following *three* processes: (*i*) spontaneous-emission, (*ii*) absorption and (*iii*) stimulated emission. The principle of laser is based on the fact that there is more stimulated emission as compared to the spantaneous-emission and absorption.

Let N_1 and N_2 be the number of atoms in the ground energy state and the excited energy state, and ρv_{12} is the radiation density in the surroundings corresponding to the frequency v_{12} of photon emitted by spontaneous emission from the excited energy state to the ground state energy, then,

rate of spontaneous emission $= A_{12} N_2$

rate of stimulated emission $= B_{12} N_2 \rho v_{12}$

and rate of absorption $= B_{12} N_1 \rho v_{12}$

where A_{12} and B_{12} are called as Einstein's A and B coefficient.

Hence,
$$\frac{\text{Rate of stimulated emimssion}}{\text{Rate of spontaneous emission}} = \frac{B_{12} N_2 \rho v_{12}}{A_{12} N_2}$$

and
$$\frac{\text{Rate of stimulated emimssion}}{\text{Rate of absorption}} = \frac{N_2}{N_1}$$

It is essential that the rate of stimulated emission should be higher than that of the rate of spontaneous emission and the rate of absorption. Hence, the following conclusions are obtained from the above relations.

(*i*) For lasing action the number of atoms in the excited energy state must be very large as compared to the number of atoms in the ground energy state, *i.e.* $N_2 >> N_1$, this is called the *population inversion.*

(*ii*) The value of v_{12} must be as high as possible, *i.e.* the radiation density in the surroundings corresponding to the frequency v_{12} of the photon emitted by spontaneous emission should be high. For this, the emitted radiation is kept enclosed inside the cavity between the two parallel reflectors so that due to the successive reflections, the number of photons corresponding to the frequency v_{12} increases and hence the energy density v_{12} increases.

(*iii*) The ratio of Einstein's A and B coefficients should be minimum. For this, the excited energy state is to be chosen in such a way that the spontaneous emission from it is forbidden. This energy state is called the *metastable state.* By doing so, the value of A becomes nearly zero. The life time of a metastable state is nearly 10^{-3} s while that of the ordinary excited state is nearly 10^{-8} s. Thus, working of laser is based on the presence of a large number of atoms in one or more metastable states. Figure 5.8 shows the laser action. Let E_0 be the ground state energy of an atom and E_1 and E_2 are the two excited states respectively, if a light is made to incident by a flash tube, many atoms absorb photons corresponding to the energy $(E_2 - E_0)$ and get excited to the energy state E_2 as shown by the transition 1 in Fig. 5.5. These excited atoms after 10^{-8} s, emit a photon of energy $(E_2 - E_1)$ by spontaneous emission and come to the ground state energy as shown by the transition 2 in Fig. 5.5.

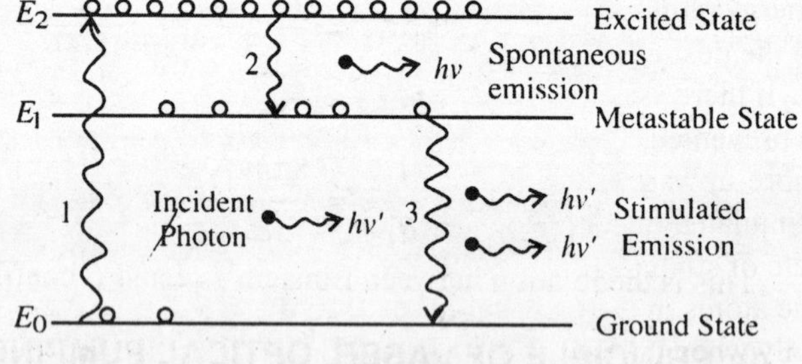

Fig. 5.5

If the energy state E_1 is the metastable state, then it is not possible for the atom to come to the ground state energy E_0 by spontaneous emission, but it is possible for the atom to come from the metastable energy state E_1 to the ground state energy state E_0 by the stimulated emission. Thus, the number of atoms in the energy state E_1 is very much increased in comparison to those in the ground state energy E_0.

Now, if photons corresponding to the energy $E_1 - E_0 = hv$ are made to incident, then the stimulated emission starts and as a result of chain process an intense light beam is obtained. The photons so emitted are reflected several times between the high reflectors so that the stimulated radiation becomes more intense as compared to the spontaneous radiation.

Thus, for a simple kind of laser, we need an assembly of atoms that have the three energy states, one the ground state of energy E_0, other a metastable state of energy E_1 and third a still higher excited state E_2 that decays to the metastable state. Hence, we need more atoms in the metastable state than in the ground state. If we can arrange this and then shine light of frequency v on the assembly, there will be more stimulated emissions from the atoms in the metastable state than the absorptions by atoms from the ground state. The result will be an amplification of the original light. *The process of increasing the number of atoms in the metastable state much larger in comparision to those in the ground state, is called the population inversion.* This is illustrated in Fig. 5.6.

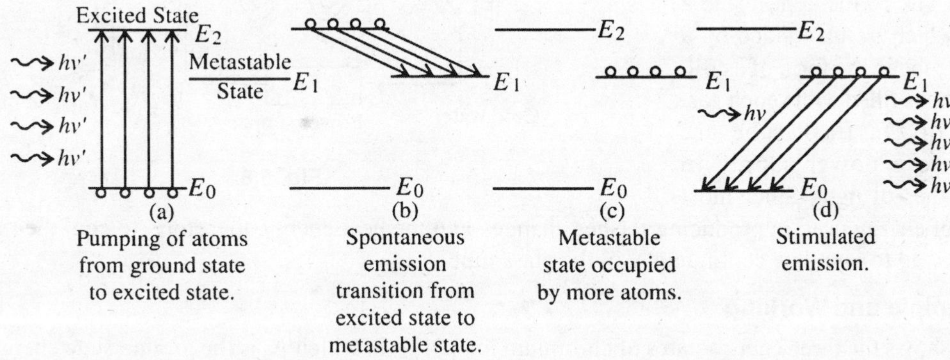

Fig. 5.6

There are several ways to produce a population inversion, one of them is the *optical pumping* in which an external light source is used which emits light of frequency such that the photons have energy just equal to the energy required to raise the ground state atoms to the excited state from where the atoms decay spontaneously to the desired metastable state.

If there are only two energy states, a ground state and metastable state above the ground state, the pumping is prevented. The reason is that more photons of frequency v which we pump into the assembly of atoms causes more upward transitions from the ground state to the metastable state and at the same time there will be stimulated transitions from the metastable state to the ground state. When half the atoms are in each state, the rate of stimulated emission will be equal to the of rate of absorption. Thus, there will never be more than half the atoms in the metastable state, so the laser amplification will not occur. A population inversion is possible only when the absorptions are more in higher energy state than the metastable state from which the stimulated emission occurs. There must be more than half the atoms present in the metastable state for the stimulated emission to predominate.

In case if there are four energy states of an assembly of atoms as shown in Fig. 5.7 the laser transition from the metastable state occurs to an unstable intermediate state rather than to the ground state. The intermediate state dacays rapidly to the ground state, so there are very few atoms in the intermediate state. Hence, in such cases even a modest amount of pumping is enough to populate the metastable state to a greater extent than the intermediate state.

E_3 — Excited state
Stimulated transition
E_2 — Metastable state
Laser transition
E_1 — Intremediate state
Pumping
E_0 — Ground state

Fig. 5.7

On the basis of different methods of pumping, we have different kinds of laser such as *ruby laser, gas laser, chemical laser and semiconductor laser*, etc.

5.5 RUBY LASER CONSTRUCTION

The construction of ruby laser is as shown in Fig. 5.8. It consists of a rod of artificial ruby crystal of length nearly 5 cm and diameter nearly 1 cm, with its ends plane. The artificial ruby is the aluminium oxide (Al_20_3) crystal mixed with 0.05 % impurity of chromium oxide. Thus, some aluminium atoms are replaced by the chromium ions (Cr^{+++}). One end of this rod is fully silvered so that this end becomes a good reflector. The other end of the rod remains translucent. The rod is kept inside a glass tube which is surrounded by a xenon flash tube in the form of a coil. The flash tube is filled with xenon gas. The ends of the flash tube are connected to a power supply to produce flashes of light. Since most of the part of energy used in producing flashes changes into the heat energy, therefore, to cool the tube, cold water is allowed to circulate continuously in the glass tube.

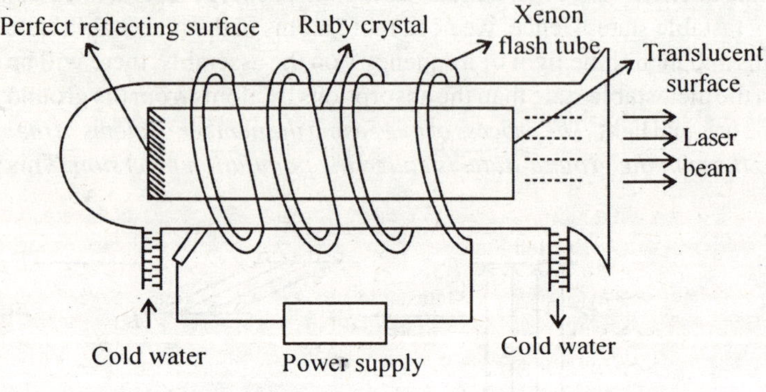

Fig. 5.8

5.5.1 Principle and Working

Figure. 5.9 shows the three energy states of chromium ion (Cr^{3+}) in which E_0 is the ground state energy and E_1 and E_2 are the two excited state energies. There is a metastable energy state M slightly below the excited energy state E_1. An ion can remain in the metastable energy state for a period of about 10^{-3} s, whereas it can remain in the excited state for a period of 10^{-8} s.

When a flash of light from the flash tube falls on the ruby rod, the chromium ions present in the rod absorbs green-yellow light of wavelength 5600 Å and get excited to the energy states E_2 and E_1 by the transitions 1 and 2 respectively as shown in Fig. 5.9. Now, these excited chromium ions lose some energy in collisions with the crystal lattice and come to the metastable state M by the transitions 3 and 4 as shown in Fig. 5.6. There is no emission of light in these transitions. Since the ions can remain in the metastable state M for a long period of about 10^{-3} s, therefore, more and more ions come to this state and the number of ions in the metastable state becomes very large as compared to that in the ground state. Now, a photon of energy corresponding to the red light of the order 6943 Å from the surrounding stimulates the ions present in the metastable state and emits other photons of energy corresponding to the red light of wavelength 6943 Å. As a result, the ions pass from the metastable state M to the ground state E_0 the transitions 5 and 6 shown in Fig. 5.9. These photons are coherent and unidirectional, and they get successively reflected at the ends of the rod. This process is repeated so that an highly intense, unidirectional and coherent beam of wave length 6943 Å is obtained parallel to the axis of rod.

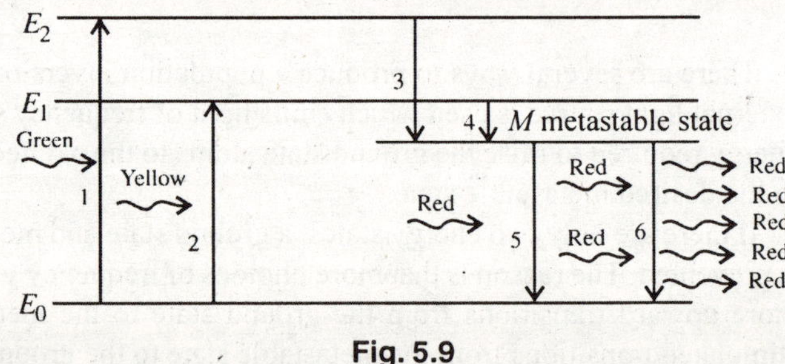

Fig. 5.9

5.6 HELIUM–NEON GAS LASER

We get a very intense laser beam from the ruby laser, but this beam is obtained in the form of pulses and therefore, it cannot be used for the practical purposes. For the practical purpose, we require a continuous laser beam. For this, a gas laser is used.

Helium-neon laser is a common gas laser in which a mixture of helium and neon gases in the ratio 10 : 1 at a low pressure (≈ 1 mm) is used. The reason is that some of the excited energy states of helium are very close to the excited energy states of neon gas. On discharge, first the helium atoms get excited. The excited helium atoms transfer some of their energy to the neon atoms during collisions with them, so that the neon atoms get excited. Then an highly intense unidirectinal monochromatic laser beam is obtained by the stimulated emission of neon atoms.

5.6.1 Construction

Figure 5.10 shows the construction of helium-neon laser. It consists of a long quartz cylindrical tube filled with a mixture of helium and neon gases. One end of the tube is provided with a perfect reflector and the other end with a partial reflector. The pumping of the gas mixture is done with the electromagnetic energy of frequency $\approx 4 \times 10^7$ Hz due to which a laser beam of nearly 40 watt and wavelength 6328 Å is obtained by the stimulated emission of radiation.

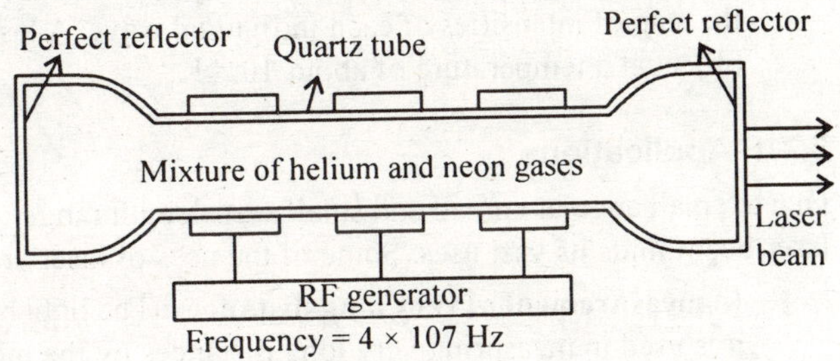

Fig. 5.10

5.6.2 Principle and Working

Figure 5.11 shows the principle of helium-neon laser by the energy level diagram. The diagram shows the energy levels of helium and neon. The excitation of the helium atoms by the electron-collisions is rapid as compared to that of the neon atoms. The life-time of excited states of helium atom is comparatively longer than that of neon. Although this laser action is due to the stimulated emission of neon atoms, but the helium atoms helps in pumping.

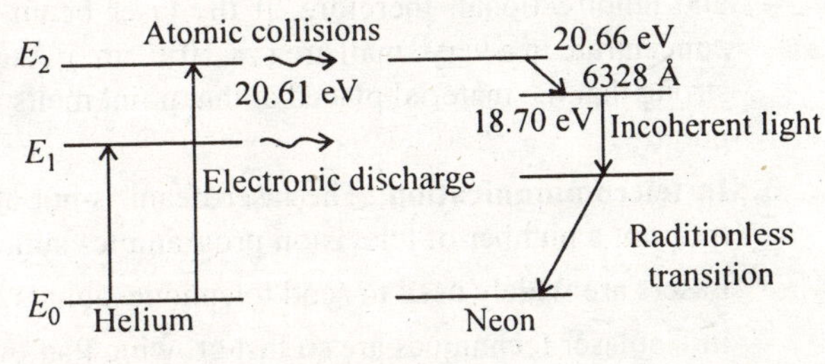

Fig. 5.11

When electric discharge is passed in the tube with the help of electromagnetic energy, the helium atoms easily reach the excited states E_1 and E_2 from the ground state which are the metastable states of He-atom.

Now, the helium atoms in the metastable state collide with the neon atoms in the ground state and transfer their energy to them with an additional energy 0.05 eV provided by the kinetic energy of the atoms. As a result, neon atoms get excited and helium atoms return back to their ground state E_0. In this process, a continuous pumping starts and the number of neon atoms in their excited states increases. Then, by the stimulated emission with the help of photons from the external source, laser transition occurs in neon from the metastable state E_3 to an excited state E_2 with the emission of photon of wavelength 6328 Å. Then another photon is spontaneously emitted in a transition to a lower metastable state E_1 which yields only the incoherent light. Now, the remaining energy is lost in collisions with the walls of the tube and the radiation–less transition occurs from the metastable state E_1 to the ground state E_0.

Since the electron impacts that excite the helium and neon atoms occur all the time, unlike the pulsed excitation from the xenon flash lamp in a ruby laser, a helium-neon laser operates continuously and a narrow red coherent beam is obtained. A He-Ne laser is used in supermarkets to read bar codes.

5.7 CHARACTERISTICS

The laser beam has the following characteristics.

1. It is a monochromatic, *i.e.* its wavelength is limited in a very small range. The reason is that all the photons emitted are of same energy.
2. It is a coherent, *i.e.* all the emitted waves are in same phase. Thus, interference pattern can be obtained not only by placing two slits in a laser beam, but also by using the two separate lasers.
3. It is unidirectional, *i.e.* its angular spread is very small. This is why a laser beam sent from the earth to a mirror on the moon by the Apollo 11 expedition remained narrow enough to be detected on its return to the earth. An ordinary light beam would have spread out too much and could not be detected on return to earth.
4. It is very intense. The reason is that all the waves are in same phase. If n waves, each of amplitude a, are in same phase, the resultant amplitude is na and the resultant intensity is $(na)^2 = n^2a^2$, which is n times the sum of intensities of each individual wave. A laser beam has an energy density same as that of a hot object at a temperature of about 10^{30} K.

5.7.1 Applications

Due to small angular spread and small wavelength range, *i.e.* unidirectional and monochromatic properties, a laser beam finds its vast uses. Some of the uses of laser are given below.

1. **In measurement of very long distances:** The light beam obtained from laser is unidirectional, therefore, it is used in measuring very long distances by the method of reflection of pulses. The distance of moon from the earth is measured by this method.
2. **In making fine holes and cutting the thick metal sheets:** The beam obtained from laser is very intense and unidirectional, therefore, if the laser beam is focussed by a convex lens, its entire energy will concentrate in a very small area ($\approx 10^{-10}$ m^2). As a result, at that point the intensity of beam becomes so strong that the material placed at that point melts and gets evaporated and thus a hole is obtained at that point.
3. **In telecommunication:** The laser beam is not absorbed by water. Hence, laser beam can be used to transmit a number of television programmes simultaneously through water.

   Lasers are widely used to send telephone signals over long distances through optical fibres.

   In fact laser techniques are so fast growing that everyday a new incredible application of laser appears.
4. **In holography:** A laser beam can be used to obtain three-dimensional images without using the lenses.
5. **In surgery:** A laser beam is used in bloodless surgery and in removing the eye tumour.
6. **In chemistry:** Laser beam is used to accelerate the various chemical reactions.
7. **In space:** The working of rockets and satellites in space is much efficiently and effectively controlled with the help of laser beam.
8. **Industrial applications:** In industrial and technical fields the laser beam is used for drilling extremely fine holes in diamonds, teeth, paper clips, hard sheets of hard metals, welding on a microscopic scale, such as integrated circuits, is taken up by means of laser beams. The laser beam is used to vaporise unwanted material during the manufacture of electronic circuit on semiconductor chips.
9. **Medical applications:** In medicine, microsurgery has become possible due to narrow angular spread of the laser beam. The laser beam can be focussed on harmful components to destroy them without seriously damaging the neighbouring regions. It can be used in the treatment of kidney stone, tumour, in cutting and sealing the small blood vessels in brain operations and retina detachment. Preliminary success has also been made to treat the human and animal cancer. The laser beams are frequently used in endoscopy. The time of completion of big surgical operations can be reduced considerably by the use of laser beam. It is also used in the delicate surgery of cornea grafting.

10. **In communication:** Narrow angular width or spread of the laser beam makes it a very useful tool for microwave communication. Communication with earth satellites, in rocketry, and in accurate range finders. The earth–moon distance has been measured with the help of lasers. In microwave communication the message in the form of signal is mounted on carrier waves through modulation. The semiconductor laser is the best light source for optical fibre communication.

11. **Chemical applications:** Lasers have a wide chemical applications. It can initiate certain chemical reactions which could not be possible in the absence of photons. It is used in the laser Raman spectroscopy.

    During war, lasers are used in laser rifles, laser pistols and laser bombs which are used in night. Lasers are also used to detect and destroy enemy missiles.

    In campact disc (CD) audio systems in place of the phonographic needle a laser beam is used. On a CD which is much smaller than a traditional record, a very large number of songs, speeches, serials a huge amount of data in a small volume such as books of large volume like dictionaries and encyclopedias are available. Using lasers, video images can also be stored on discs and can be displayed on a Television with the help of a laser disc player. Nowadays this technique is used in computers for data retrieval and storage.

12. **In spectroscopy:** Introduction of laser as a source of excitation has revolutionised the field of spectroscopy. The development of tunable dye lasers and the frequency doubling techniques have added a new dimension to the studies in Raman effect. The investigation of Raman scattering with the aid of lasers led the discovery of new Raman scattering phenomena of fundamental interest. The three new phenomena which arises from the non-linear interaction of a system with intense monochromatic radiation are (*i*) The stimulated Raman effect, (*ii*) The hyper-Raman effect and (*iii*) Coherent anti-stoke Raman scattering (CARS).

    The stimulated Raman effect has been observed in a very large number of Raman laser material, which provides us nearly hundreds of coherent sources of light from ultraviolet to infrared regions. Stimulated Raman scattering has been observed in optical fibre wave guides.

13. **Coherent anti-stoke Raman scattering:** Is widely used in high resolution spectroscopy. This technique is frequently used as an effective spatial probe of concentration of gases in flames. This technique allows powerful fluorescence-free analysis of biological samples.

    Saturation absorption spectroscopy is used to study the fine details of the spectra of free atoms and molecules.

5.8 MASERS

The term maser stands for "molecular (previously microwave) amplification by simulated emission of radiation". It is extremely useful as a source for carrier of microwaves.

The working of maser is similar to that of laser. Actually the principle of stimulated emission on which laser is based was first employed in microwave region (in frequency range $\sim 10^{10}$ Hz).

In the same way, the principle of maser can be stated as under.

The maser action is based on the principle of 'population inversion' followed by 'stimulated emission'. For stimulated emission to predominate, most of the atoms should be in the excited state. So population of atoms in higher energy level is increased by pumping the matter by photons of apropriate size. If during the excited state of the atom, it is struck by a matching photon having precisely the energy as that emitted by the atom is spontaneous emission, then the excited atom emits photons that fall precisely in phase with the stimulating photons. So the excited atom emits waves in step with incoming wave and increases its intensity. Finally resulting emergent wave is extremely intense and diverges the least.

5.9 DIFFERENCE BETWEEN ORDINARY LIGHT AND LASER RADIATION

Basically light from a laser and from any ordinary conventional source, such as a flame or an incandescent lamp are both electromagnetic in nature but they differ enormously in coherency, monochromacity, directionality, intensity and other more detailed properties.

1. The light emitting from a conventional source is incoherent whereas those emitted by laser has a very high degree of coherence and is almost perfectly parallel.
2. Spatially coherent beam of light from an ordinary source can be produced by passing it through a pin hole, whereas a laser beam is already highly spatially coherent. The high coherence of laser emission makes it possible to realize a tremendous spatial concentration of light power such as 10^{13} W in a space with linear dimension of only 1 μm. Radiation of such high intensity is used to cut metal and micro welding.
3. The coherence time for a laser beam is much larger in comparison to ordinary light source.
4. The conventional light sources emit radiation in all directions, while the laser emit radiation only along one direction, that is, laser is highly directional.
5. The light emitted by a laser is vastly more monochromatic than that of any conventional monochromatic source. The line emitted by ordinary source is never perfectly sharp but it spreads over a wide frequency range, whereas light from a laser has a very small spreading.
6. Light from a lamp spread out more or less uniformly in all directions, whereas laser gives out light into a narrow beam and its energy is concentrated in very small regions. For example if we look at a 100 W lamp filament at a distance of 30 cm, the power entering the eye is less than a thousand of a watt. On the other hand, 1 W laser would appear many thousand times more intense than 100 W ordinary lamp.

5.9.1 Laser Welding

In performing the task of welding laser welding is superior than the other types of welding such as are welding, or electron beam welding.

In laser welding the laser beam is focussed on to the spot to be welded as shown in Fig. 5.12. Due to the heat generated, the material melts over a tiny area on which the beam is focussed. The heat produced by the beam is so intense that, impurities in the materials such as oxides float up on the surface, and upon cooling, the material within, becomes homogeneous solid structure, which makes it a stronger joint. In this type of welding, virtually no destruction occurs in the shape of work piece, since the total amount of heat input is very small compared to the regular welding (it is the focussing of high laser energy over a small area which does the job), and the heat is also dissipated immediately. Since the laser beam could be controlled to a great precision, the welding spots could also be located precisely.

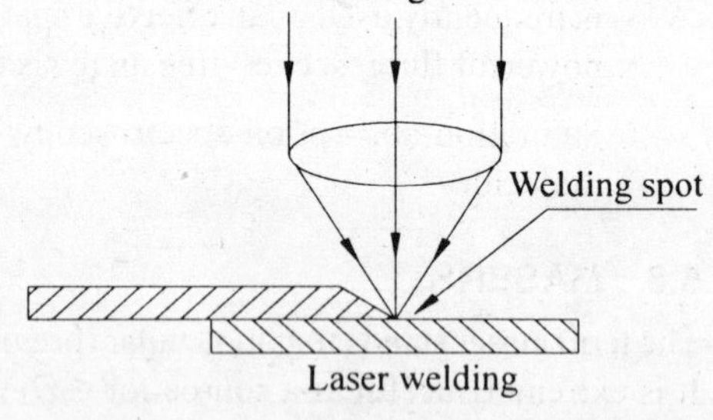

Fig. 5.12

Laser welding is a contact-less process, and thus no foreign material has a chance to get into the welded joint. Since the heat affected zones are very small, laser welding is ideal at places which are surrounded by heat sensitive components, such as the ones we encounter in microelectronics. Carbon dioxide lasers are the most popular ones in this particular application.

5.9.2 Laser Cutting

Laser cutting of metals is generally done assisted by gas blowing. The laser focussing part is surrounded by a nozzle into which oxygen gas is fed. A jet of the gas is issued through the nozzle right at the spot where laser beam is focussed. The cobustion of the gas burns the metal thus reducing the laser power requirement for cutting. The tiny splinters along with the molten part of the metal will be blown away by the oxygen jet. The blowing action increase the depth and also the speed of cutting. But it is the laser, and not the burning gas, which controls the accuracy of cutting. Thus the cut edges will be of very high quality as shown in Fig. 5.13.

Laser cutting is employed in tailoring industry where mass production is employed. A number of layers of the cloth material are stockpiled. A laser beam is focussed on the pile and moved in a path along which cutting is required. The movements path details are programmed in an NC machine. The beam cuts the entire stock pile right through along the path.

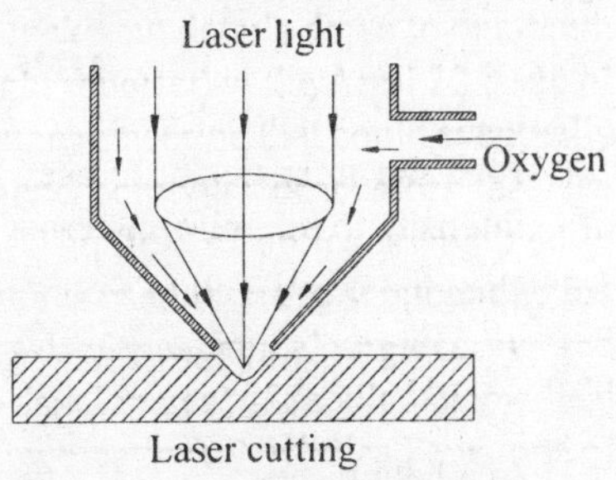

Fig. 5.13

Advantages of Laser Cutting

(*a*) The cutting process could be programmed which results in high production rates.

(*b*) The quality of cutting is very high.

(*c*) There will be no thermal damage and chemical change when cutting is done in inert atmosphere.

(*d*) Cutting a complicated profile even in 3-dimensions is possible.

5.9.3 Laser Drilling

Drilling of holes is achieved by subjecting the material to powerful light pulses. The pulses will be of 10^{-4} to 10^{-3} second duration. The intense heat generated over a short duration by the pulses evaporates the material locally, thus leaving a hole, Nd–YAG laser is used to drill holes in metals, whereas a CO_2 laser is used in case of both metallic and non-metallic materials.

Advantages of Laser Drilling

(*a*) The tools wear out while drilling by conventional method, whereas that problem doesn't exist with laser set-up.

(*b*) Drilling can be achieved at any oblique angle through the material whereas it could be done only to a limited extent in conventional methods.

(*c*) Very fine holes to the dimensions of even 0.2 to 0.5 mm diameter could be drilled with a laser beam. The holes may be even adjacent to each other.

(*d*) Very hard materials or brittle materials could be subjected to laser drilling since there is no problem of mechanical stresses with a laser beam.

5.9.4 Pollutants in Atmosphere

Lasers are used in pollution measuring system to measure the pollution caused by the pollutants like carbon dioxide, sulphur dioxide, freon, nitrous oxide, ethylene and vinyl chloride and automotive exhaust, industries. The schematic diagram is as shown in Fig. 5.14.

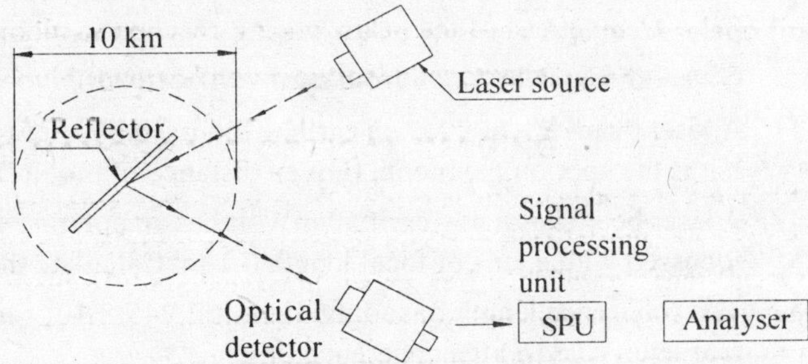

Fig. 5.14

It consist of a laser source, optical unit, SPU and reflector. The laser beams are projected through the atmosphere of the area in which the pollution to be measured, the measurement is based on the spectral absorption. The laser beam of different energy and wave length will be absorbed by the different gases in the atmosphere. The laser beam of different frequencies can be projected through the atmosphere to a reflector placed at far off distances, the reflected laser beam is received by optical detector and analyser. This laser beam under go attenuation due to the absorption by the pollutants in the atmosphere. A signal generated by the differences in the wavelength absorbed is analysed converted into digital signal on the computer gives the concentration of the pollutants.

UNSOLVED NUMERICAL PROBLEMS

1. Find the intensity of a laser beam of 20 mW power and having a diameter of 1.3 mm. Assume the intensity to be uniform across the beam.
2. Imagine that we chop a continuous laser beam assumed to be perfectly monochromatic of wavelength λ_o = 623.8 nm into 0.05 ns pulses using some sort of shutter. Compute the resultant line width $\Delta\lambda$ bandwidth and coherence length.
3. In a helium-neon laser the two plane mirrors forming the resonant cavity are at a distance 0.5 m. What is the mode separation of longitudinal cavity in terms of frequency?
   This is the expected frequency of consecutive modes.
4. Calculate the energy difference in eV between the two energy levels of the Ne atoms of a He–Ne gas laser, the transitions between which results in the emission of a light of wavelength 632.8 nm. Also, calculate the number of photons emitted/second, if the optical power output is 2 mW.
5. The ratio of population of two energy levels out of which upper one corresponds to a metastable state is 1.059×10^{-30}. Find the wavelength of light emitted at 330 K.
6. Find the ratio of population of the two energy states of the ruby laser, the transition between which is responsible for the emission of photons of wavelength 6928 Å. Assume the ambient Temperature as 18 K.
7. A pulsed laser emits photons of wavelength 780 nm with 20 mW average power/pulse. Calculate the number of photons contained in each pulse if the pulse duration is 10 ns.
8. Calculate the energy and momentum of a photon of a laser beam of wavelength 6328 Å.
9. A laser beam of intensity 500 w/m^2 falls on a perfectly reflection plane mirror for an hour. The area of the mirror is 0.5 cm^2. Calculate the momenum imparted to the mirror in this time and the average force acting on this mirror.
10. A laser beam having a wavelength of 8000 Å and aperture 0.5 cm is sent to moon. Calculate
    (*i*) angular spread of the beam
    (*ii*) areal spread of the beam when it reached to moon.
    Given: distance of moon from earth = 4×10^8 m).
11. A laser beam λ = 6000 Å on earth is focussed by a lens of diameter 2 m on to the surface of moon. How big is the spot on the moon. (Given distance of moon from earth = 4×10^5 km)
12. A laser beam has a power of 50 mW. It has an aperture of 5×10^{-3} m and wavelength 7200 Å. A beam is focussed with a lens of focal length 0.1 m. Calculate the area and the intensity of the image.
13. The coherence length of sodium light is 2.945×10^{-2} m and its wavelength is 5890 Å. Calculate (*i*) the frequeney and (*ii*) the coherence time.
14. A laser beam λ = 6000 Å on earth is focused by a lens of diameter 2 m on to a crater on the moon. The distance of the moon is $4 \cdot 10^8$ m. How big is the spot on the moon?
15. Assume that we chop a continuous perfectly monochromatic laser beam of wavelength λ = 6238 Å into 10^{-10} sec pulses using some sort of shutter. Calculate the resultant line width, bandwidth and coherence length.

16. A laser beam has a bandwidth of 2800 Hz. obtain the coherence length.
17. Determine which emission predominates for a cavity temperature of 100 K and for one micron wavelength.
18. Calculate the difference in energy between metastable and ground state electron for the ruby laser. Given the wavelength of the laser beam from a ruby is 6943 Å.
19. A helium – neon laser is generating a laser beam with an average power of 3.5 mW. Determine the number of photons emitted per second by the laser. Given the wavelength of the emitted radiation is 6238 Å.
20. A pulsed laser has an average power output of 1.5 mW per pulse and the pulse duration is 20 ns. The number of photons emitted per pulse is estimated to be 1.0472×10^8. Find the wavelength of the emitted lasers.
21. A ruby laser is emitting photons of wavelength 694.3 nm and the temperature of operation is 27° C. Find the ratio of population of the two energy states responsible for the release of photons.
22. A ruby laser emmits 1.0 J pulses of light of wavelength 6940 Å. Calculate the minimum number of Cr^{3+} ions in the ruby.
23. Assuming the excited state of water at 100° C to be the steam at 100° C, calculate the wavelength of laser beam obtained if in transition from steam to water, the energy lost per molecule appears as a photon. Given : Latent heat of steam = 2260 kJ kg^{-1}.
24. The energy flux of a laser beam is 10 W cm^{-2}. It falls on 1 cm^2 of the surface of an ideal reflecting mirror for 1 h. Calculate the momentum transferred and the force exerted on the mirror.

MULTIPLE CHOICE QUESTIONS

1. Laser was inveted by
   (*a*) Einstein (*b*) Town and Shallow (*c*) Sir. CV Raman (*d*) Niels Bohr.
2. Laser beam is not
   (*a*) monochromatic (*b*) unidirectional (*c*) coherent (*d*) non-coherent.
3. The wavelength of Ruby laser is
   (a) 6328 Å (*b*) 5000 Å (*c*) 6943 Å (*d*) 6000 Å
4. The wavelength of helium-neon laser is
   (*a*) 6943 Å (*b*) 6328 Å (*c*) 6534 Å (*d*) 6845 Å
5. Emission of photon when an electron jumps from higher energy state to lower energy state due to interaction of external energy called
   (*a*) spontaneous emission (*b*) stimulated emission
   (*c*) induced emission (*d*) amplified emission
6. The population of the various energy levels of a system in thermal equilibrium is given by
   (*a*) Boltzmann distribution law (*b*) Einstein relations
   (*c*) Planck's law (*d*) Beer's law
7. A three level laser system will be
   (*a*) always continious wave
   (*b*) either continious wave or pulsed
   (*c*) always pulsed
   (*d*) made continious wave as well as pulsed by temprature control.

8. Choose the correct statement
   (*a*) Four level laser system will be always a pulsed system
   (*b*) Three level laser system will be always a CW system
   (*c*) Resonators help in increasing the brightness of laser beam only
   (*d*) Resonators act as frequency selectors and also give rise to directionality to the output beam.
9. T Maiman invented
   (*a*) He–Ne laser (*b*) CO_2 laser (*c*) Ruby laser (*d*) Nd: YAG laser
10. The wavelength of emission of CO_2 laser is
   (*a*) 632.8 nm (*b*) 10.6 mm (*c*) 1.064 mm (*d*) 694.3 nm
11. The ratio of pressure of $CO_2 : N_2 :$ He used in CO_2 laser in
   (*a*) 5 : 4 : 1 (*b*) 4 : 5 : 1 (*c*) 1 : 4 : 5 (*d*) 1 : 5 : 4
12. CO_2 laser was invented by
   (*a*) T Maiman (*b*) Ali Jawan (*c*) Einstein (*d*) CKN Patel
13. In ruby lasing material the percentage of chromium ions in aluminium oxide is
   (*a*) 0.5 (*b*) 0.05 (*c*) 5 (*d*) 0.005
14. The colour of the laser output from a ruby laser is
   (*a*) green (*b*) blue (*c*) red (*d*) violet
15. The wavelength of radiation from Nd : YAG/Nd: glass is
   (*a*) 1.064 mm (*b*) 10.6 mm (*c*) 694.3 mm (*d*) 1.064 nm
16. Measurement of variation of divergence of laser beam with distance is used to determine
   (*a*) coherence (*b*) monochromacity (*c*) brightness (*d*) directionality
17. Coherence of light if measured from
   (*a*) variation in spot size with distance (*b*) visibility of interference fringes it produces
   (*c*) brightness of the beam (*d*) wavelength of the beam
18. In He–Ne lasers, the ratio of He–Ne is in the order
   (*a*) 1 : 10 (*b*) 1 : 1 (*c*) 100 : 1 (*d*) 10 : 1

EXERCISE

1. What is an optical pumping?
2. Explain the meaning of population inversion.
3. The ground state energy and the excited state energy of an atom are E_1 and E_2 respectively. For stimulated emission, photons of what frequency will be absorbed?
4. In which region of electromagnetic spectrum, the probability for stimulated emission is maximum and why?
5. Explain the Acronym of laser.
6. Write down the wavelength of laser beam obtained from ruby laser.
7. What is the wavelength of laser beam obtained from the He–Ne laser?
8. Differentiate between the spontaneous emission and stimulated emission.

9. State the necessary conditions for the strong stimulated emission and in this reference explain the Einstein's A and B coefficients.
10. Explain the meaning of Einstein's A and B coefficients.
11. What is laser? Explain the principle of laser. How is laser radiation different from the ordinary light?
12. Explain the meaning of population inversion.
13. For laser action to occur, the medium used must have at least three energy states. What must be the nature of each of these states? Why is three the minimum number?
14. Explain the construction and working of ruby laser.
15. Explain the construction and working of helium-neon laser. Write two characteristics of the beam obtained from the helium-neon laser.
16. State some uses of chemical and semiconductor lasers.
17. State the important characteristics and uses of laser beam.
18. Explain the terms

    (*i*) Optical pumping (*ii*) Metastable energy state (*iii*) Population inversion
19. Write short notes on

    (*a*) Laser (*b*) Ruby laser (*c*) He–Ne laser (*d*) Semiconductor lasers
20. Mention any two applications of laser each in the field of scientific research, engineering and medicine.
21 (*a*) What do you understand by population inversion? How it is achieved?

    (*b*) Derive the relation between the probabilities of spontaneous emission and stimulated emimssion in terms of Einstein's coefficient.
22. Explain the purpose of an active medium in a gas laser.
23. Explain the need of a cavity resonator in a laser.
24. Describe a semiconductor laser. Give the applications of lasers.
25. Explain the basic principles for producing laser beam. Write the medical application of lasers.
26. What is population inversion in lasers? How is it achieved? What are the advantages of lasers in communications?
27. Describe the construction and working of a semiconductor laser. List out its merits and limitations.
28. What are Einstein's coefficients? Explain the three main components of any laser system.

Holography

5.10 HOLOGRAM

The process in which the image of an object can be recorded by wave front reconstruction, which does not record the image of the object, but records the light waves, this photographic record of light waves is called a hologram.

The technique in which the photographic process of recording and reproducing the three dimensional image of an object on two dimensional light sensitive surface is called holography.

When an object is ordinarly photographed it records only the intensity distribution in a particular plane. In holography both intensity and the phase distribution are recorded in any plane the resultant photograph is called as hologram.

Coherent sources are necessary for studing holography.

5.11 REQUIREMENTS

(*i*) The light source used must be monochromatic. If a broad spectrum is used each wave length will give rise to number of fringe patterns, which leads to the over lapping and blurring of the image.

(*ii*) The light source should have spatial coherence if not the resultant effect of the fringe pattern will average to some uniform value and some fringe will be missing.

5.12 PROPERTIES

(*i*) The remarkable property of hologram is the three dimensional imaging and highly exact reproduction.

(*ii*) The three dimensional holograms will contain a fag like suspension particles.

(*iii*2) Each part of the hologram reproduces entire image, because hologram receives light from all parts of the object.

5.13 APPLICATIONS

(*i*) Since the holograms contains the deep information of the object, they are used in the study of transient microscopic events.

(*ii*) It is used in the field of interferometry to find out the strain suffered by the body. (Young's modulus of the material).

(*iii*) It is used in image storage and re-examination.

5.14 PRINCIPLE AND CONSTRUCTION OF HOLOGRAM

If the object is illuminated by a light from a coherent source, each point of the object act as a source of secondary waves, which spreads out in the surrounding space with intensity characteristic of the object as shown in Fig. 5.15.

When the object is illuminated by coherence light the obtained spherical waves corresponds to each point on the object. Since the source of light is coherent the wave will also be coherent and they produces interference or diffraction pattern on the plate the pattern is characteristic of the object and will be different for different object.

If the intensity pattern is recorded in a suitable manner and the recording is again illuminated with light from a coherent source the original intensity distribution is reproduced giving rise to the image of the original object.

Let us consider two identical coherent beam P and Q from the source incidenting at different angles on photosensitive surface as shown in Fig. 5.16.

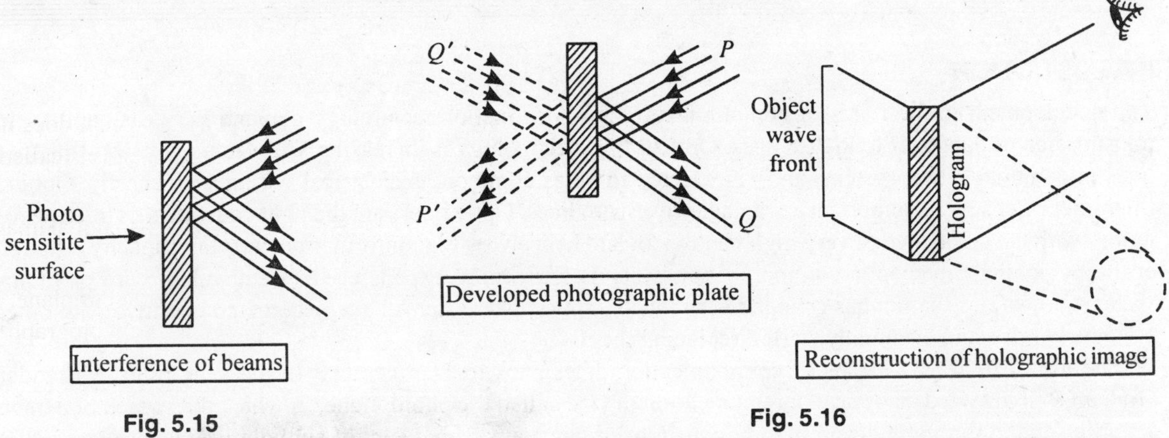

Fig. 5.15 **Fig. 5.16**

Due to the interference effect by the two beams on the photosensitive surface results in the recording of interference fringes on it.

On developing the plate it can be observed that the two incident beams P and Q from sources pass through the fringe patterns. It can seen on the developed photographic plate that, if the emergent beam is viewed from the transmission side, then the beam P appears as the continuation of the beam P. Similarly, the beam Q appears as the continuation of Q.

Due to the constructive interference of the emerging light waves, the incident light wave will suffer diffraction on the photographic plate and the part of the incident wave deflected from the incident direction which produces the beam P. The part of the beam which deflected occurs precisely to form the beam Q, *i.e* the transmission part of the beam Q which will be observed from the incident direction but the same can be viewed from transmitted direction and it appears as though the beam Q itself is transmitted.

In holography an arrangement has done to get one of the object beam reflected from the three-dimensional object on the photographic plate where as the reference beam incident directly on the plate. When the incident, reflected and the position of the developed plate are set identical to the above explanation the three dimensional view of the object appears when viewed from the transmission direction of the object beam.

The interference fringes obtained as recorded on the photographic plate are called as hologram.

Optical Fibers

INTRODUCTION

The development in the fields of communication and information technology demand very easy and rapid transmission of data over long distances. Optical fiber technology is increasingly replacing wire transmission lines in communication systems and is expected to be as common as electrical wiring very shortly. Optical fiber lines offer several important advantages over wire lines. Optic fibers are the light equivalent of microwave guides with the advantage of very high bandwidth and hence very high information carrying capacity. Though at the beginning, fiber optic communication systems were more expensive than equivalent wire or radio-systems, now the situation has changed very much. Optic fiber systems have become competitive with other systems in price and eventually started replacing them.

The use of light as a means of communication dates back to 18th century. In the year 1880, Alexander Graham Bell invented an optical communication device called the photophone, in which the speech of person was transmitted through a beam of sunlight. In his device, a narrow beam of sunlight was focused on a thin mirror, the sound waves from a speaker's voice were allowed to strike the mirror and hence vibrate it, which in turn would vary the intensity of the light in proportion to the strength of the sound waves. The reflected sunlight was made to fall on a receiver, the selenium detector whose resistance varies with changing intensity of the incident light beam, thus converting the light beam into an electronic signal to recreate the speech. Using his photophone device, Alexander Graham Bell was able to transmit the voice to a distance of about 100 m.

However, Bell's optical system was not practically a viable communication system due to the non-availability of a suitable light source, and the inefficiency of air as a medium of transmission. But, the invention of lasers and light emitting diode's in 1960's, and that of a high purity, low loss optical fibre in 1970 by Corning Glass Works, USA, changed the trends in optical communication systems. A practical optic fibre communication channel was established at the Bell laboratories, USA in 1977, which connected two central offices and was successfully used for transfer of audio/video data. By the end of 1980 s, many functional optic fibre communication systems were installed and operated throughout the world. By the turn of the 20th century, optical communication has become a great success.

5.15 PROPAGATION MECHANISM

John Tyndall in the year 1870 demonstrated that light could be guided within a water-jet based on the phenomenon of total internal reflection. But it was not until the mid 1960 s that the idea of communication system based on the propagation of light within the water jet like circular wave guide called optical fibers was considered.

The light transmitted at one end of the fiber has to travel through the entire length and reach the other and without much loss. Initially only single circular dielectric rods were considered as waveguides but the light launched through its one end penetrated into the air surrounding the rod thereby causing the losses to be very high. Moreover they had to be made very thin to accommodate a single electromagnetic mode. These major problems were overcome with the development of cladded dielectric waveguides in the year 1954. Optical fiber is a very thin and flexible medium having a cylindrical shape consisting of three sections.

1. **Core:** The core is the inner most section and is made of glass or plastic. The remarkable property of core is to conduct the optical beam.
2. **Cladding:** The core is surrounded by its own cladding a glass or plastic coating, that has optical properties which are different from core.
3. **Jacket:** The outer section is called the jacket made of plastic or polymer which is provided for the protection against moisture, absorption crushing and other environmental danger.

The structure of an optical fiber is as shown in Fig. 5.17. The fiber has a core surrounded with a cladding with refractive index slightly less than that of the core to satisfy the condition for total internal reflection. To give mechanical protection to the fiber, a protective skin called outer jacket is used.

The light transmitted inside the core through its one end propagates to the other end due to total internal reflection at the core and cladding interface.

Total internal reflection at the fiber wall can occur only when the following conditions are satisfied.

1. The refractive index of the core material m_1 must be slightly higher than that of the cladding m_2 which is surrounding it.
2. At the core-cladding interface the angle of incidence q between the ray and the normal to the interface must be greater than critical angle and is defined as

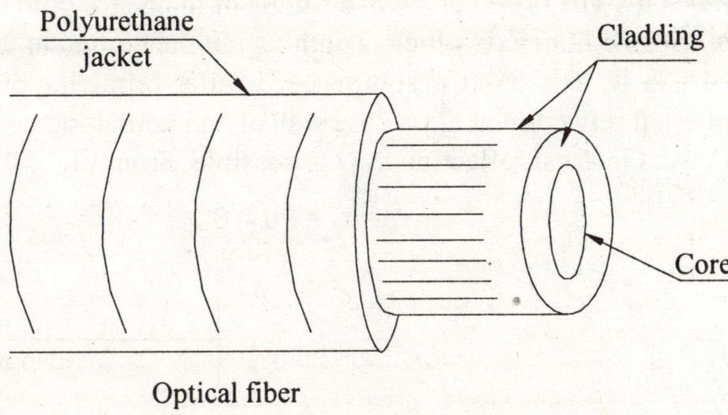

Fig. 5.17

$$\sin \theta_c = \frac{\mu_2}{\mu_1} \tag{1}$$

These conditions are shown in Fig. 5.18. When light ray travels from core of refractive index μ_1 to cladding of refractive index μ_2, refraction occurs. Since it travels from denser to rarer medium the angle of refraction is greater than the angle of incidence as shown in Fig. 5.18 (a). With increase in angle of incidence the angle of refraction also increases and for a particular angle of incidence the refracted ray just glazes the interface between the core and cladding. This angle of incidence is known as critical angle θ_c shown in Fig. 5.18 (b). When angle of incidence is further increased, the ray is reflected back into the core at the interface obeying the law of reflection. This phenomenon is called as total internal reflection and is as shown in Fig. 5.18 (c).

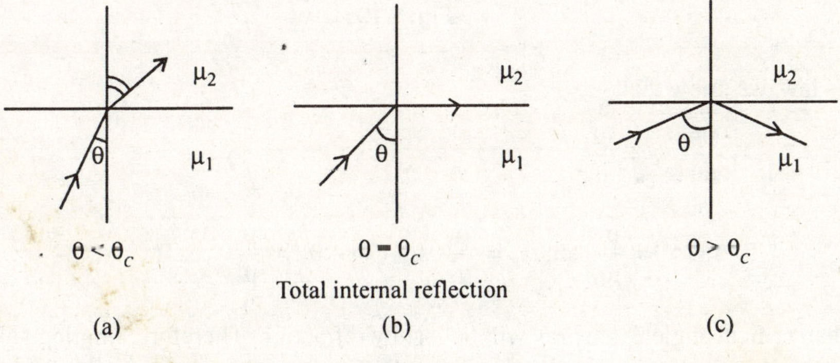

Fig. 5.18

5.16 ACCEPTANCE ANGLE AND NUMERICAL APERTURE

When the light beam is transmit into a fiber at its one end using a focussing lens, the entire light may not pass through the core and propagate. Only the rays which make the angle of incidence greater than critical angle at the core-claddin interface undergo total internal reflection and propagate through the core. The other rays are refracted to the cladding and are lost. Hence, it is very essential to know at what angle called acceptance angle we have to transmit the beam at its end to enable the entire light to pass through the core.

Condition for propagation shows a longitudinal cross-section of the transmission end of a fiber with a ray entering it. The light transmission from a medicum of refractive index $\mu_o = 1$ for air into core of refractive index μ_1. The ray enters with an angle of incidence α_i to the fiber end face, *i.e.* the incident ray makes an angle α with the fiber axis which is nothing but the normal to the end face at its centre. This particular ray enters the core at its axis point P and proceeds after refraction at an angle α_r from the axis. It then undergoes total internal reflection at Q on core wall at an internal incidence angle θ. Let us now find up to what value of θ_i at P, total internal reflection at Q is possible. From Fig. 5.19, in triangle PQR,

$$\alpha_r = 90 - \theta \tag{2}$$

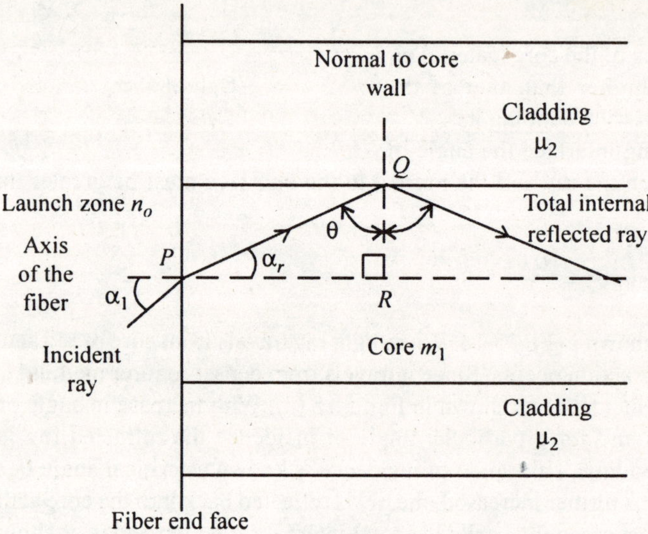

Fig. 5.19

From Snell's law, we know that,

$$\frac{\sin \alpha_i}{\sin \alpha_r} = \frac{\mu_1}{\mu_0} \tag{3}$$

$$\sin \alpha_i = \frac{\mu_1}{\mu_0} \sin \alpha_r = \frac{\mu_1}{\mu_0} \sin (90 - \theta) = \frac{\mu_1}{\mu_0} \cos \theta \tag{4}$$

If θ is less than critical angle θ_c, the ray will be lost by refraction. Therefore, limiting value for containing the beam inside the core, by total internal reflection, is θ_c. Let θ_i be the maximum possible angle of incidence at the fiber end face at A for which θ is equal to θ_c. If for a ray α_i exceeds α_i (max), then θ will be less than θ_c and hence at B the ray will be refracted.

Hence, the above equation can be written as $\sin \alpha_i \text{ (max)} = \frac{\mu_1}{\mu_0} \cos \theta_c$ (5)

From Eq. (1) $$\cos \theta_c = \sqrt{1 - \sin^2 \theta_c} = \sqrt{1 - \frac{\mu_2^2}{\mu_1^2}} = \frac{\sqrt{\mu_1^2 - \mu_2^2}}{\mu_1} \tag{6}$$

Therefore, $\sin \alpha_i(\max) = \dfrac{\sqrt{\mu_1^2 - \mu_2^2}}{\mu_0}$ or $\alpha_i(\max) = \sin^{-1}\left(\dfrac{\sqrt{\mu_1^2 - \mu_2^2}}{\mu_0}\right)$ (7)

This maximum angle is called the acceptance angle or the acceptance cone half-angle rotating acceptance angle about the fiber axis as shown in Fig. 5.20 explains the acceptance cone of the fiber. Light transmitted at the fiber end within this acceptance cone alone will be accepted and propagated to the other end of the fiber by total internal reflection. Larger acceptance angles make launching easier. Light collecting capacity of the fiber is expressed in terms of acceptance angle and is termed as 'numerical aperture'.

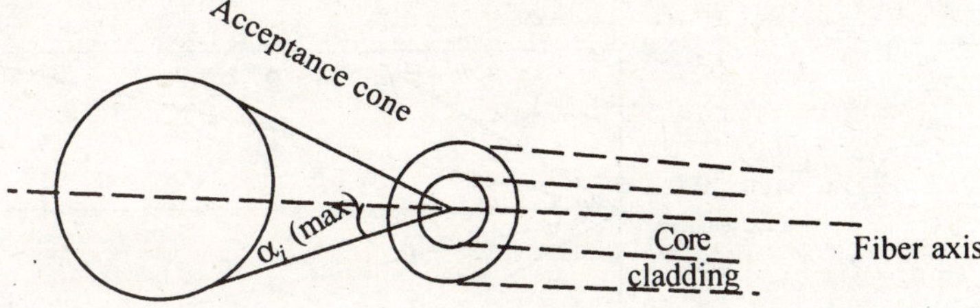

Fig. 5.20

5.17 NUMERICAL APERTURE

Sine of the maximum acceptance angle is called the numerical aperture of the fiber and is denoted as *NA* and is given by,

$$NA = \sin \alpha_i\ (\max) = \frac{\sqrt{\mu_1^2 - \mu_2^2}}{\mu_0} \tag{8}$$

It is often convenient to introduce the quantity Δ known as fractional index change and can be written as

$$\Delta = \frac{\mu_1^2 - \mu_2^2}{2\mu_1^2} \tag{9}$$

For most of the fibers $\mu_1 \approx \mu_2$

Hence,

$$\Delta = \frac{(\mu_1 + \mu_2)(\mu_1 - \mu_2)}{2\mu_1^2} = \frac{2\mu_1(\mu_1 - \mu_2)}{2\mu_1^2}$$

i.e. $\Delta = \dfrac{\mu_1 - \mu_2}{\mu_1}$ known as fractional difference in refractive indices (10)

from Eq. (9), we can write

$$\mu_1^2 - \mu_2^2 = 2\mu_1^2\Delta$$

Hence, $\mu_1^2 - \mu_2^2 = \mu_1\sqrt{2\Delta}$

Substituting this in Eq. (8), the numerical aperture becomes

$$NA \approx \frac{\mu_1\sqrt{2\Delta}}{\mu_0} \tag{11}$$

When the light transmmited into the fiber from air ($\mu_0 = 1$). Then,

$$NA = \mu_1\sqrt{2\Delta} \tag{12}$$

It should also be noted that the numerical aperture of a fiber, *i.e.* its light collecting capacity is effectively dependent only on the refractive indices of the core and cladding materials and is not a function of the fiber dimensions.

5.18 TYPES OF OPTICAL FIBERS

When the light is transmitted at one end of the fiber within the acceptance cone alone propagates through the fiber by total internal reflection not all the light within the acceptance cone propagates but only specified directions are allowed which satisfy the conditions for constructive interference. The rays travelling in these specified directions are called as modes and is as shown in Fig. 5.21.

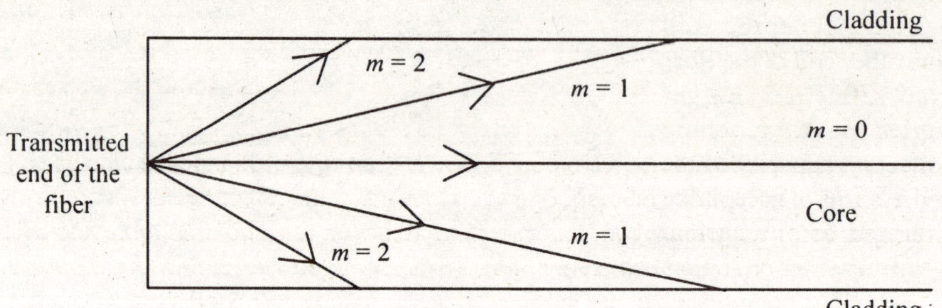

Internal guide angles corresponding to the possible modes inside the core of the

Fig. 5.21

The number of modes that the fiber supports depends on the dimension of the fiber. If the thickness of the fiber is so small that it supports only one mode then the fiber is called monomode or single mode fiber. If it supports more than one mode then it is called multimode fiber. The monomode fiber has small core diameter of the order of 2 to 8 μm. The core diameter of multimode fiber is of the order of 50 μm.

5.19 SINGLE MODE STEP INDEX FIBER

In these fibers the entire core has uniform refractive index μ_1 slightly greater than the refractive index of the cladding μ_2 as shown in Fig. 5.22. Since the index profile is in the form of a step, these fibers are called step index fibers.

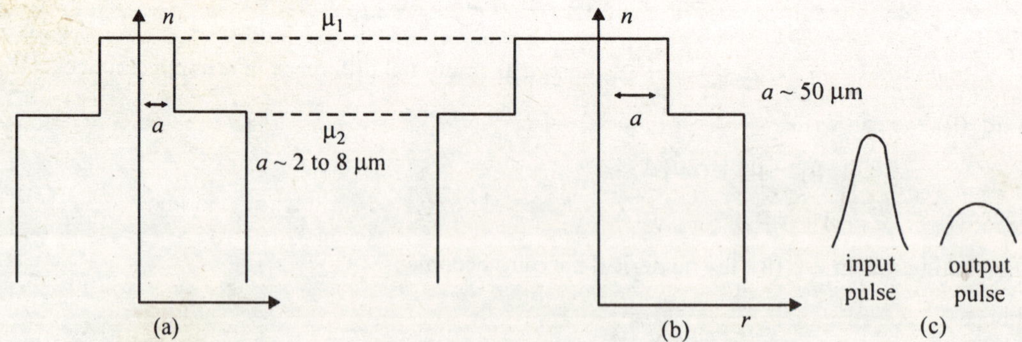

Refractive index profile for (a) monomode and (b) multimode stem index fibers

Fig. 5.22

5.19.1 Transmission of Signal in Step Index Fiber

Usually the signal is sent through the fiber in digital form, *i.e.* in the form of pulses representing 0 s and 1 s. Let us consider the propagation of one such pulse through the multimode fiber. The same pulsed signal travels in different paths represented by multimodes. Hence, at the receiving end only the ray (1) which travels along the fiber axis reaches first while the rays taking zigzag paths (2) reach after some time as shown in Fig. 5.23. Hence, the pulsed signal received at the other end is broadened. This is called intermodel dispersion. This imposes limitation on the seperation between pulses thereby reducing the transmission rate and capacity.

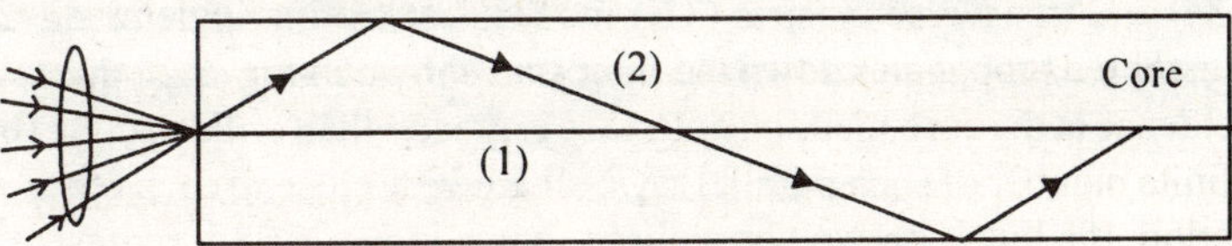

Signal transmission in step index fiber

Fig. 5.23

5.19.2 Graded Index Fiber

Graded index multimode fiber is one in which the refractive index of the core varies radially as shown in Fig. 5.24. It has maximum refractive index at its centre, which gradually falls with increase of radius and at the core — cladding interface matches with the refractive index of the cladding. The variation of refractive index of the core (μ) with radius (r), measured from the centre of the core, is given by

$$\mu(x) = \mu_1 \left[1 - 2\Delta\left(\frac{x}{a}\right)^p\right]^2 \tag{13}$$

where μ_1 is refractive index at the centre of the core α is radius of the core

$$\Delta = \frac{\mu_1 - \mu_2}{\mu_1} \quad p \text{ is grading profile index number.}$$

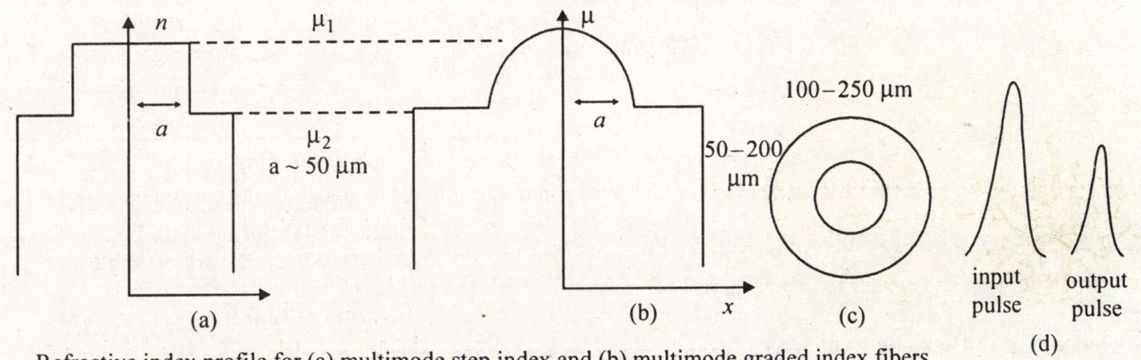

Refractive index profile for (a) multimode step index and (b) multimode graded index fibers

Fig. 5.24

5.19.3 Transmission of Signal in Graded Index Fiber

Let us consider a signal pulse travelling through graded index fiber in two different paths 1 and 2. The pulse 1, travelling along the axis of the fiber, though travels along shorter route it travels through a medium of higher refractive index. The other pulse 2, travelling away from the axis undergo refraction and bend as shown in the Fig. 5.25. Though it travels longer distance, it travels along the path possessing relatively lesser refractive index and hence both the pulses reach the other end simultaneously. Thus, the inter modal dispersion can be overcome by using graded index fiber.

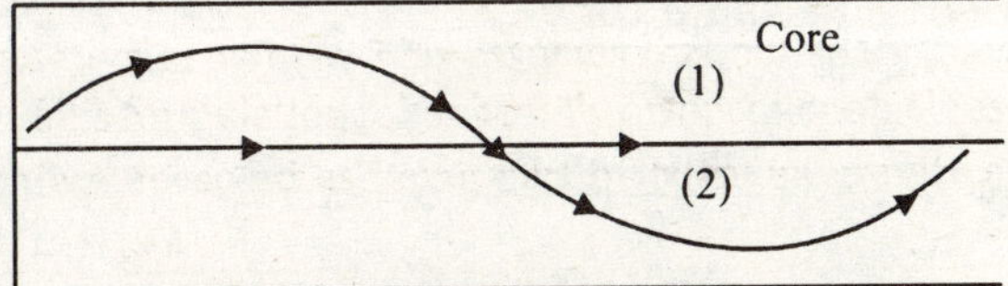

Propagation of different model rays inside graded index fiber

Fig. 5.25

5.20 MODES OF PROPAGATION

Light transmitted through one end of the fiber propagates down the fiber in the form of transverse magnetic (TM) and transverse electric (TE) modes. Let us first understand how the modes are formed. A ray of light transmitted propagates down the fiber is as shown in Fig. 5.26 due to total internal reflection provided angle of incidence at the core-cladding interface is greater than critical angle ($\theta > \theta_c$). A ray indicated in reality represents infinite number of such parallel rays, all are very close to each other since the beam transmitted has some cross section. If a line is drawn normally as shown in Fig. 5.27 it represents plane wavefront and all points along the same wavefront must be in phase with each other. The ray *PQ* after reflection at *Q* travels to *R* and then travels parallel to *PQ* after reflection at *R*. Normal drawn to *PQ* from *R*, *i.e. UTPR* represents a plane wavefront and hence the points *P* and *R* must be in phase with each other. If they are out of phase with each other destructive interference will take place and propagation is not possible. Now moving from *P* to *R* along the ray the phase change is given by

$$(PQ + QR)\,\frac{2\pi\mu_1}{\lambda_0} - 2\phi \tag{14}$$

where λ_0 is the wavelength of light in vacuum and ϕ is the phase change due to reflection. It should be remembered that whenever light is reflected at the interface of denser to rarer media additional phase change occurs and the magnitude of the change ϕ depends on angle of incidence θ. Such change is absent if reflection takes place at the interface of rarer to denser medium. Since reflection takes place both at *Q* and *R*, 2ϕ appears in the Eq. (14).

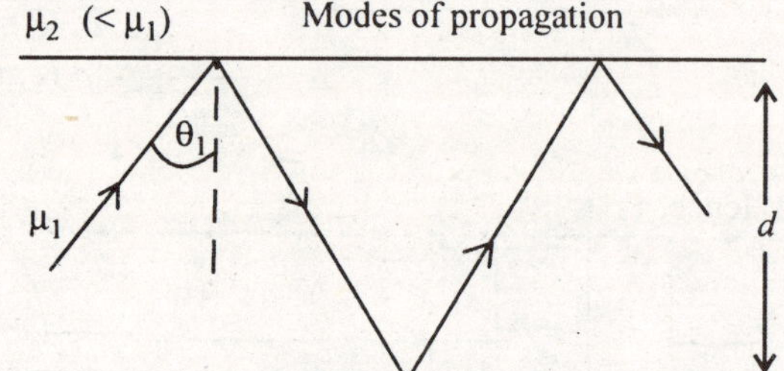

Path of a light ray down a planar dielectric waveguide that results when the angle of incidence at the boundary q_i is greater than the critical angle q_c

Fig. 5.26

From Triangle *PQR*, $\cos 2\theta = PQ/QR$

or
$$PQ = QR\cos 2\theta$$

$\therefore$
$$PQ + QR = QR\,(1 + \cos 2\theta)$$
$$= 2QR\cos^2\theta$$

From Triangle *QSR*

$$\cos\theta = QS/QR$$

$$QS = d = QR\cos\theta$$

$\therefore$
$$PQ + QR = 2d\cos\theta$$

Hence, Eq. (14) can be written as

$$\frac{4\pi\mu_1 d\cos\theta}{\lambda_0} - 2\phi$$

Now the condition for the mode to propagate is

$$\frac{4\pi\mu_1 d\cos\theta}{\lambda_0} - 2\phi = 2\pi m$$

or
$$\frac{2\pi\mu_1 d\cos\theta}{\lambda_0} - \phi = \pi m \tag{15}$$

where *m* is an integer.

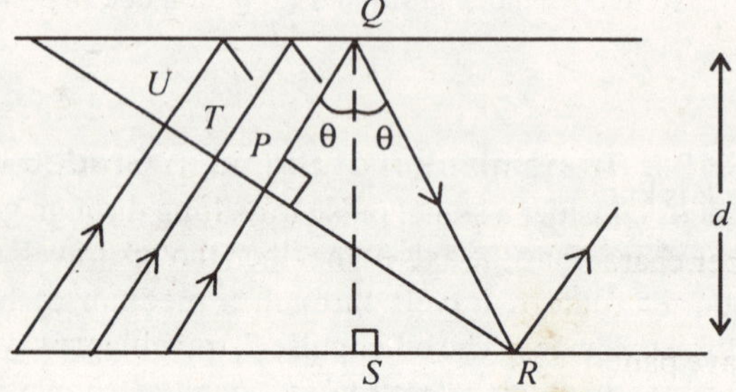

Ray paths having the same internal angle *q* within a planar waveguide. A wavefront is shown connecting the points *UTPR* must therefore have the same phase

Fig. 5.27

For each value of *m*, there will be a corresponding value of θ, *i.e.* θ_m. Rearranging Eq. (15), we get

$$\cos\theta_m = \frac{(m\pi + \phi)\,\lambda_0}{2\pi d \mu_1} \tag{16}$$

It is not possible to obtain an explicit expression for θ in terms of m (m can have only few specific integer values) since θ_m can take values in the range θ_c to $\pi/2$.

Rearranging Eq. (15), we get

$$m = \frac{2d\mu_1 \cos\theta_m}{\lambda_0} - \frac{\phi}{\pi} \tag{17}$$

from Eq. (1)

$$\sin\theta_c = \frac{\mu_2}{\mu_1}$$

Since $\theta_m \geq \theta_c$ to maintain total internal reflection.

$$\sin\theta_m \geq \frac{\mu_2}{\mu_1}$$

$$\therefore \quad \cos\theta_m = \left(1 - \sin^2\theta_m\right)^{1/2} \leq \left[1 - \left(\frac{\mu_2}{\mu_1}\right)^2\right]^{1/2} \leq \frac{\left(\mu_1^2 - \mu_2^2\right)^{1/2}}{\mu_1}$$

Hence, Eq. (17) can be written as

$$m \leq \frac{2d\mu_1}{\lambda_0}\,\frac{\left(\mu_1^2 - \mu_2^2\right)^{1/2}}{\mu_1} - \frac{\phi}{\pi} \tag{18}$$

or

$$m \leq \frac{2V}{\pi} - \frac{\phi}{\pi} \tag{19}$$

where,

$$V = \frac{\pi d}{\lambda_0}\left(\mu_1^2 - \mu_2^2\right)^{1/2} \tag{20}$$

$$V = \frac{\pi d}{\lambda_0} NA \tag{21}$$

V is known as the normalized frequency or cutoff parameter or V number. This number depends only on the fiber characteristics and the wavelength of the light propagating. Each value of m is associated with a distinct wave pattern or mode within the waveguide. If $m > \left(\frac{2V}{\pi} - \frac{\phi}{\pi}\right)$, the condition for total internal reflection will not be met. When the total internal reflection condition is not met, the mode is said to be beyond "cut-off". All guided modes are below cut-off. From Eq. (19) we find that if $2V < \phi$, no mode can be propagated. However, for any value of V it is always possible to find an angle θ such that the corresponding value of ϕ is less than $2V$. Consequently, at least one mode can always be propagated. In general, the light launched at the fiber end within acceptance cone meets the core-cladding interface at an angle θ varying between θ_c and $\pi/2$. Of all these rays only those which satisfy Eq. (19) alone propagate forming modes as shown in Fig. 5.4.

5.20.1 Transverse Electric and Transverse Magnetic Modes

In free space, electromagnetic wave travels as a TEM wave. When it travels through a fiber, when light is incident on an interface, two different situations are possible.

(*i*) electric field vector E being in the plane of incidence and

(*ii*) electric field vector E being perpendicular to the plane of incidence

When total internal reflection takes place, depending on whether electric vector E is parallel or perpendicular to the plane of incidence, two different phase shift (ϕ) values are observed. They give rise to two independent sets of modes. These modes are called transverse magnetic (TM) and transverse electric (TE) corresponding to E parallel and E perpendicular situations respectively. To specify the modes completely two integers, l and m, are required rather than one (m). If Z is the direction of propagation through the fiber, the guide is bounded in two dimensions rather than one. Thus we refer to the TE_{lm} and TM_{lm} modes.

A ray that propagates through the core of a fiber undergoing total internal reflection is called meridional ray. Meridional rays are not be only type that correspond to guided modes through the fibers. There is another category of rays called as skew rays that describe angular helical path as they progress along the fiber as shown in Fig. 5.28. Consequently, the modes originating from skew rays are designated as either HE_{lm} or EH_{lm}, depending onwhether their magnetic or electric character dominates.

Path of skew ray in a circular index fiber

Fig. 5.28

In most practical waveguides, the full set of modes, *i.e.* *EH, HE* and *TE* can be approximated to a single set called linearly polarized (LP_{lm}) modes. Figure 5.29 illustrates the electric field intensity profiles of three such modes $LP_{0,1}$, $LP_{2,1}$ and $LP_{4,2}$. An LP_{lm} mode has m field maxima along a radius vector and 2l field maxima around a circumference.

When $V >> 1$, the number of modes, N, that can propagate is given approximately by

$$N \approx \frac{V^2}{2} \tag{22}$$

Cut off V numbers for the lowest order modes that a fiber can support are given in Table 5.1.

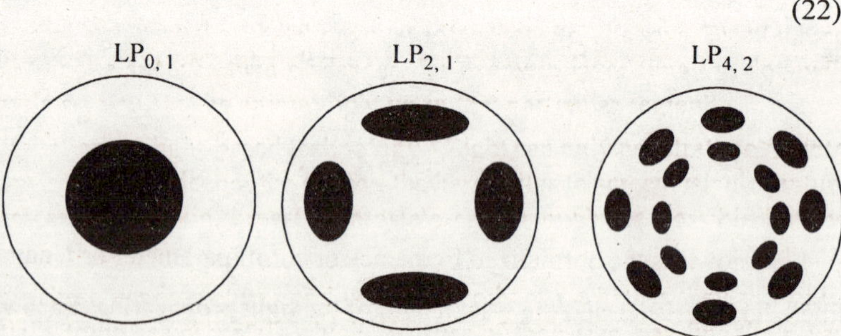

Electrical field distribution is three circular wave guide modes

Fig. 5.29

If the normalized frequency V is less than 2.405, only HE_{11} mode is supported by the fiber. HE_{11} mode represents the propagation of the light along the axial length of the fiber and the fiber becomes single mode fiber. If V number is greater than 2.405 but less than 2.42, multi modes viz HE_{11}, TE_{01} and TM_{01} are supported.

Thus for single mode transmission in a step index fiber, V must be less than 2.405. The wavelength at which the fiber becomes single mode is called cutoff wavelength λ_c.

From Eq. (21) λ_c can be expressed as

Table 5.1 Cut off *V* numbers for the lowest order modes

Mode	*Cut off V number*
HE_{11}	0
TE_{01}, TM_{01}	2.405
HE_{21}	2.42
HE_{12}, EH_{11}	3.83
HE_{31}	3.86
EH_{21}	5.14
HE_{41}	5.16
TE_{02}, TM_{02}	5.52
HE_{22}	5.53

$$\lambda_c = \frac{\pi d}{2.405} NA \tag{23}$$

5.21 DISPERSION

In an optical fiber communication, the signal to be transmitted is transmitted in the form of pulses of light with a given width, amplitude and spacing between pulses. During transmission several effects result in spreading of pulse width, which is called as dispersion and it tend to widen out and flatten it reducing its amplitude. The widening of the output pulse may result in overlapping of adjacent pulses at the receiving end of the fiber. This forces to increase the pulse to pulse seperation larger inorder to avoid overlapping of pulses. This reduces the upper limit on the pulse transmission rate. The maximum allowable transmission rate is called bandwidth. The product of bandwidth and dispersion, or bandwidth-dispersion product (BDP), is used as a quality factor of the fiber.

They are three separate dispersion mechanisms in an optic fiber.

(*i*) intermodal dispersion,

(*ii*) material or chromatic dispersion, and

(*iii*) waveguide dispersion.

5.21.1 Intermodal Dispersion

In a step-index fiber when a pulse of light is sent, depending on the *V*-number of the fiber, the same signal will travel along different paths, each representing one possible mode as shown in Fig. 5.30. The shortest path is for the HE_{11} mode which travels along the axial length of the fiber, while the longest path occurs for the mode nearest to the cut-off limit. Thus a pulse coupled into several modes at the transmitting end travel different paths and arrive at the receiving end at slightly different times. The shortest path mode (HE_{11}) pulse arrives first while the longest path mode pulse arrives last. Hence, the received pulse is the summation of these pulse modes, each delayed by a different time.

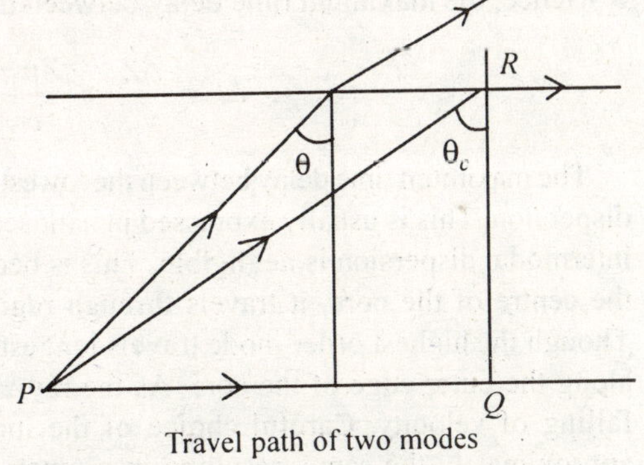

Travel path of two modes

Fig. 5.30

Let Z be the total length of the fiber. Figure 5.30 *PQ* represent lowest order mode (HE_{11}) travelling parallel to the axial length and *PR* represent higher order mode for which θ is almost critical angle θ_c. We know $\sin \theta_c = \mu_2/\mu_1$.

Now *AC* is the longest path while *AB* is the shortest.

In Triangle PQR $\sin\theta_c = PQ/PR$

or
$$PR = \frac{PQ}{\sin\theta_c} = PQ\,\frac{\mu_1}{\mu_2}$$

The path difference between PQ and PR though they cover the same length PQ of the fiber.

$$PQ\,\frac{\mu_1}{\mu_2} - PQ = PQ\,\frac{(\mu_1-\mu_2)}{\mu_2}$$

The path difference when these two modal rays travel a length Z of the fiber is given by

$$\Delta Z = Z\left(\frac{\mu_1-\mu_2}{\mu_2}\right)$$

We know from Eq. (10)

$$\Delta = \frac{\mu_1-\mu_2}{\mu_1} = 1-\frac{\mu_2}{\mu_1}$$

Hence,
$$\frac{\mu_2}{\mu_1} = 1-\Delta$$

or
$$\frac{\mu_1}{\mu_2} = \frac{1}{1-\Delta}$$

i.e.
$$\frac{\mu_1}{\mu_2} - 1 = \frac{1}{1-\Delta} - 1$$

$$\frac{\mu_1-\mu_2}{\mu_2} = \frac{\Delta}{1-\Delta}$$

$\therefore$
$$\Delta Z = Z\,\frac{\Delta}{1-\Delta} \tag{24}$$

Since μ_1 is the refractive index of core the phase velocity $v_p = c/\mu_1$.
Hence, the maximum time delay between the modes is

$$\Delta t = \frac{\Delta Z}{v_p} = \frac{Z\mu_1}{c}\left(\frac{\Delta}{1-\Delta}\right) \tag{25}$$

The maximum time delay between the lowest and highest order modes in a step index fiber is called intermodal dispersion. This is usually expressed in nanoseconds per kilometer (ns/km). In multimode graded index fibers intermodal dispersion is negligible. This is because though the lowest order mode travels shortest path along the centre of the core, it travels through region of higher refractive index and hence with lesser velocity. Though the highest order mode travels longest zigzag path, it passes through a zone of lower refractive index along the outer edge of the core. As the ray zigzags across the core it experiences an alternative rising and falling of velocity. Careful choice of the index profile causes the average velocity for all modes to be approximately the same, resulting in a much lower dispersion than in a step-index fiber. In an intermodal dispersion the following points are to be remembered.

1. Intermodal dispersion occurs only in step-index fibers.
2. Intermodal dispersion cannot occur in single mode fibers.
3. Intermodal dispersion is a characteristic of the fiber and is not affected by the wavelength of light used.

5.21.2 Material (or Chromatic) Dispersion

We know that, the refractive index is a function of wavelength. Material dispersion is due to the dependence of wavelength on the refractive index. It is expressed by

$$M = \frac{\lambda}{c}\frac{d^2\mu}{d\lambda^2} \quad (26)$$

where c is the velocity of light, M is a function of glass composition. Wavelength corresponding to $d^2\mu/d\lambda^2 = 0$ is called zero dispersion wavelength λ_0. If $\Delta\lambda$ is the spectral width of the light source the material dispersion that gives pulse spreading can be expressed as

$$\Delta_{mat} = \frac{\lambda\,(\Delta\lambda)}{c}\left(\frac{d^2\mu_1}{d\lambda^2}\right)Z \quad (27)$$

where λ = is the average wavelength of the transmitted light
μ_1 = is the refractive index of the core material
Z = is the length of the fiber

Typical material dispersion pulse spreading values for glass fibers using *LED*s in the 820 nm region are 3.0 – 3.5 ns/km ($\Delta\lambda$ = 380 Å). For lasers operating in the 1040 nm these values are 0.2 – 0.3 ns/km ($\Delta\lambda$ = 20 Å).

5.21.3 Waveguide Dispersion

In a single mode fibers near the wave length λ_0 = 1300 nm the intermodal and material dispersion disappear. But another type of dispersion called waveguide dispersion arises. The waveguide dispersion is a function of core radius, Δ and λ. It is due to the variation in group velocity of the fundamental mode v_g with wavelength, *i.e.* $dv_g/d\lambda$. This shift varies in a complex way with the wavelength, and results in broadening of the received pulses. This effect is the result of the guiding properties of the fiber, and hence this name. It has been shown that for an ideal single mode step index fiber, the waveguide dispersion has the peak value Δ_{wg} = 6.6 ps/nm-km.

5.21.4 Total Dispersion

Each of these three dispersions contributes to pulse broadening during signal transmmission to a greater or lesser extent. Independent dispersion effects combine in a root mean square manner and the total dispersion is given by

$$\Delta_t\ (\text{total}) = \sqrt{\Delta_t^2\,(imd) + \Delta_t^2\,(md) + \Delta_t^2\,(wg)} \quad (28)$$

At short wavelength the material dispersion is dominant, but near λ_0 the contribution of waveguide dispersion approaches that of material dispersion. At wavelengths greater than λ_0, the two are opposite in sign. At a particular longer wavelength, the waveguide dispersion and material dispersion counterbalance reducing the net dispersion to zero. Thus by balancing these two effects, single mode fibers can be designed to have λ_0 over a fairly wide wavelength range.

5.22 LOSSES IN FIBERS (ATTENUATION)

Another very important parameter of an optical fiber is losses in fibers is also known as attenuation. The power of the light at the output end is found to be always less than the power launched at the input end. The loss is dependent on fiber material, wavelength of light and length of the fiber.

Losses intrinsic to fibers result in decrease of light transmittance. These losses have three main sources (*i*) scattering losses, (*ii*) absorption losses, and (*iii*) bending losses.

5.22.1 Scattering Losses

The glass used in optical fiber is an amorphous solid that is formed by allowing the glass to cool from its molten state at high temperature until it freezes. While it is still plastic, the glass is drawn in the form of a very thin fiber under proper tension. During this process, submicroscopic variations in the density of the glass are frozen into the glass. Dopants added to silica to modify the refractive index also cause fluctuation in refractive index. These inhomogenities act as reflecting and refracting facets to scatter a small portion of the light passing through the glass, contributing for the losses. If the scale of these fulctuations is of the order of $\lambda/10$ or less, each irregularity acts as a point source scattering centre. This type of scattering is known as Rayleigh scattering. The losses induced because of the scattering vary inversely with the fourth power of the light wavelength used. Hence, at a wavelength 1.3 μm scattering losses are less than about 0.3 dB/km while at a wavelength 0.7 μm they are about 5 dB/km. The intrinsic Rayleigh scattering losses in silica fiber are shown in the Fig. 5.31.

5.22.2 Absorption Losses

Three different mechanisms contribute to absorption losses in glass fibers. These are ultraviolet absorption, infrared absorption, and ion resonance absorption. In pure fused silica, absorption of ultraviolet radiation around 0.14 μm results in ionization of valence electrons into conduction band. Thus, there is a loss of light due to ionization. Also during the fabrication of fiber GeO_2 is doped to change the refractive index of the glass to any desired value. This causes shift in the UV absorption band towards longer wavelength region. Absorption of infrared photons by atoms within the glass molecules results in increase of random mechanical vibrations and hence heating the molecules. This infrared absorption also exhibits a main spectral peak corresponding to silicon at 8 μm, with minor peaks at 3.2, 3.8 and 4.4 μm. The peaks are broad and tail off into the visible part of the spectrum. OH^- ions arepresent in the material due to trapping of minute quantities of water molecules during manufacturing. These ions absorb energy at peaks of 0.95, 1.25 and 1.39 μm with the main peak at 1.39 μm and is as shown in Fig. 5.32.

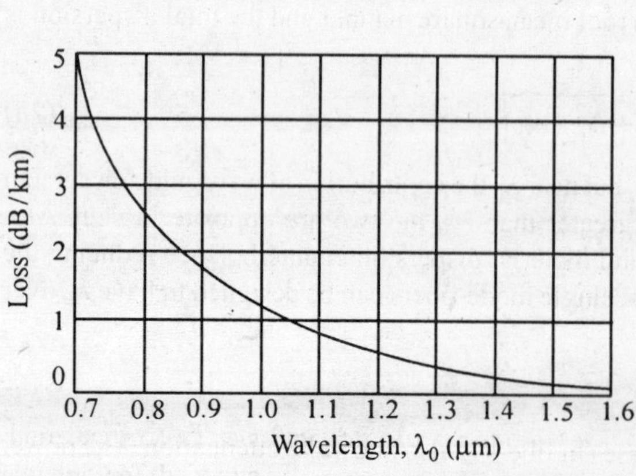

Fig. 5.31

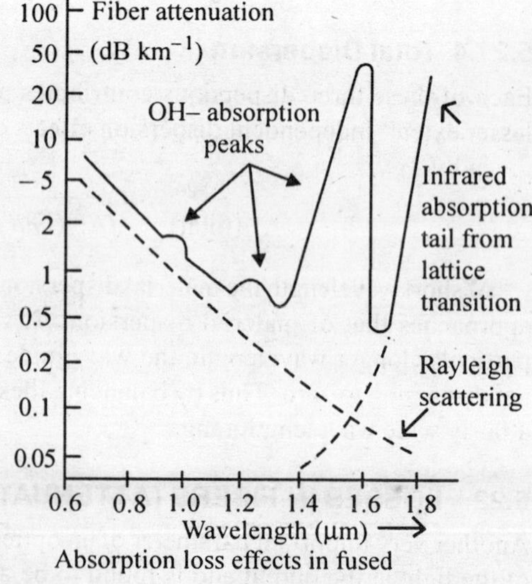

Absorption loss effects in fused

Fig. 5.32

The presence of other impurities such as iron, copper and chromium may also create unacceptable losses within the usable portion of the spectrum and hence extra care has to be taken while purification of silicon to avoid the impurities.

5.22.3 Bending Losses (Radiation Losses)

The distortion of the fiber from the ideal straight-line configuration may also result in fiber losses. Consider a wavefront which is travelling perpendicular to the direction of propagation. In order to maintain this, the part of the mode which is on the outside of the bend has to travel faster than that on the inside as shown in Fig. 5.33. As per the theory, each mode extends an infinite distance into the cladding though the intensity falls exponentially.

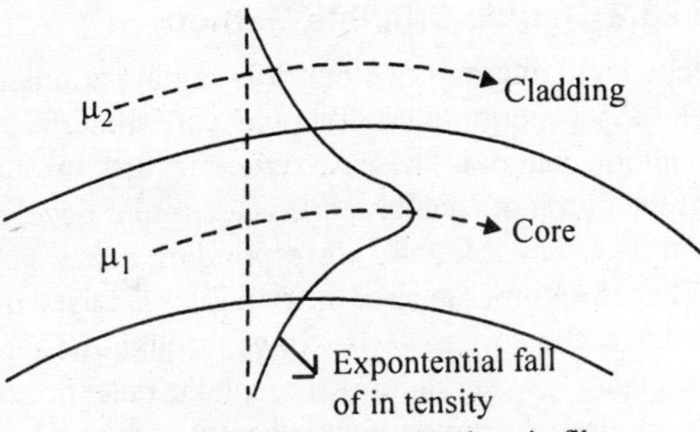

The mechanism of radiation loss in fiber

Fig. 5.33

Since the refractive index of the cladding is less than that of core, the part of the mode travelling in the cladding will attempt to travel faster. As per Einstein's theory of relativity since part of the mode cannot travel faster, the energy associated with this particular part of the mode is lost by radiation. The loss can be represented by an absorption coefficient (α_B) and is given by

$$\alpha_B = Ce^{\frac{-R}{R_c}}$$

where C is a constant, R is the radius of curvature of the fiber bend and R_c is given by $R_c = r/(NA)^2$ where r is the radius of the fiber. Since bends with radii of curvature of the order of magnitude of the fiber radius give rise to heavy losses, it has to be avoided.

5.22.4 Loss in Decibel

The loss in optical fiber is defined differently depending on the type of loss described and the design of the fiber. Attenuation loss is generally measured in terms of the decibel (dB) which is a logarithmic unit. The decibel loss of optical power in a fiber is calculated through the formula

$$dB = -\ 10 \log (P_{\text{out}}/P_{\text{in}})$$

where,

P_{out} = is the power coming out of the fiber
P_{in} = is the power launched into the fiber

Most fiber manufacturers characterize attenuation loss by the number of decibles loss per kilometer of fiber. This value can be çalculated by

$$dB/km = -\ (10/L) \log (P_{\text{out}}/P_{\text{in}})$$

where,

P_{out} = is the power coming out of the fiber
P_{in} = is the power launched into the fiber
L = is the length of the fiber tested

The loss per kilometer (or dB/km) is a standard unit for describing attenuation loss in all fiber designs.

5.23 FIBER MATERIALS

Glasses and plastics are mainly used in the manufacture of optical waveguides. Plastic fibers are easy to manufacture at low cost but because of their high losses they are used only for short distances.

5.23.1 Glass Fibers

Though the glass is fiber, silica is used in manufacture of these fibers than glass. An obvious requirement of the material used is that it must be possible to vary the refractive index. At 1 μm wavelength pure silica has a refractive index of 1.45. B_2O_3 is added to lower its refractive index, while addition of GeO_2 raises it. A typical fiber consists of an $SiO_2 : GeO_2$ core with pure SiO_2 cladding. Though glass fibers can be easily made with a relatively wide range of refractive indices, control of the impurity content is very difficult.

Low loss fibers are mainly manufactured using double crucible method and vapour phase deposition method.

5.23.2 Double Crucible Method

Schematic diagram of fiber drawing apparatus using the double crucible technique is as shown in Fig. 5.34. The inner platinum crucible has core material in molten form while the outer platinum crucible has molten cladding material. These two crucibles are surrounded by furnace.

At the bottm of each crucible is a circular nozzle with both inner and outer crucibles and corresponding nozzles being concentric. When the temperature of the crucibles is raised using the external furnace, the core material flows through the inner nozzle while the cladding material flows through the outer nozzle. The emerging drum, passes through plastic coating bath and curing over. This provides a protective coating of plastic over the drawn fiber. This results in step index fiber. If the glasses used have the property of interdiffusion, graded index fibers can be obtained.

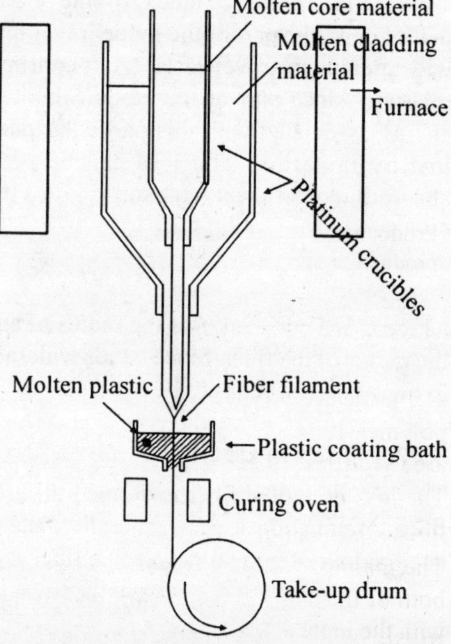

Double crucible method of fiber drawing

Fig. 5.34

5.23.3 Chemical Vapour Deposition (CVD) Method

A number of vapour phase deposition techniques have been developed for producing fibers with minimal impurity content. One of the most popular is the modified chemical vapour deposition (MCVD) method. In this method onto the inner surface of a pure silica tube a doped silica layer is deposited. As a result of chemical reaction taking place between the vapour constituents that are being passed down the tube, deposition occurs. Typical vapours used are $SiCl_4$, $GeCl_4$ and O_2. The reactions taking place are as follows.

$$SiCl_4 + O_2 \rightarrow SiO_2 + 2Cl_2$$

and

$$SiCl_4 + O_2 \rightarrow SiO_2 + 2Cl_2$$

A traversing oxy-hydrogen flame heats the reaction zone to a temperature about 1600° C as shown in Fig. 5.35 (a). If the process is repeated with different concentrations of the deposit vapours, one can produce any desired graded index profile. Once the deposition is over, the tube is collapsed into a solid rod called a preform by heating the tube to its softening temperature ($\approx$ 2000° C) as shown in Fig. 5.35 (b). Conversion of preform rod into a fiber is done by heating the end of the preform rod to its softening point and drawing off the fiber from its tip as shown in Fig. 5.35 (c). The fiber is coated with a layer of plastic for protection before being wound on the drum.

5.24 PLASTIC FIBERS

Plastic fibers are made either entirely from plastics or with silica cores and plastic claddings. Full plastic fibers suffer from very high attenuation because of a large Rayleigh scattering. Hence, such fibers are of any practical use only for short distances, low band width systems, preferably in the wavelength range 600 – 700 nm. These fibers can be made with large diameters because they are more flexible than glass.

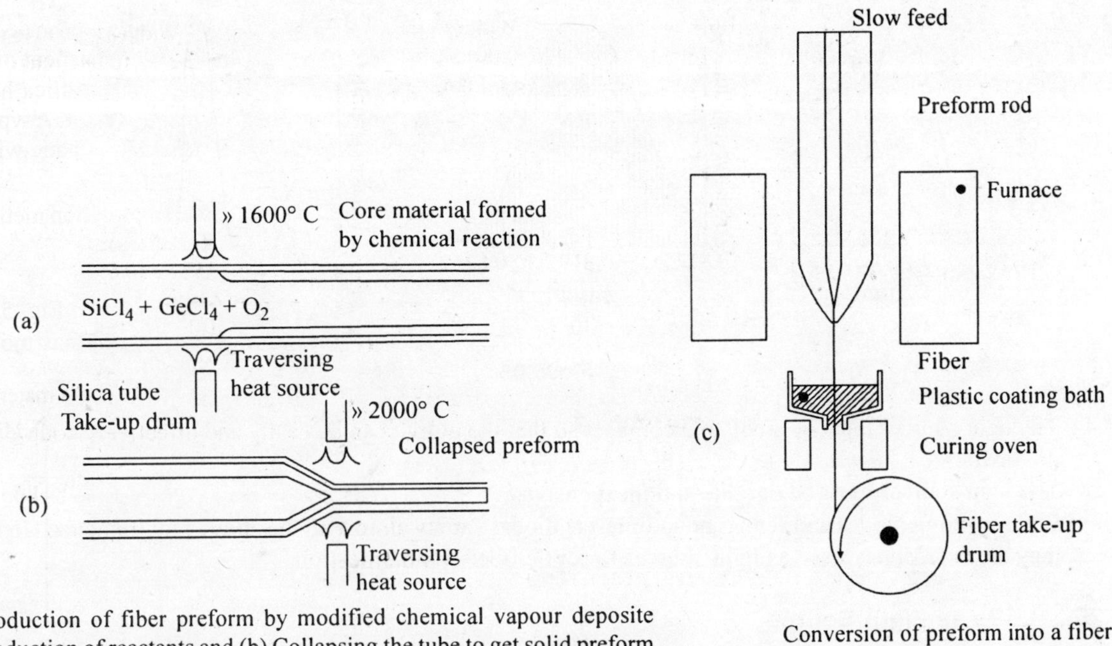

Production of fiber preform by modified chemical vapour deposite production of reactants and (b) Collapsing the tube to get solid preform

Conversion of preform into a fiber

Fig. 5.35

Plastic coated silica fibers are easy to manufacture. The silica fiber core is drawn through a bath of suitable polymer. It is subsequently cured by heating to a higher temperature to provide a solid cladding. These fibers are used for medium distance, moderate bandwidth communications systems.

5.25 OPTICAL FIBER COMMUNICATION

The main parts of the optical fiber communication system are transmitters, receivers and fibers connecting both of them. The transmitter has light source. The output from the light source is modulated in accordance with the input signal to be transmitted. This modulated light is coupled to the fiber, which carries the signal. At the other end of the fiber the received light signal is detected, demodulated and fed to the display unit.

5.25.1 Light Sources for Optical Fibers

Light sources for fiber optics act as signal emitters. The following are the requirements of light source.

Light emission spectra for typical light transmitters. (a) *LED* operating at 0.65 mm. (b) Broad-spectrum laser diode at 0.8197 mm. (c) Narrow-spectrum laser diode at 0.85275 mm.

1. The light emitted must be monochromatic, *i.e.* emission bandwidth must be as narrow as possible to minimize the chromatic dispersion increase the transmission bandwidth. Emission bandwidth of different light sources are shown in Fig. 5.36. A light source emitting at much narrower bandwidth has to be selected.
2. The light emitted must be intense enough so that transmission over relatively longer distances is possible inspite of the intrinsic losses.
3. The light sources must be capable of being easily modulated. Light emitting diodes and diode lasers can be easily modulated by modulating the current flow through them. Other light sources such as gas ionization lamps, lasers except diode lasers cannot be easily modulated. For such sources external modulation techniques are to be adopted.

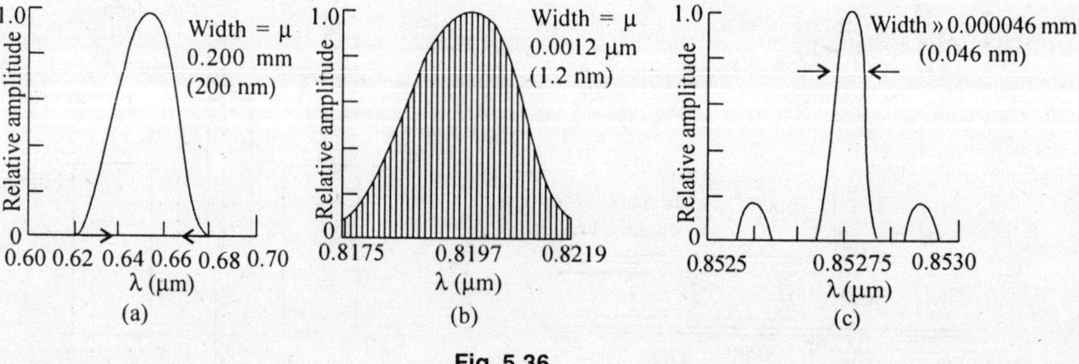

Fig. 5.36

4. The light sources must be small and compact so that the output can be easily and effectively coupled to the fibers.
5. The light sources must be durable and inexpensive.

Light emitting diodes and semiconductor lasers diodes satisfy almost all the above requirements. Hence, they are extensively used as light sources for optic fiber communication.

5.25.2 LEDs as Light Source

The two main types of *LED* most often used in optical fiber systems are the surface etched well emitter (*S-LED*) and the edge emitter (*E-LED*). In the *S-LED*, a well is etched into the top of a planer *LED* structure. This enables the fiber end to be kept as close as possible to the light emitting region. In *S-LED* the current is confined to a smaller circular region of the surface. This area is typically 20 to 50 μm in diameter. The fiber is held in position by the use of of a transparent epoxy resin as shown in Fig. 5.37 (a).

In *E-LED*, the radiation is confined to a narrow light guiding layer with a structure very similar to that of the double heterostructure laser as shown in Fig. 5.37 (b). Total optical powers from the edge emitters are typically several times smaller than from surface emitters. Narrower beam divergence gives rise to more coupled power. In general, *E-LED*s are preferred to use with small *NA* fibers whereas *S-LED*s are better for large *NA* fibers.

When we consider the output characteristics of *LED*s, they have an almost linear relationship between drive current and light ouput. This makes the *LED* more suitable for amplitude modulation.

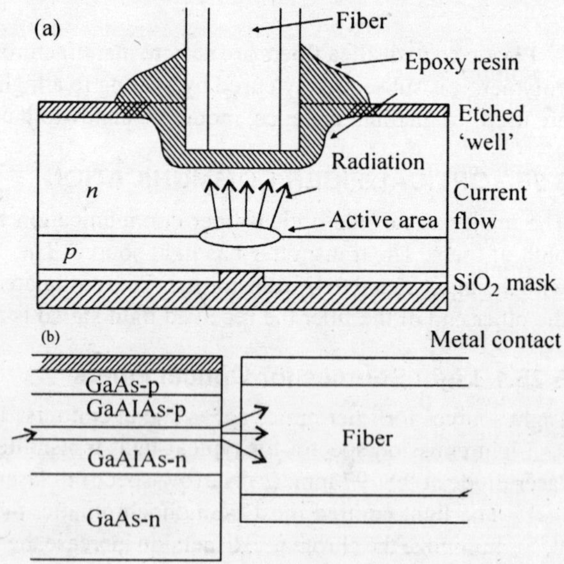

Two types of emitter designed for more efficient coupling into optical fibers. (a) the etched well type and (b) an edge emitter

Fig. 5.37

5.25.3 Lasers as Light Source

Lasers are highly monochromatic and intense than *LED*s. Semiconductor lasers exhibit much higher powers, narrower beam divergence and a smaller emitting area. Hence, lasers couple significantly more power into fibers than *LED*s. They are useful for small diameter low numerical aperture fibers.

5.25.4 Emitter Design

Efficiency of coupling of light from a flat emitting surface into a fiber optical waveguide is given by the relation.

$$\text{Coupling efficiency} \qquad \eta_c = \frac{\text{Energy coupled into fiber}}{\text{Energy emitted by LED}}$$

$$\text{It can be shown that} \qquad \eta_c = \frac{(NA)^2}{\mu_0^2}$$

In order to maximize the coupling efficiency, one has to select a fiber with as large *NA* as possible and a source with an area not larger than the fiber core area. By increasing the *NA* mode dispersion increases also and the signal bandwidth decreases.

5.26 MODULATION

Optical carrier modulation differs from radiofrequency carrier modulation because of the characteristics and limitations of the devices used for modulation. There are three methods of direct modulation, namely, the analog, pulse and digital forms of modulation. Both *LED*s and laser diodes are most suitable sources for the above modulation methods. Other light sources, such as gas laser cannot be modulated internally. This makes external modulators attractive.

5.26.1 Analog and Digital Modulation

There are many different ways of modulation and we may divide these into three categories namely analog, pulse and digital. With analog modulation, the primary information signal continuously varies the appropriate wave parameter of carrier wave. At any one time, there is one-to-one relationship between the original signal amplitude and the magnitude of the wave parameter. In the other two methods, the signal amplitude is only sampled at regular intervals and this information is then conveyed by means of a series of 'pulses'. In pulse modulation if the pulse width is varied in proportion to the required signal amplitude, then it is called pulse width modulation; alternatively if a fixed pulse width pulse is used and the information is transmitted as a series of discrete pulses (0 and 1) it is called binary digital modulation.

Figure 5.38 represnets three possible schemes of modulating an optical light beam. In (*a*) the amplitude is continuously modulated in time. The envelope of the waveform gives the required signal. In (*b*) the signal is sampled at the beginning of each 'time slot' and a pulse is emitted during the time slot whose length is proportional to the signal amplitude. In (*c*) the signal is sampled at regular intervals and then a series of pulses is emitted. The presence of a pulse during a time slot indicates a 'one', while the absence of a pulse indicates zero.

Most of the devices used are external modulators which are basically amplitude and phase modulators in both digital and analog modulation schemes. In this scheme of modulation the CW light from the source is incident on the modulator in which the signal is impressed through the bias circuit. The output of the device is modulated light.

5.27 PHOTODETECTORS

At the receiver end of the fiber link, conversion of the signal received from optical to electrical is being done by photodetectors. The photodetectors for this purpose are required to have

1. High quantum efficiency (photon to electronic conversion efficiency) at the appropriate spectral range.
2. Adequate frequency response (refers to speed of response as it should respond to the modulation frequency).
3. Low dark current (dark current refers to the leakage current in a reverse biased diode in the absence of illumination and is due to the thermal generation of electron — hole pairs).

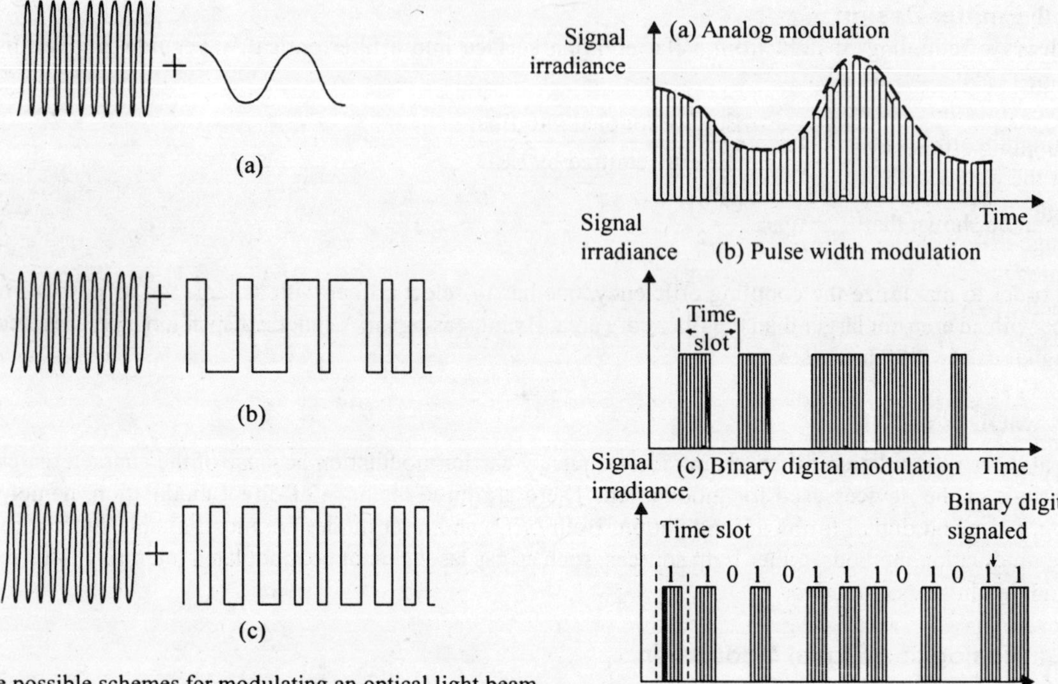

Three possible schemes for modulating an optical light beam

Fig. 5.38

4. Low signal dependent noise: Photoconductive and photoemissive types of detectors have poor noise performance and sensitivity and hence are not used for optical fiber communications. PIN diodes and the avalanche diodes (APD) have the speed of response and high sensitivity needed for useful communication applications.

5.28 OPTICAL FIBER COMMUNICATION LINK

Optical fibers are very attractive alternative to twisted wire or coaxial cables in communication links. Technological development has led to the manufacturing of fibers with losses less than 1 dB km^{-1} and with high information carrying capacity (bandwidths of the order of 10^{14} Hz). A schematic diagram of an optical transmission system is shown in Fig. 5.39.

At the transmitting end, the emitter is usually an *LED* or semiconductor laser. Transmitter includes modulators, switches and drive circuits to lasers. It also has couplers to couple the modulated light signals to be transmitted with the fiber. Fibers may also require connectors to increase the length of the fiber medium. At low information

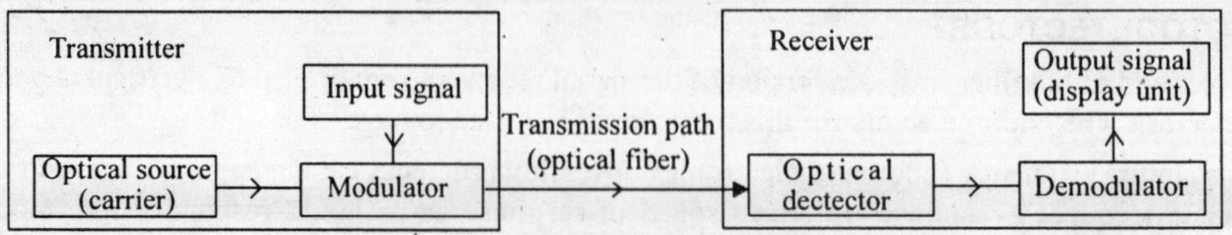

Schematic diagram of a optical communication system

Fig. 5.39

rates the length of a optic fiber transmission line is limited almost entirely by its losses. At higher rates, the length is limited by the amount of pulse dispersion. Repeaters must be included with fibers if either the loss limit or the dispersion limit is exceeded. At the repeaters an avalanche photodiode (APD) or a PIN diode receives the attenuated and dispersed train of pulses from one link, detects and amplifies. After reshaping and retiming, they are retransmitted into the next link.

At the receiving end, the optical receiver has three functions, namely (*a*) the conversion of optical signal into electrical signal, (*b*) amplification and (*c*) demodulation of the signal received to retrieve the original message transmitted. The first function of the receiver, namely the conversion from optical to electrical signal is achieved by the use of photodetectors together with their associated electronic circuits. The amplifiers must be such as to introduce minimum amount of noise and distortion. The detected, amplified and regenerated pulses are demodulated in order that the original signal waveforms are recovered.

5.29 ADVANTAGES AND DISADVANTAGES OF OPTICAL FIBER COMMUNICATION

Advantages

(a) **Large bandwidth:** In the coaxial cable transmission the bandwidth is up to around 500 MHz only whereas in fiber optical communication it is as large as 10^5 GHz. Thus, the information carrying capacity of optical fiber system is far superior to the best copper cable system.

(b) **Electrical isolation/hazards:** Since fiber optic materials are insulators, unlike their metallic counterparts, they do not exhibit earth loop and interface problems. Hence communication through fiber even in electrically hazardous environment do not cause any fear of spark hazards.

(c) **Immunity to interference and cross talk (electromagnetic interface):** Since optical fibers are dielectric waveguides, they are free from any electromagnetic interference (EMI) and radiofrequency interference (RFI). Hence, fiber cables do not require special shielding from EMI. Since optical interference among different fibers is not possible, unlike communication using electrical conductors, cross talk is negligible even when many fibers are cabled together.

(d) **Signal security:** Unlike the situation with copper cables a transmitted optical signal cannot be drawn from a fiber without tampering it. Such an attempt will affect the original signal and hence can be easily detected.

(e) **Small size and weight:** Since fibers are very very small in diameter the space occupied by the fiber cable is negligibly small compared to metallic cables. Optical cables are light in weight; these merits make them more useful in aircrafts and satellites.

(f) **Low transmission loss:** Since the loss in fibers is now as low as 0.2 dB km^{-1}, transmission loss is very less compared to copper conductors. Hence, for long distance communication fibers are prefered. Number of repeaters required is reduced.

(g) **Ruggedness and flexibility:** Fiber cable structures are flexible, compact and extremely rugged.

(h) **Low cost:** Since fibers are made of silica which is available in abundance, optical fibers offer the potential for low cost line communication.

Disadvantages

(a) **Bend losses:** If the optical fibre is bent severally, a part of the light escapes out through the cladding (without undergoing total internal reflection) causing signal loss. In copper cables bending has no effect on the signal strength.

(b) **Cable splicing:** The only time consuming process in laying the optical fibres is the cable splicing (joining of the ends of two optical fibres to correctly transmit the light wave). Effective splicing operation costs both time and money.

(c) **Thermal stability:** The expansion and contraction of optical fibres under varying environmental conditions, may affect the alignment at the splices and cause transmission losses.

(d) **Cost:** Compared to conventional cables, the initial cost of optical fibres is higher actually it is not the raw material cost, but the manufacturing process cost of the optical fibres, that makes it expensive.

Also, the maintenance, like the joining of broken fibre is time consuming as well as expensive compared to a conventional copper cable.

5.30 OPTICAL FIBER IN SENSING APPLICATIONS

Sensors are devices used to measure/monitor quantities such as displacement, pressure, temperature, flow rate, liquid level, chemical composition, etc. The measurement techniques developed using fibers are more sensitive and reliable. They can be developed so that they are compatible with any device/machine. Optical fiber sensors can be divided into two main categories, namely, active and passive. In active sensors, the quantity to be measured acts directly on the fiber itself to modify the radiation passing down the fiber. In passive sensors, the modulation takes place outside the fiber. The fiber merely acts as a convenient transmission channel for the radiation. There is no limit on sensor design. Any innovative design can be implemented.

5.30.1 Passive Optical Fiber Sensors

When two fiber ends are moved out of alignment, the coupling loss depends on the displacement. In a similar type of sensor shown in Fig. 5.40 (a), a shutter moves between two fiber ends thereby modulating the light coupled between two longitudinally displaced fibers. In order to improve the sensitivity a pair of gratings can be placed between the fiber ends while one grating is fixed, the other is movable as shown in Fig. 5.40 (b). Though the sensitivity has increased, the output is going to be periodic thereby reducing the range. To increase

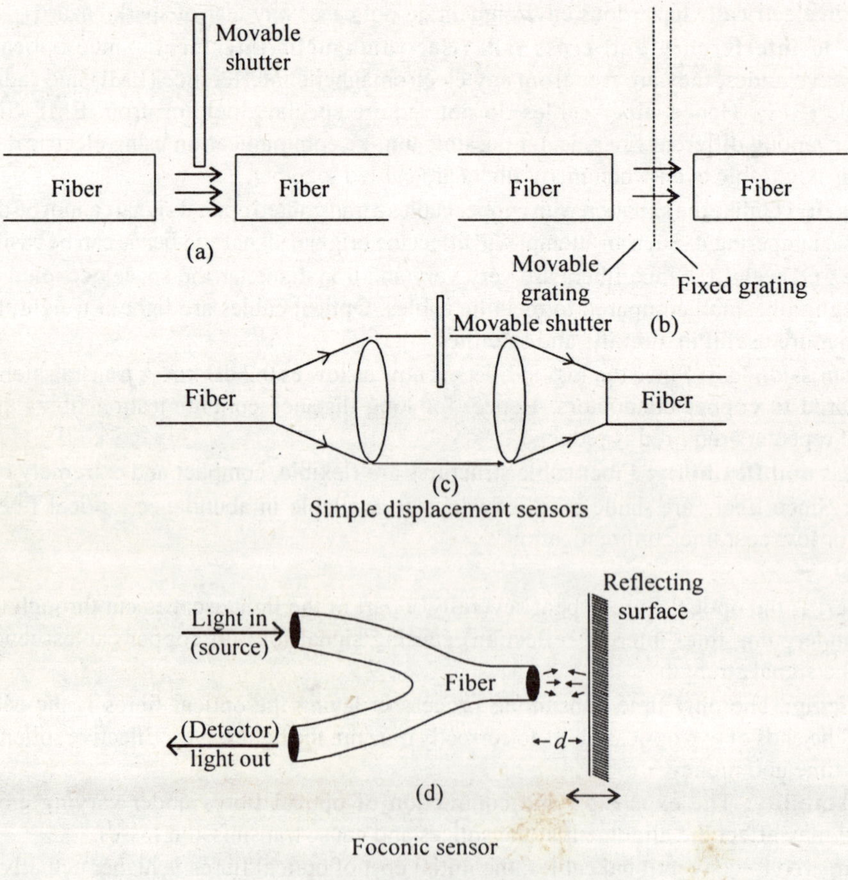

Fig. 5.40

the mea.urable range of displacement a beam expansion system can be employed as shown in Fig. 5.40 (c). Fotonic sensor shown in Fig. 5.40 (d) was the first commercially available sensor. This uses a bundle of fibers, half of which are connected to a source of radiation, the other half to a detector. A reflecting surface is placed at certain distance d from the bundle end. Now depending on the distance *d*, the light coupled to the fiber end after reflection at the reflecting surface varies. Hence, this can be used as a deflection sensor.

5.30.2 Active Optical Fiber Sensors

One of the simplest method of measuring displacement or pressure is changing the amount of light flowing down a fiber by means of microbending loss. In a typical device, the fiber passes between a pair of ridged plates which impart a periodic perturbation to the fiber as shown in Fig. 5.41. As a result the intensity of the light output varies. This method can therefore be used as displacement or pressure sensor. Some sensors involve polarisation modulation, some involve phase change, etc.

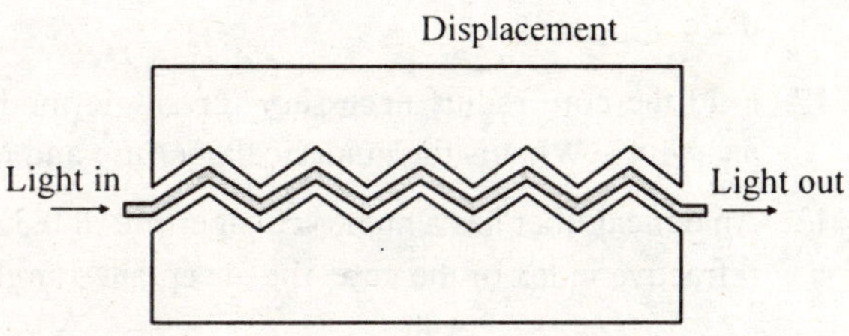

Measurement of microbending loss in fiber

Fig. 5.41

5.31 OTHER APPLICATIONS OF OPTICAL FIBERS

Other applications of optical fibers include its use in medicine to see inside the human body known as endoscopy and in road signs where one light bulb and a set of fibers are used to illuminate different parts of the sign thus saving electrical energy. A further recent application is in security fences. The metal strands of the fence contain a piece of optic fiber material down which a beam of light passes. If the strand is cut the light beam is interrupted and an alarm sounds. It is thought that this type of system is impossible to bypass.

UNSOLVED NUMERICAL PROBLEMS

1. An optical fibre has a numerical aperature of 0.39. The refractive index of its clad is 1.50. Find the refractive index of the core, the acceptance angle of the fibre, and the fractional index change.
2. Calculate the numerical aperture and the angle of acceptance of a given optical fibre, if the refractive indices of the core and cladding are 1.563 and 1.498 respectively.
3. A single mode step index optical fibre used in communication has a core with refractive index 1.45, refractive index change of 5×10^{-3}, and a core diameter of 6 mm. If the operating wavelength of the communication system is 1.2 mm, determine the *V*-parameter of the cable.
4. The attenuation of light in an optical fibre is estimated to be 2.0 dB/km. What fraction of the initial intensity remains after 1 km and after 8 km ?
5. A step index optical fibre has a core index of 1.46 and the cladding index of 1.409. If the core diameter is 80 mm and the wavelength of the light source is 1.2 mm, determine the number of modes present in the fibre.
6. A fibre sample 500 m long has an input power of 8.6 mW and an output power of 7.5 mW. What is the loss specification for the cable sample?
7. An optical fiber has a NA of 0.20 and a cladding refractive index of 1.59. Determine the acceptance angle for the fiber in water which has a refractive index of 1.33.
8. A glass clad fiber is made with core glass of refractive index 1.5 and the cladding is doped to give a fractional index difference of 0.0005. Find (*a*) the cladding index, (*b*) the critical internal reflection angle, (*c*) the external critical angle, acceptance angle and (*d*) the numerical aperature.

9. Calculate the numerical aperture and angle of acceptance of a given optical fiber if the refractive indices of the core and cladding are 1.563 and 1.498 respectively.
10. A He-Ne laser is generating a laser beam of power 4 mW. Calculate the number of photons emimtted by the laser. Given the wavelength of the emitted radiation is 680 nm.
11. A step index fiber in air has a NA of 0.12, a core refractive index of 1.42 and a core of diameter of 20 cm. Determine the normalised frequency for the fiber when light at a wavelength 0.8 mm is transmitted. $v = 9.43 \times 10^4$.
12. Find the core radius necessary for single mode operation at 850 mm of fiber with $m_1 = 1.480$ and $m_2 = 1.47$. What is the numerical aperture and maximum acceptance angle of this fiber?
13. An optical fiber has a numerical aperture of 0.32. The refractive index of cladding is 1.48. Calculate the refractive index of the core, the acceptance angle of the fiber and the fractional index change.
14. A step index optical fiber has diameter of 60 μm, a core index of 1.48 and the cladding index of 1.41. If the wavelength of the light source is 0.8 μm, determine the number of modes present in the fiber.
15. The attenuation of light in an optical fiber is estimated to be 2.2 dB/km. What fractional of the initial intensity remains after 2 km and after 6 km?
16. An optical signal has lost 85% of its power after traversing 500 m in a fiber. What is the loss in dB/km of this fiber?
17. Calculate the numerical aperture and angle of acceptance for an optical fiber having refractive indices 1.563 and 1.498 for the core and cladding respectively.
18. An optical fiber has a core material with refractive index 1.55, and its cladding material has a refractive index of 1.50. The light is launched into it in air. Calculate its numerical aperture, the acceptance angle and also the fractional index change.
19. Calculate the *V*-number for a fiber of core diameter 40 mm and with refractive indices of 1.55 and 1.50 respectively for core and cladding when the wavelength of the propagating wave is 1400 nm. Also calculate the number of modes that the fiber can support for proporgation.
20. Calculate the numerical aperture and acceptance angle of a fiber with a core index of 1.54 and a cladding index of 1.50.
21. In a fiber if $m_1 = 1.53$, $m_2 = 1.50$ and $l_0 = 1$ mm, calculate value of V and the number of modes N possible if the radius of the core is 50 mm.
22. A signal of 100 mW is injected into a fiber. The outcoming signal from the other end is 40 mW. What is the loss in dB?
23. For a fiber of core diameter 50 μm calculate the cut-off parameter and also the number of modes. $\mu_1 = 1.47$ (core), $\mu_2 = 1.45$ (cladding), operating wavelength = 0.85 μm.
24. Compute the cut-off parameter and the number of modes supported by a fiber μ_1 (core) = 1.54 and μ_2 (cladding) = 1.5; core radius 25 μm and operating wavelength is 1300 nm.
25. Compute the dispersion per km of length and the total dispersion in 10 km length of a step index fiber from the following data. Refractive index of the core = 1.55; $D = 0.026$.
26. Compute the maximum dispersion for an optically graded fiber from the following data. Refractive index at the centre of the core = 1.5; $D = 0.026$. The length of the fiber is 1 km.

EXERCISE

1. Describe the step index fiber and explain the transmission of a signal through it.
2. Explain briefly 'Numerical aperture of a fiber'.
3. Explain how it is made possible to transmit light through bent fibers.
4. Briefly explain, acceptance angle of a fiber.
5. Describe graded index optical fiber and explain the transmission of signal through it.
6. Explain briefly 'Basic principle of an optical fiber'.
7. Explain the advantages of optical fibers in communication.
8. Explain step index optical fibers.
9. Explain the propagation of light waves through optical fiber.
10. Explain how modes are formed in optical fibers. Derive an expression for V number.
11. What is intermodal dispersion? Derive an expression for the time delay between the lowest and highest order modes in a step index fiber.
12. What are different losses in optical fibers? Write a brief note on each.
13. Write a notes on

    (*i*) fiber materials, (*ii*) light sources for fiber optics, (*iii*) photodetectors for fiber optics.
14. Draw the block diagram of fiber optic communication system and explain the function of each element in the system.
15. Describe the structures of different types of optical fibers with ray paths.
16. Explain the terms numerical aperture and accpetance angle of a fiber. Derive expression for them.

6 Crystal Structures

6.1 CRYSTALLOGRAPHY

The study of geometric form and other physical properties of a crystalline solids by using X-rays, electron beams and neutron beams, etc. is called crystallography.

The solids in which the atoms or molecules are arranged in a regular and orderly manner in a three-dimensional pattern are known as crystalline solids. Each atom or molecule is fixed at a definite point in space, at a definite distance and in a definite angular orientation to all other surrounding. The internal spatial symmetry of atomic or molecular orientation is an essential feature of crystalline solids. The angular arrangement of the space positions of the atoms in a crystal is called lattice space.

6.2 SPACE LATTICE

A space lattice is defined as an infinite array of points in three dimensions in which every point has identical surroundings to that of every other point in the array.

Let us consider a finite two-dimensional array of points as shown in Fig. 6.1. If we locate any lattice point and look around in the plane of the paper, the condition is same irrespective of where are we. Consider a point P, if we look due north or due east from this point, we see another point at a distance of 1 unit along north-east, we see the nearest point at a distance of $\sqrt{2}$ units and along north-north-east, the nearest point is at a distance of $\sqrt{5}$ units. The surrounding would remain unchanged even if we shift our position to another point in the array.

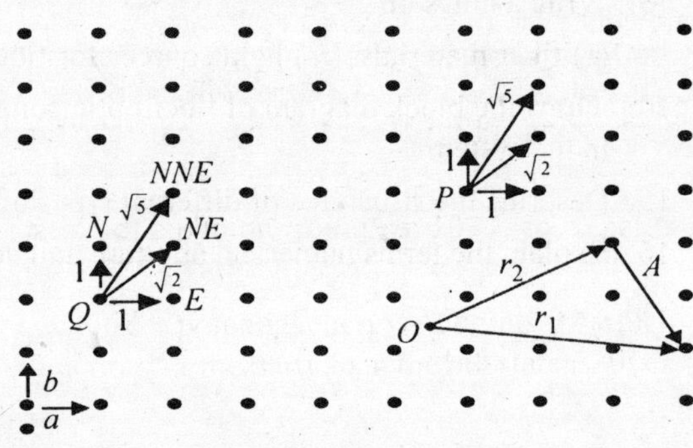

Fig. 6.1

Since this is true for every point in the array, if this array is extended infinitely satisfies the definition given above and can be called as two-dimensional square lattice. The space lattice may be defined in a more rigorous mathematical form which is as follows.

Let O be an arbitary point as origin and vectors r_1 and r_2 join any two lattice points to this origin. The difference A of the vectors should satisfy the following relation.

$$A = n_1 a + n_2 b$$

where a and b are fundamental translation vectors charateristic of the array and n_1 and n_2 are intergers. When the above relation is true for all values of r_1 and r_2, the array of points is a two dimensional lattice.

For a position vector P for a given point O is given by such a vector is called as lattice vector.

6.3 CRYSTAL LATTICE AND UNIT CELL

The complete lattice structure of a crystal is found to consist of identical blocks known as unit cells. The unit cell is the smallest block or geometric Figure from which the crystal is built up by repetition in three dimensions. The unit cell is a parallelepiped by moving which, in the direction of each of the coordinate axis x, y and z, are arranged parallel to edges of the figure, the crystal lattice can be constructed. As seen, this process resembles brick laying. It is also known as fundamental elementary pattern because it is repeated again and again to form the lattice structure of a crystal

∴ *A* unit cell may be defined as the volume of a solid from which the complete crystal can be constructed by translational repetition in three dimensions as shown in Fig. 6.2. The unit cell is the smallest part of a crystal having all the structural properties of the given lattice.

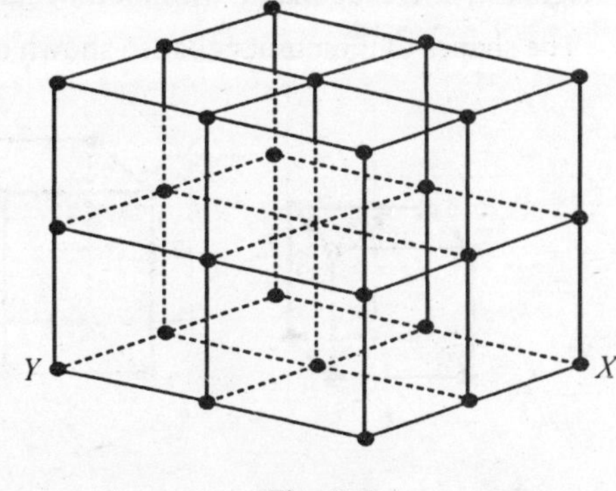

Fig. 6.2

If each atom in a lattice is replaced by a point, then each such point is called a lattice point and the arrangement of the points is referred to as the three-dimensional lattice array.

The length of the side of a unit cell is the distance between the atoms. In the case of pure metals whose crystals have simple cubic structure as shown in Fig. 6.3, it is equal to the basic distance *a*. However, in the case of crystals of chemical compounds like the cubic crystal of sodium chloride (NaCl), it is twice the distance *d* between two adjacent atoms as shown in Fig. 6.4.

6.4 LATTICE PARAMETERS OF A UNIT CELL

The three sides of a unit cell are called as crystallographic axes and are chosen in such a way that they pass regularly through the lattice points associated with like-atoms. Let us consider an arbitary arrangement of crystallographic axes *X*, *Y* and *Z*, defining a unit cell is as shown in Fig. 6.5. The angles between the three axes are called interfacial angles. The intercepts *a*, *b* and *c* define the dimensions of a unit cell and are known as its primitives or characteristic intercepts on the axes. The primitives and interfacial angles constitute the lattice parameters of the unit cell. These are also called as geometrical constants of a given crystal.

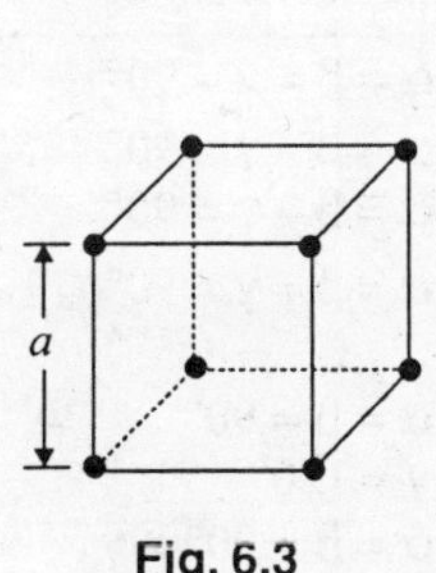

Fig. 6.3

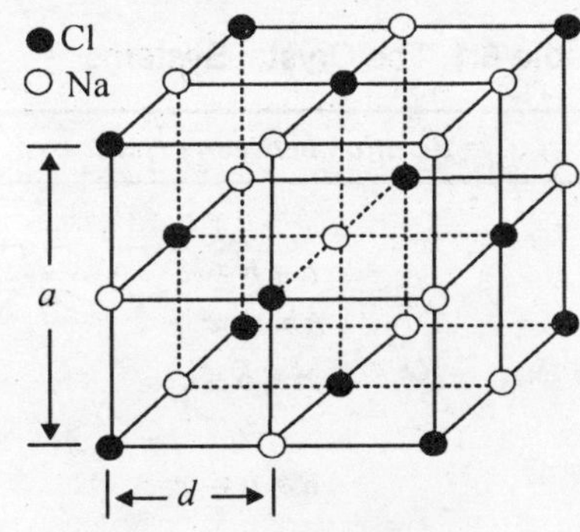

Fig. 6.4

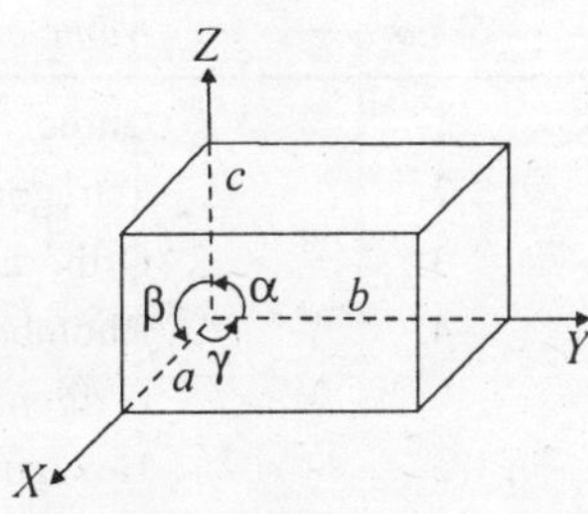

Fig. 6.5

If we know the linear dimensions *a*, *b* and *c* and the interfacial angles α, β and λ, provided the form and actual size of the unit cell. But if we do not know actual values of primitives but only their ratio and the values of interaxial angles, then we can determine only the form of the unit cell and not its actual size.

In a space lattice, all lattice points may be included in a set of parallel and equally spaced planes known as lattice planes. These planes give the orientations of the possible faces of a crystal. In theory, there are 320 such lattices groups divided into 32 crystal classes. These can be arranged in seven crystal systems which are characterised by three symmetry elements (*i*) the plane of symmetry, (*ii*) the axes of symmetry, and (*iii*) the centre of symmetry.

6.5 THE SEVEN CRYSTAL SYSTEMS

There are seven crystal systems corresponding to seven distinct types of unit cells. These unit cells are all parallel pipeds whose shapes are determined by the lengths of the three sides *a*, *b*, *c* and the values of the three interfacial angles α, β and γ.

The names of the seven crystal systems are: 1. Cubic, 2. Tetragonal, 3. Orthorhombic, 4. Rhombohedral (Trigonal), 5. Hexagonal, 6. Monoclinic, and 7. Triclinic.

The shapes of their unit cells are shown in Fig. 6.6 and their lattice parameters are summarised in Table 6.1.

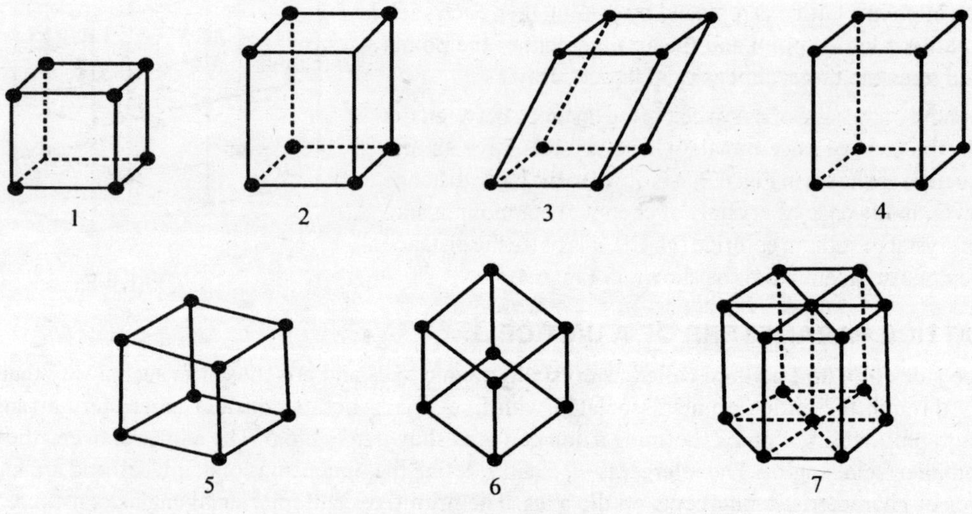

Fig. 6.6

Table 6.1 The Crystal Systems

S. no.	*Name of system*	*Relation between primitives*	*Interfacial angles*
1.	Cubic	$a = b = c$	$\alpha = \beta = \gamma = 90°$
2.	Tetragonal	$a = b \neq c$	$\alpha = \beta = \gamma = 90°$
3.	Orthorhombic	$a \neq b \neq c$	$\alpha = \beta = \gamma = 90°$
4.	Rhombohedral (Trigonal)	$a = b = c$	$\alpha = \beta = \gamma = 90°$
5.	Hexagonal	$a = b \neq c$	$\alpha = \beta = 90°$ $\gamma = 120°$
6.	Monoclinic	$a \neq b \neq c$	$\alpha = \beta = 90° \neq \gamma$
7.	Triclinic	$a \neq b \neq c$	$\alpha \neq \beta \neq \gamma \neq 90°$

1. In simple *cubic* structure: Ex: NaCl, CaF_2, $NaClO_2$
   (*i*) The three crystal axes are at right angles to the other two but
   (*ii*) The repetitive intervals along the three axes are equal.
2. A simple *tetragonal* structure has the following features: Ex: $NiSO_4$, SnO_2
   (*i*) The three crystal axes are at right angles to each other
   (*ii*) The repetitive intervals along two of the axes are equal
   (*iii*) The interval in the third direction is different from that in the other two.
3. *Orthorhombic* structure has the following features: Ex: KNO_3, $BaSO_4$, $MgSO_4$
   (*i*) The three crystal axes are at right angles
   (*ii*) The repetitive interval along each axis is different.

4. In *rhombohedral* structure: Ex : $CaSO_4$
   (*i*) The three crystal axes are equally inclined to each other at angles other than a right angle.
   (*ii*) The intervals along the three axes are different.
5. In *hexagonal* arrangement: Ex: SiO_2, AgI
   (*i*) There are three axes in one plane 120° from each other and a fourth axis normal to this plane.
   (*ii*) The repetition intervals along the three coplanar axes are equal but the interval along the fouth axis is unique.
6. In a simple *monoclinic* structure: Ex: Na_2SO_4, $FeSO_4$
   (*i*) One of the axes is at right angles to the other two but
   (*ii*) The other two axes are not at right angles to each other and
   (*iii*) Repetitive interval along each axis is different.
7. In Triclinic structure: Ex: $CuSO_4$, $K_2Cr_2O_7$
   (*i*) The characteristic feature of *triclinic* structure is a complete lack of symmetry, *i.e.*
   (*ii*) No two of the three axes are at right angles to each other
   (*iii*) Repetitive interval along each axis is different.

It will be seen that cubic system possesses the greatest symmetry whereas triclinic system has the least symmetry.

6.6 BRAVAIS' LATTICE

According to Bravais there are 14 possible types of space lattices in these, seven systems of crystals as illustrated in Fig. 6.7 are: 1. three cubic, 2. two tetragonal, 3. four orthorhombic, 4. one rhombohedral 5. one hexagonal, 6. two monoclinic, and 7. one triclinic, it has been proved mathermatically that there are only fourteen independent ways of arranging points in three dimensional space such that each arrangement conforms to the definition of a space lattice.

The cubic system is the simplest and most common crystal system that will be considered for discussion.

6.7 SYMMETRY ELEMENTS OF A CRYSTALLINE SOLID

The main symmetry elements of a crystalline solid are: (*i*) Axes of symmetry, (*ii*) Plane of symmetry, and (*iii*) Centre of symmetry.

Let us consider a simple geometrical solid body such as a cube. Since it has a regularity in the arrangement of its faces, it is also called as symmetrical solid. A description of symmetries present in a solid is a way to describing this regularity of an arrangement of its faces as shown in Fig. 6.7.

When a cube is placed on a table its four side faces become vertical. The intersections of these four faces are four vertical lines which are parallel to each other which are as shown in Fig. 6.8. It is obvious that if one of the intersection edges is made vertical, then the other four face leads to the conclusion that these faces are related to each other in same way.

These four faces are said to constitute a zone and any direction parallel to the intersection edges is called the zone axis.

6.7.1 Axis of Symmetry

If a cube is rotated about a vertical axis as shown in Fig. 6.9 parallel to the intersection edges and passing through the centre of the horizontal face, then in one complete of rotation, there are found to be four positions of the cube which are coincident with its original position. When the cube is rotated through 90°, it is not

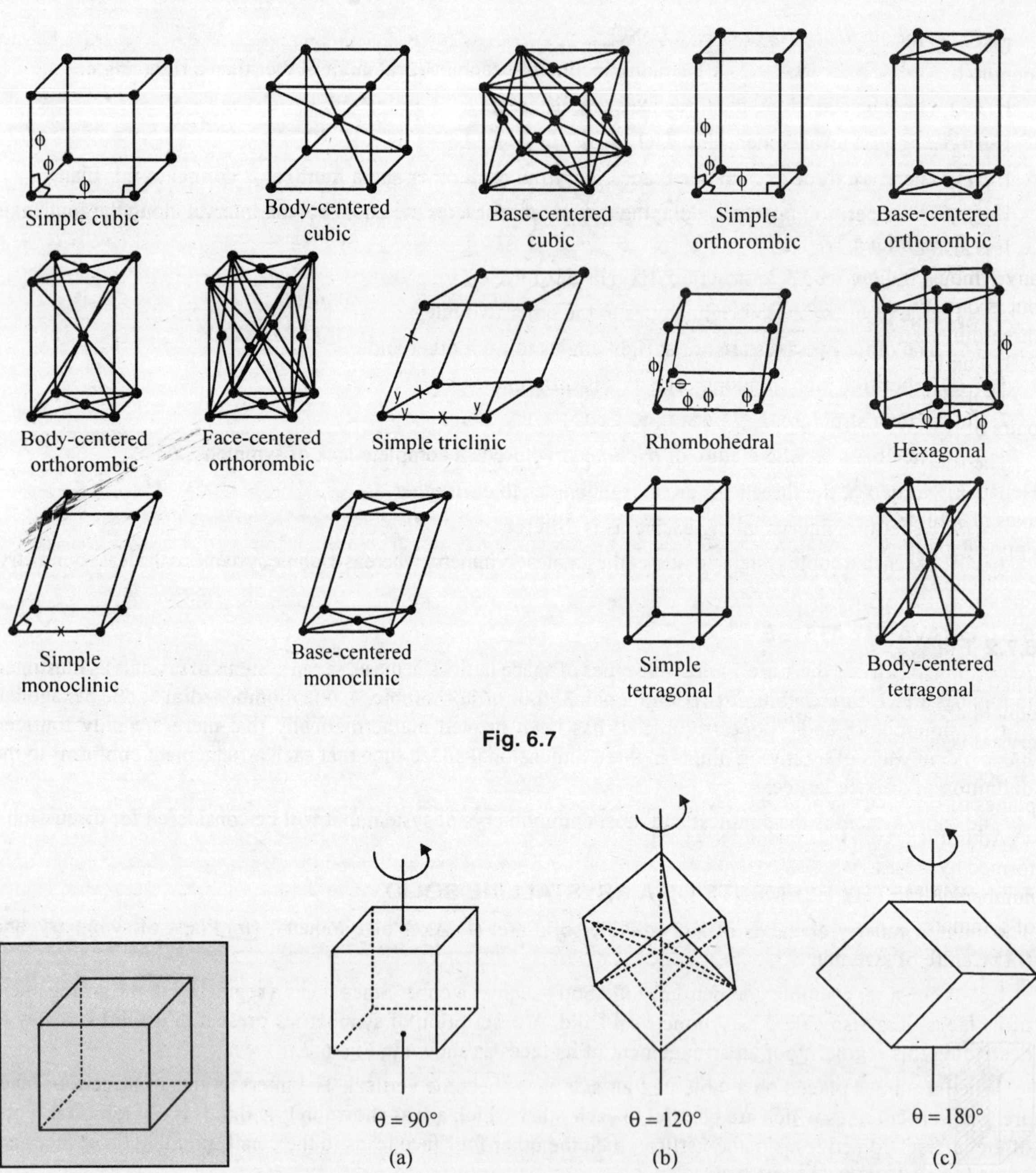

Fig. 6.7

Fig. 6.8

Fig. 6.9

possible to distinguish it from its original position. In other words, each rotation of 90° brings the cube into self-coincidence position.

This rotation axis is known as an axis of symmetry and since there are four congruent positions in one complete revolution, it is called a four-fold axis of symmetry. In general, if a rotation through an angle of radians or 360/n degrees about an axis, brings a structure into congruent position, then, the axis is called an n-fold axis of symmetry.

If $n = 1$, the crystal has to be rotated through 360° to achive self-coincidence. Such an axis is called an identity axis. Each crystal possesses an infinite number of such axes.

If $n = 2$, the crystal has to be rotated through 180° to achieve self-coincidence and the axis is termed a diad axis.

If $n = 3$, the angle of rotation has to be 120° and the axis is called a triad.

If $n = 4$, the angle of rotation has to be 90° and the axis is termed as tetrad.

If $n = 6$, the corresonding angle of rotation is 60° and axis is called as hexad.

It is also found that crystalline solids show only 1, 2, 3, 4 and 6- fold symmetry. They do not show 5-fold or any symmetry axis higher than 6. The reason for this fact is well-known, that is a crystal is not just a solid body, but is one in which the atoms or molecules are internally arranged in very regular and periodic manner in a three-dimensional pattern. Consequently, identical repetition of a unit can take place only when we consider 1, 2, 3, 4 and 6-fold axes.

Therefore, it is clear that a cube possesses three axes of four-fold symmetry, one normal to each of the three pairs of parallel faces. Other additional axes of symmetry possessed by it are as follows.

The rotation of the cube about a solid diagonal through 120° brings it into a position of self-coincidence. Hence, it is an axis of 3-fold symmetry, *i.e.* a triad. In all, these are four solid diagonals in a cube and hence four axes of 3-fold symmetry, *i.e.* four triads. As shown in Figure 6.9 (c), a line joining the middle points of a pair of opposite parallel edges provides a diad axis. Since there are 12 edges in a cube, the number of diad axes is 6.

Obviously, a cube has 13 axes of symmetry : 3 tetrad axes; 4 triad axes and 6 diad axes.

6.7.2 Planes of Symmetry

The other kind of symmetry shown by a cube is known as bilateral symmetry. A crystal can be cut into two equal halves such that one half is the reflection of the other half. There is a certain plane along which if the crystal is cut and then placed on a mirror, the image reproduces the other half of the crystal.

Such a plane is called as a plane of symmetry and is represented by the symbol m. In a cube, there are three planes of symmetry parallel to the faces of the cube.

Additionally, there are 6 diagonal planes of symmetry one of which is shown in Fig. 6.10 (a). This plane is formed by a pair of opposite parallel edges. Since there are 6 such pairs of edges which are shown by the same numbers in Fig. 6.10 (b), the number of diagonal planes of symmetry is six. Hence, the total numbers of planes of symmetry in a cubic crystal is = (3 + 6) = 9.

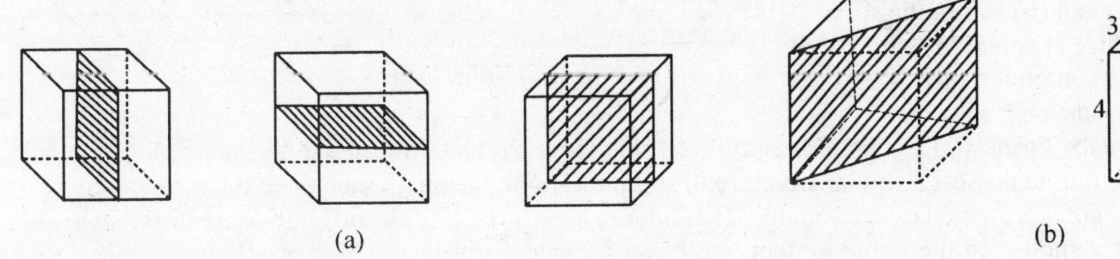

Fig. 6.10

6.7.3 Centre of Symmetry

It is also known as the centre of inversion (I). It is defined as a point in a crystal in which if a line is drawn from any point on the crystal through this point and produced an equal distance on the other side of the centre, it meets an identical point.

Hence, it is obvious that the centre of symmetry leads to parallel pairs of faces on opposite sides of the crystal. In fact, centre of symmetry really represents a reflection through a point instead of reflection in a plane.

It may also be noted that the whole crystallographic symmetry of a cubic crystal comprises of the 23 elements of symmetry and are as follows.

(*i*) 13 axes of symmetry-3 tetrads, 4 triads and 6 diads.

(*ii*) 9 planes of symmetry-3 planes parallel to the faces of the cube and 6 diagonal planes.

(*iii*) one centre of symmetry.

The unit cells belonging to the seven crystal systems have the following symmetry axes

(*i*) triclinic system has no axis of symmetry.

(*ii*) the monoclinic system has one diad axis only. Moreover, there are no axes of higher degree.

(*iii*) the orthorhombic system has three diad axes only.

(*iv*) the rhombohedral system has one 4-fold axis. This unique axis is called as the *z*-axis.

(*v*) the cubic system has four triad axes.

(*vi*) the tetragonal system has one triad axis. This unique axis is equally inclined to the three coordinate axes.

(*vii*) the hexagonal system has one hexad. This unique axis is at right angles to the other coordinate axes.

6.8. SPACE LATTICES OF CUBIC SYSTEM

Cubic structure is the simplest type of array in which the atoms take positions at the corners of the cube. However, there are four types of lattices possible in this system depending on the positions of the lattice points in the unit cell and are shown in Fig. 6.11.

(*i*) simple cubic (SC) lattice—Fig. 6.11 (a).

(*ii*) body-centred cubic (BCC) lattice—Fig. 6.11 (b).

(*iii*) face-centred cubic (FCC) lattice—Fig. 6.11 (c).

Some elements and compounds possess FCC lattice whereas many others crystalline into BCC lattice. These are listed in Table 6.2 which also gives the value of primitive *a* in angstrom units (1 Å = 10^{-10} m).

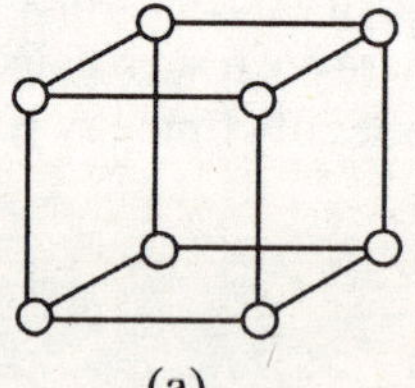

(a)

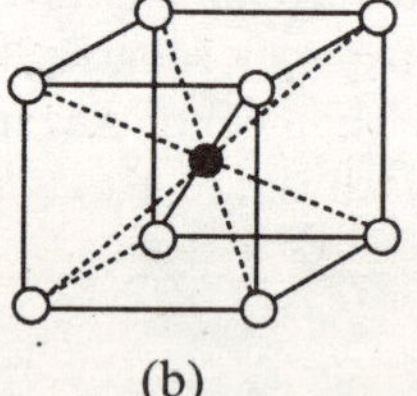

(b)

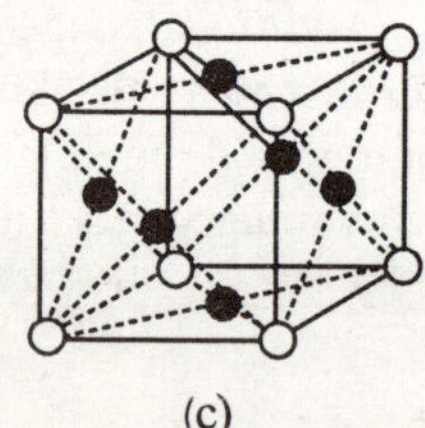

(c)

Fig. 6.11

Figure 6.12 (a) and (b) shows the cubical crystal lattice of common salt (NaCl) which is a compound of two elements, *i.e.* sodium and chlorine. Here, Na and Cl each forms an FCC pattern and the two patterns alternate in three dimensions. As seen, an FCC structure of chlorine possess a similar structure of sodium. The above mentioned three types of lattice, *i.e.* SC BCC and FCC are represented by the sumbols *P, I* and *F* respectively. Not all such lattice are possible in all crystal systems. Example: In the cubic system, we have *P, I* and *F* respectively. Not all such lattices are possible in all crystal systems. For example, in the cubic system, we have *P, I* and *F* whereas *P* lattice can be expected in triclinic and hexagonal.

6.9 CALCULATIONS OF DIMENSIONS OF UNIT CELL

The dimensions of the unit cell or the intratomic distance in a crystal lattice can be computed from the knowledge of molecular weight of the crystalline compound, Avogadro's number *N*, density of the material and its crystalline form. Example:

Table 6.2 Elements of Cubic Crystal System

Face-centred		*Body-centred*	
Element	*a (in Å)*	*Element*	*a (in Å)*
Nickel	3.524	α-iron	2.867
γ-iron	3.591	Tungsten	3.115
Copper	3.615	Titanium	3.306
Aluminium	4.049	Uranium	3.474
Gold	4.079	Zirconium	3.620
Platinum	4.086	Sodium	4.291
Silver	4.086	Barium	5.025
Lead	4.951	Rubidium	5.630

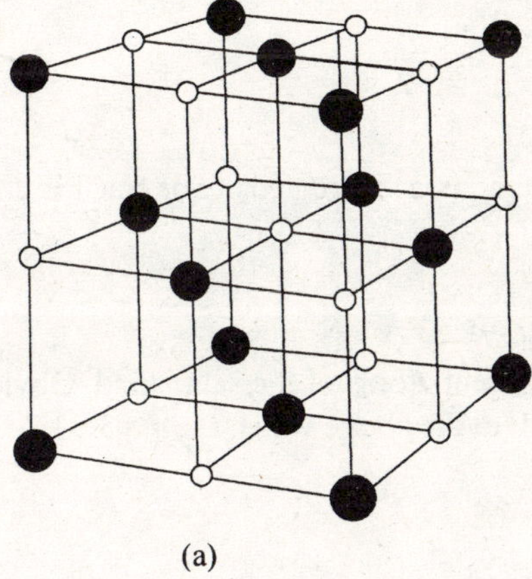

(a)

Na Cl

(b)

Fig. 6.12

(*i*) *Body-centred Cubic* (*BCC*) *Lattice:* The lattice which is afforded by α-iron crystal whose each unit cell as shown in Fig. 6.13 consists of 8 corner iron atoms and one atom is at the centre of the body. The atom at the body centre belongs entirely to the unit cell. However, each of the other 8 corner atoms is shared by 8 adjacent unit cells so that each contributes one-eighth of an atom to the cell under consideration. Hence, total number of iron atoms associated with each unit cell = 1 + (8/8) = 2.

For iron, the density is 7.86 g/cm^3 and the atomic weight is 55.85, hence, we get

$$a^3 \times 7.86 = 2 \times 55.85/6.02 \times 10^{23}$$

$$a = 2.87 \times 10^{-8} \text{ cm} = 2.87 \text{ Å}, \; a = 2.87 \text{ Å}$$

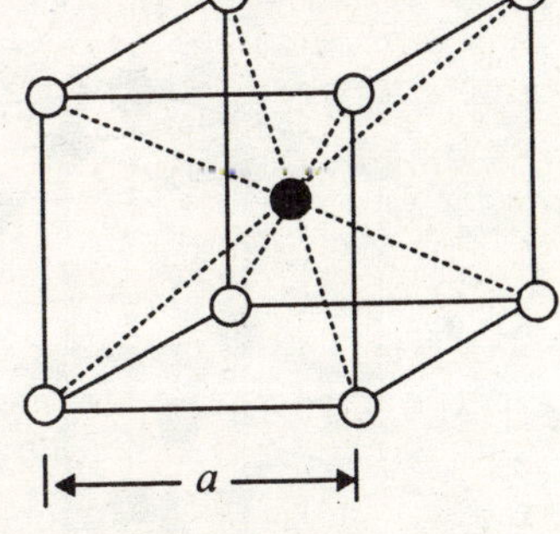

Fig. 6.13

we can also use the following formula

$$\rho a^3 = n^{A/N}$$

where A = atomic weight.

(*ii*) *Face-centred Cubic* (*FCC*) *Lattice:* It can be proved that each unit cell in sodium chloride has four Na^+ atoms and four C1 atoms associated with it. As shown in Fig. 6.13, the sodium atom in the centre belongs to entire cell. Each of the other twelve sodim atoms is shared with *three* adjacent unit cells so that each contributes one-fourth of an atom to the cell under consideration. Hence, total number of sodium atoms in each unit cell is

$$= 1 + \frac{1}{4}(12) = 4$$

Now, if we consider chlorine atoms. There are 8 corner atoms and 6 face atoms, *i.e.* one in each face. Out of the fourteen chlorine atoms, only four are associated with the unit cell. It is easy to visualize that each one of the corner chlorine atoms is shared by 8 adjacent unit cells so that each contributes one-eighth of an atom to the cell under consideration. Hence, number of corner atoms per unit cell is = 8/8 = 1.

If we consider the face-centred chlorine atoms. Each of these atoms is shared by *two* adjacent unit cells. Hence, each contributes one-half of an atom to the cell under consideration. Number of face centred chlorine atoms associated with each unit cell = 6/2 = 3. Hence, total number of chlorine atoms belonging to each unit cell of NaCl is = 1 + 3 = 4. Taken together, it means that there are 4 *molecules* of NaCl per unit cell.

∴ The molecular weight of NaCl = 23 + 35.5 = 58.5

Therefore the mass of each molecule = $58.5/6.02 \times 10^{23}$ g

Then the mass of 4 molecules = $4 \times 58.5/6.02 \times 10^{23}$ g

If a cm is the length of the edge of the unit cell, then its volume is a^3 since density of NaCl is 2.18 g/cm^3, mass of each unit cell = $a^3 \times 2.18$ gram.

$$\therefore \quad a^3 \times 2.18 = 4 \times 58.5/6.02 \times 10^{23}$$

$$\therefore \quad a = 5.63 \times 10^{-8} \text{ cm} = 5.63 \text{ Å}, \; a = 5.63 \text{ Å}$$

It is shown in Fig. 6.14, this is the distance between two adjacent atoms of the same kind. Obviously, the distance between two adjacent atoms, *i.e.* Na and Cl atoms is half of the above value, *i.e.* $d = 5.63/2 = 2.815$ Å.

6.10 CRYSTAL STRUCTURES

Many commercial importance materials have relatively simple crystal structures.

(*i*) **The hexagonal close-packed structure:** This structure is built-up by stacking close-packed planes in a simple sequence. The method of stacking can be understood with the help of Fig. 6.15 which shows that, two close-packed planes lying one above the other. It is also seen that each atom of one plane rests *in a depression* against three atoms of the other. Also the plane Q has two possible positions, **the one as indicated and the other** given by position R. For either choice, one atom of plane Q rests against three

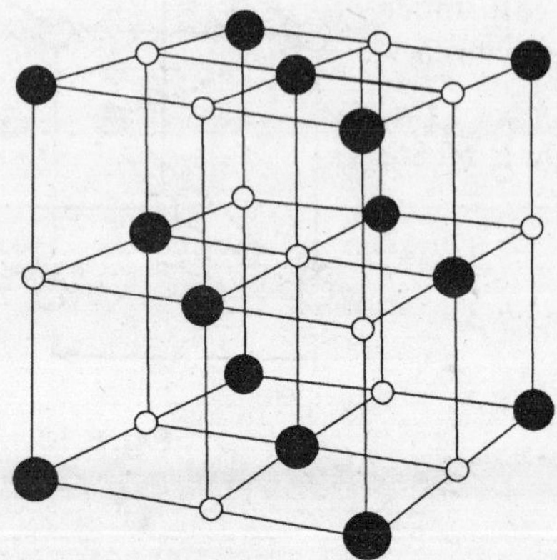

Fig. 6.14

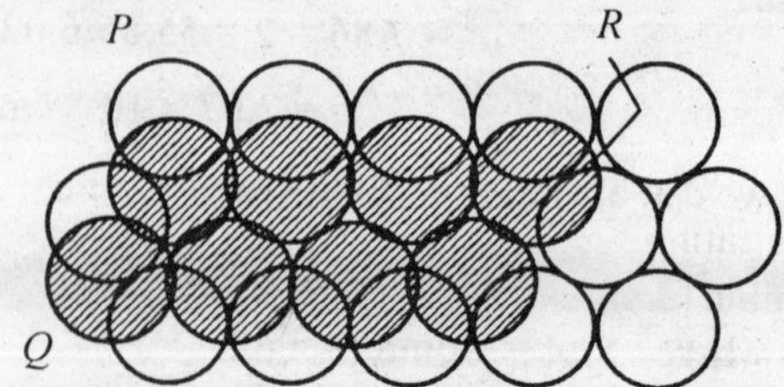

Fig. 6.15

atoms of lower plane *P* and *vice versa.* However, the only important choice is that for the third plane which also has two choices. Let the first two planes have the sequence *PQ*. Then the third plane may be added over the second in position *P.* In that case, the atoms of plane 3 would be directly above those of plane 1.

Alternatively, plane 3 may be stacked in position *R* (Fig. 6.15) so that atoms of plane 3 are directly above *interstices* of plane 1. This structure has the first stacking sequences of *PQPQ* ····· as shown in Fig. 6.16 (b). Heavy lines in Fig. 6.16 (a) show the unit cell of the crystal.

This structure is called close-packed because it possesses the maximum density of packing of hard spheres. Each atom has 12 *nearest* neighbours, and the coordination number is 12.

(*ii*) **The face-centred cubic (FCC) or cubic close-packed structure:** The unit cell of this structure is shown in Fig. 6.17.

It has the stacking sequence of *PQRPQR* ····· This structure is also close-packed because each atom has 12 *nearest* neighbours. As shown in Fig. 6.17 (b), the atoms are in contact with the diagonal atoms.

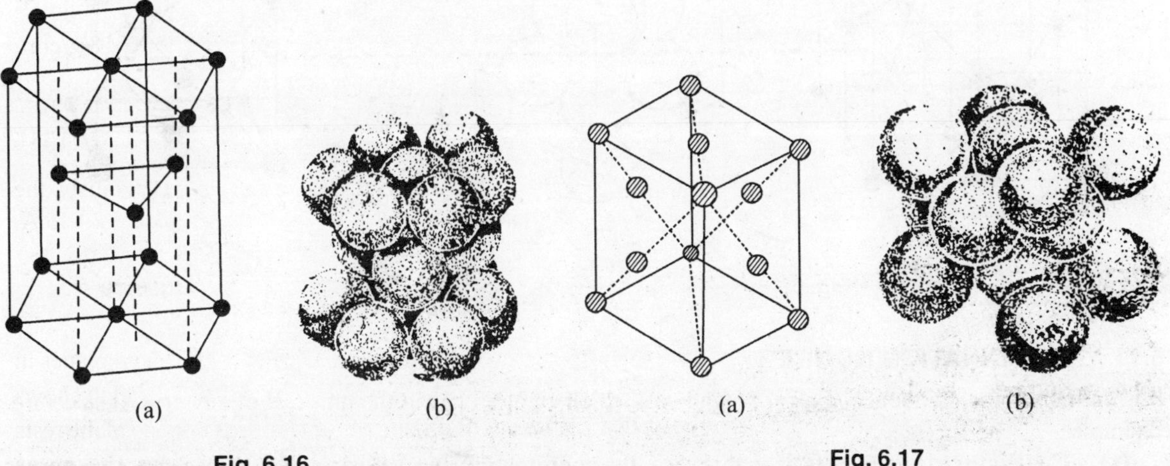

(a) (b)

Fig. 6.16

(a) (b)

Fig. 6.17

(*iii*) **The body centred cubic structure:** The unit cell of such a structure is shown in Fig. 6.14. It has one atom at each corner and one atom at the centre of the cube. Such a structure is not formed of close-packed planes as the above two since each atom has only eight nearest neighbours. However, the atoms are *in contact along body diagonals.*

(*iv*) **The simple cubic structure:** The structure is as shown in Fig. 6.12 (a), the unit cell of this structure has atoms only at the corners of the cube. Consequently, these atoms touch along cube edges. This structure is loosely-packed because each atom has only six *nearest* neighbours. Only one element, *i.e.* polonium (Po) exhibits this structure.

Many crystal structures are a variation of this simple structure. In fact, many of them consist of sets of interpenetrating simple cubic sub-lattices. For example, BCC lattice can be considered to consist of two identical simple cubic lattices, one consisting of corner atoms and the other at the centre atoms. Similarly, an FCC lattice can be considered to contain four identical simple cubic lattices properly disposed with respect to each other.

Many elements solidify in one of the structures as described above, several other elements and all compounds possess other structures. Some of these structures are derived from or combination of the basic structures.

Diamond cubic structure. Germanium (Ge), silicon (Si) and carbon (C) possess this structure which is a combination of two interpenetrating FCC sub-lattices. As shown Fig. 6.18 (a), one sub-lattice has its origin at (0, 0, 0) and the other at a point one quarter of the way along the body diagonal, *i.e.* at the ($a/4$, $a/4$, $a/4$) point. This structure is loosely-packed since each atom has only four nearest neighbours.

Zinc blende or sphalerite cubic structure: This structure Fig. 6.16 results if the two sub-lattices are occupied by two different elements. Some of the important compounds which possess this structure are semiconductors like InSb and GaAs; also ZnS, CuCl and many others.

NaCl structure: The structure is shown in Fig. 6.19 and consists of two FCC sublattices, one of the Cl ions having its origin at the (0, 0, 0) point and the other of the Na ions having its origin midway along a cube-edge, say, at ($a/2$, 0, 0) point.

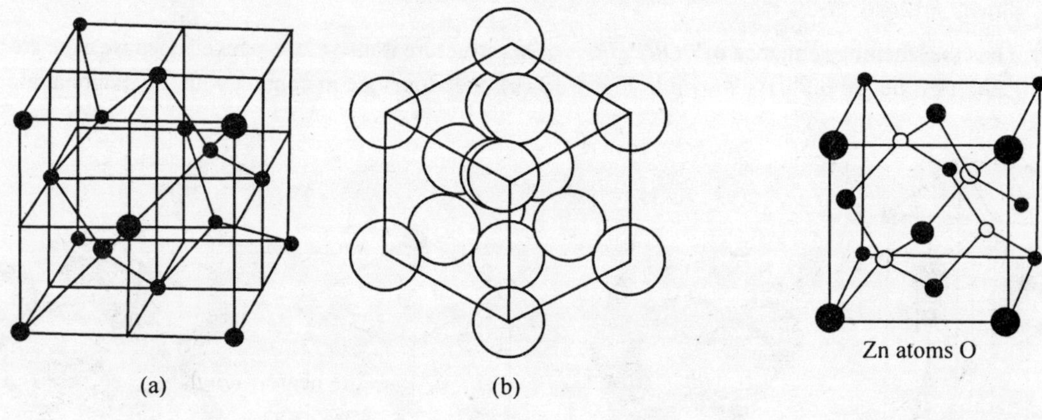

(a) (b)

Fig. 6.18

Fig. 6.19

6.11 COORDINATION NUMBER

It is defined as the number of *nearest* neighbours which an atom has in the unit cell of any crystal structure. Example:

(*i*) **SC structure** [Fig. 6.11 (a)]: In this case, the coordination number is *six*. It is so because each corner atom is linked with seven other unit cells that can be imagined to be built around the unit cell containing the atom. In that case, each corner atom has four neighbours in the same plane, one vertically above and one immediately below, giving a total of six *nearest* neighbouring atoms.

(*ii*) **BCC structure** [Fig. 6.11 (b)]: Its coordination number is *eight.* It should be noted that in this case, the *nearest* neighbours of any corner atom are the body-centred atoms. Since, there are eight surrounding unit cells for any corner atom, their eight body-centred atoms form the *nearest* neighbours for any corner atom.

(*iii*) **FCC structure** [Figure 6.11 (c)]: The coordination number of such a structure is *twelve.* In this case, the *nearest* neighbours of any corner atom are the face-centred atoms of the surrounding unit cells. Any corner atom has four such atoms in the own plane, four in a plane above it and four in a plane below it.

6.12 ATOMIC RADIUS

It is defined as *half the distance between nearest neighbours in a crystal of a pure element.* It is expressed in terms of the cube edge *a*. It must be remembered that *any two nearest neighbouring atoms touch each other.*

(*i*) **SC structure:** By definition, *AB* represents the atomic radius. It is obvious from Fig. 6.20 (a) and 6.20 (b) that $r = a/2$.

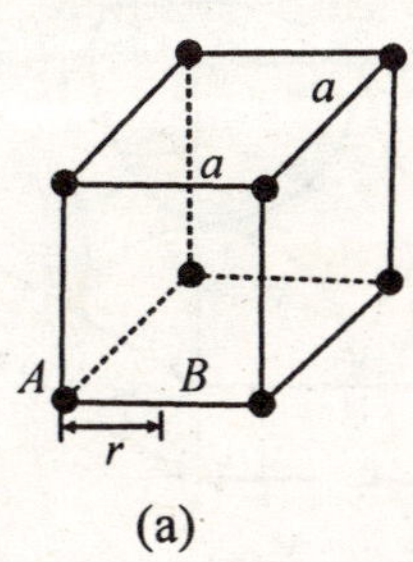

(a)

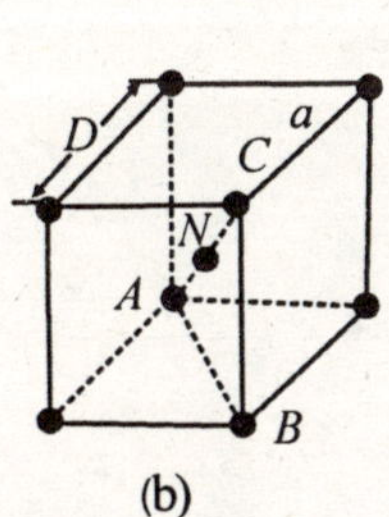

(b)

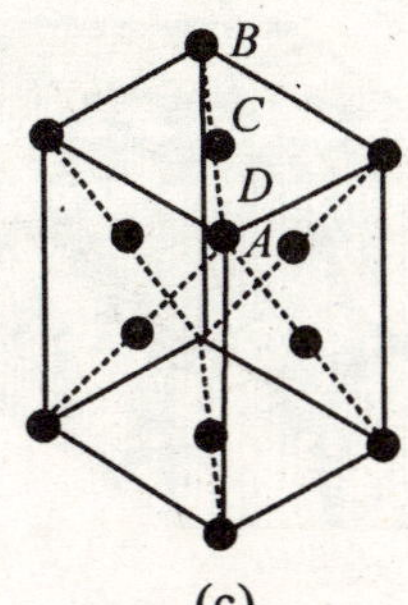

(c)

Fig. 6.20

(*ii*) **BCC structure:** From Fig. 6.22 (b), atoms *C* and *N* are the nearest neighbours. By definition $r = CN/2$. Now

$$AC = \sqrt{AB^2 + BC^2} = \sqrt{\left(\sqrt{2a}\right)^2 + a^2} = \sqrt{3 \cdot a}$$

$$r = CN/2 = AC/4 = \sqrt{3 \cdot a/4}$$

$$\therefore \qquad a = 4r/\sqrt{3}$$

A more detailed view is shown in Fig. 6.22 (b) where solid diagonal *AC* equals four times the radius of a single atom.

(*iii*) **FCC structure:** From Fig. 6.20 (c), here atoms *A* and *C* are the nearest neighbours.

Hence, $\qquad r = AD = AC/2 = AB/4$

Now, $\qquad AB = \sqrt{2 \cdot a} \qquad \therefore r = \sqrt{2 \cdot a/4}$

$$\therefore \qquad a = 4r/\sqrt{2}$$

Figure 6.22 (b) shows more detailed view of the same. Here, *surface* diagonal *AC* equals four atomic radii.

The lattice constant or side of a cubic unit cell as well its face and body diagonals in terms of atomic radius are shown in Table 6.3.

6.13 NUMBER OF ATOMS PER UNIT CELL

(*i*) **Simple Cube:** Since each corner atom is shared by *eight* surrounding cubes, share of each cube comes to one-eighth of an atom as shown in Fig. 6.23 (a). As there are eight corner atoms in all, each cube has $8 \times 1/8 = 1$ atom.

(*ii*) **Body-centred Cube:** It has eight corner atoms and one at the centre of an atom Fig. 6.23 (b). The number of atoms which belong to this cube are

(*a*) one centre atom and

(*b*) $8 \times 1/8 = 1$ corner atom.

Hence, number of atoms per cube $= 1 + 1 = 2$.

(*iii*) **Face-centred Cube:** It has 6 face-centred and 8 corner atoms. Figure 6.23 (c) each of the 6 face-centred atoms is shared by the two adjoining cubes. Hence, a total of $6/2 = 3$ such atoms belong to the cube. As before, one corner atom comes to the share of this cube. Hence, the total number of atoms per cube is $(3 + 1) = 4$.

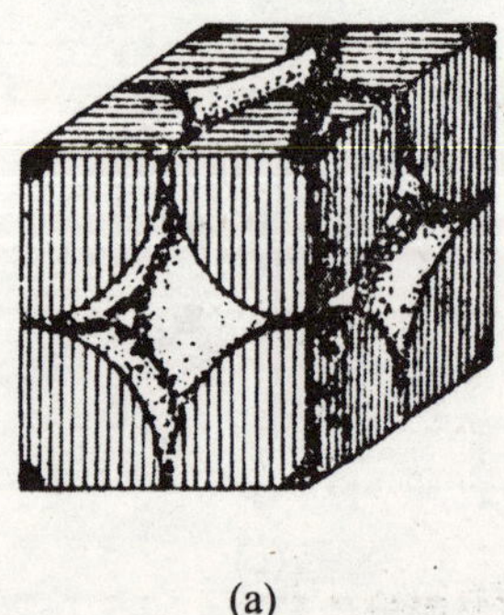

(a)

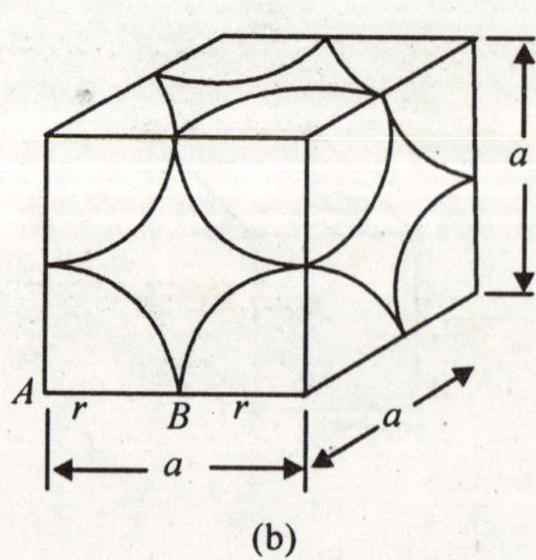

(b)

Fig. 6.21

(a)

C
a
4r
B
a
A
a

(b)

Fig. 6.22

Table 6.3

	SC	*BCC*	*FCC*
Side	$2r$	$4r/\sqrt{3}$	$4r/\sqrt{2}$
Face diagonal	$\sqrt{2}\ (2r)$	$\sqrt{\frac{2}{3}}\ (4r)$	$4r$
Body diagonal	$\sqrt{3}\ (2r)$	$4r$	$\sqrt{\frac{2}{3}}\ (4r)$

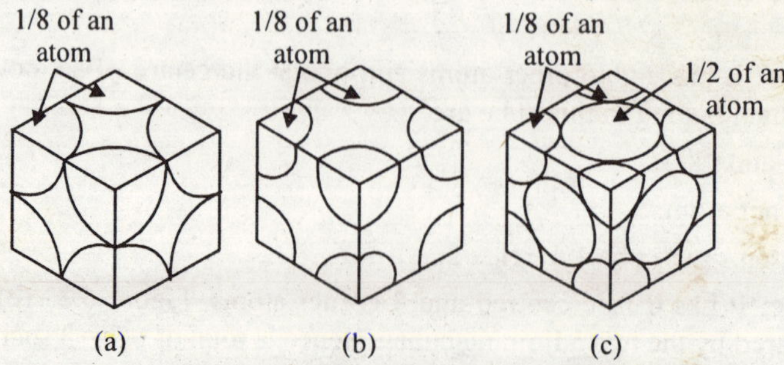

Fig. 6.23

6.14 DENSITY OF PACKING

The relative density of packing is defined as the fraction of volume occupied by spherical atom as compared to the total available volume of the structure.

***(i)* Simple Cube:** Since the cube has one atom.

Volume of one atom $v = \frac{4}{3}\pi r^3$

The, atomic radius is given by $r = \frac{a}{2}$

$\therefore$ Volume of the atom, $v = \frac{4}{3}\pi\left(\frac{a}{2}\right)^3 = \frac{\pi a^2}{6}$

Volume of the cube, $V = a^3$

$\therefore$ relative density of packing is $= \frac{v}{V} = \frac{\pi a^3/6}{a^3} = \frac{\pi}{6}$

***(ii)* Body-centred Cube:** In this case, number of atoms per cube is 2

Therefore, the volume occupied by *two atoms* is

$$v = 2 \times \frac{4}{3}\pi r^3$$

Now $$r = \sqrt{2}\cdot a/4$$

$$\therefore \quad v = 2 \times \frac{4}{3}\pi\left(\frac{\sqrt{3}\cdot a}{4}\right)^3 = \frac{\sqrt{3}\pi}{8}\cdot a^3$$

Volume of the cube, $V = a^3$

$$\therefore \quad \frac{v}{V} = \frac{\sqrt{3}\cdot\pi}{8} \times \frac{a^3}{a^3} = \frac{\sqrt{3}\cdot\pi}{8}$$

$\therefore$ relative density of packing $= \frac{\sqrt{3}\cdot\pi}{8}$

***(iii)* Face-centred Cube:** In this case, there are 4 atoms per cube.

$\therefore$ volume occupied by *four* atoms is

$$v = 4 \times \frac{4}{3}\pi r^3$$

Now $$r = \sqrt{2}\cdot a/4$$

$$\therefore \quad v = 4 \times \frac{4}{3}\cdot\pi\left(\frac{\sqrt{2}\cdot a}{4}\right)^3 = \frac{\sqrt{2}\cdot\pi}{6}\cdot a^3$$

$$V = a^3$$

$$\therefore \quad \frac{v}{V} = \frac{\sqrt{2}\cdot\pi}{6}\cdot\frac{a^3}{a^3} = \frac{\sqrt{2}\cdot\pi}{6}$$

$\therefore$ relative density of packing $= \frac{\sqrt{2}\cdot\pi}{6}$

6.15 IMPORTANT PLANE SYSTEMS IN A CUBIC CRYSTAL

In a cubic crystal, there are three sets of planes which are very rich in atoms. Consequently, Bragg reflection from these planes are more intense than from others. These three sets of planes which have different spacings are shown in Fig. 6.24.

(*i*) The first set of planes consists of *surface* planes of the cube and those parallel to them such as *QUVR, PTWS, PQRS, TUVW*, etc. as shown in Fig. 6.24 (a). Let the distance between consecutive parallel surface planes be d_1.

(*ii*) The second set of planes consists of parallel planes like *PUVS* which are inclined at an angle of 45° to the planes mentioned above as shown in Fig. 6.24 (b). Their spacing, are shown in Fig. 6.25 (a), is d_2.

(*iii*) The third set of planes consists of those planes which are parallel to the plane *PUW* as shown in Fig. 6.25 (c). Let the spacing of such planes be d_3 as shown in Fig. 6.25 (b).

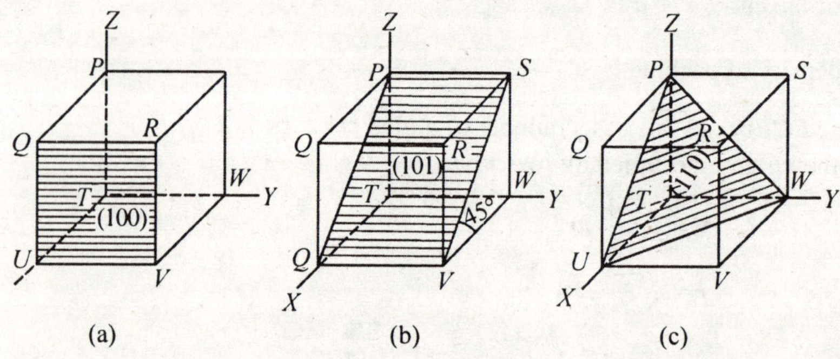

Fig. 6.24

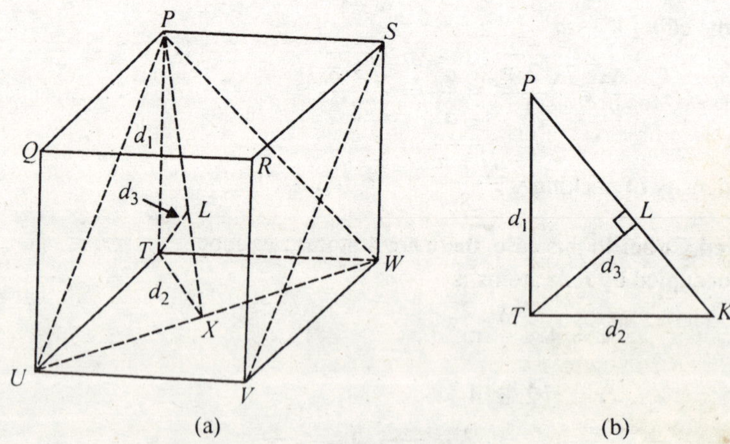

Fig. 6.25

Spacing of planes: The knowledge of spacing of these three sets of planes is essential for identifying different types of crystals. It is so because for each basic crystal structure, a definite ratio exists between the spacings of planes which are rich in atoms.

Let us Consider a simple cubic crystal with atoms at the corners of the cube as shown in Fig. 6.24 (a). The spacing of face planes like *PQRS* and *TUVW* is d_1.

The spacing of planes like *PUVS* which are inclined at 45° to the first set of planes is $d_2 = d_1/\sqrt{2}$.

The triangle PUW represents the third set of planes whose spacing d_3 may be found by drawing the triangle PTX in same shape as in Fig. 6.25 (b).

In Fig. 6.25 (a), TX is perpendicular to VW and $TZ = d_3$ is perpendicular to PX.

From the two similar triangles PZX and PTX of Fig. 6.25 (b), we have,

$$\frac{TZ}{TX} = \frac{PT}{PZ}$$

or
$$\frac{d_3}{d_2} = \frac{d_1}{\sqrt{d_1^2 + d_2^2}}$$

or
$$d_3 = \frac{d_1 d_2}{\sqrt{d_1^2 + d_2^2}}$$

Now, putting $d_2 = d_1 / \sqrt{2}$, we get

$$d_3 = \frac{d_1^2}{\sqrt{2} \cdot \sqrt{d_1^2 + \frac{1}{2} d_2^2}} = \frac{d_1}{\sqrt{3}}$$

$$\therefore \quad \frac{1}{d_1} : \frac{1}{d_2} : \frac{1}{d_3} :: 1 : \sqrt{2} : \sqrt{3}$$

Similarly, it can be shown that for a body-centred lattice structure

$$\frac{1}{d_1} : \frac{1}{d_2} : \frac{1}{d_3} :: 1 : \frac{1}{\sqrt{2}} : \sqrt{3}$$

Also, for face-centred lattice structure

$$\frac{1}{d_1} : \frac{1}{d_2} : \frac{1}{d_3} :: 1 : \sqrt{2} : \frac{\sqrt{3}}{2}$$

Bragg by using X-ray spectrometer not only verified the above ratios but also used them to recognise whether the crystal was simple, cubic or a body-centred type.

6.16 DIRECTIONS, PLANES AND MILLER INDICES

The basic necessity in crystal analysis is that, the ability to describe relative orientations of lines and planes. It can be achieved with the help of a suitable frame of reference. Latitude and longitude serve this function on a map whereas x-and y-axes are used on the graph paper as shown in Fig. 6.26 and two directions have been shown by arrows starting from the origin and ending at points A and B. These directions can be satisfactorily described by giving the coordinates of the *first* whole-numbered point (x, y) through which each passes. For direction OA, it is (2, 1) and for OB, it is (1, 1).

Three-dimensional graph paper of Fig. 6.27 can be used for describing a direction in space. As before, these directions are described by giving the coordinates of the *first* whole-numbered point (x, y, z) through which they pass. The directions of various lines of Fig. 6.27 are as follows:

OA (110)	OR (100)
OB (010)	OD (½, ½, 1) or (112)
OC (111)	OE (1, 0, ½) or (201)

The square brackets are used to indicate a *direction*. The digits in square brackets indicate the indices of that direction. In general, the indices of a direction are written as (uvw). A negative index is indicated by a bar over the digit.

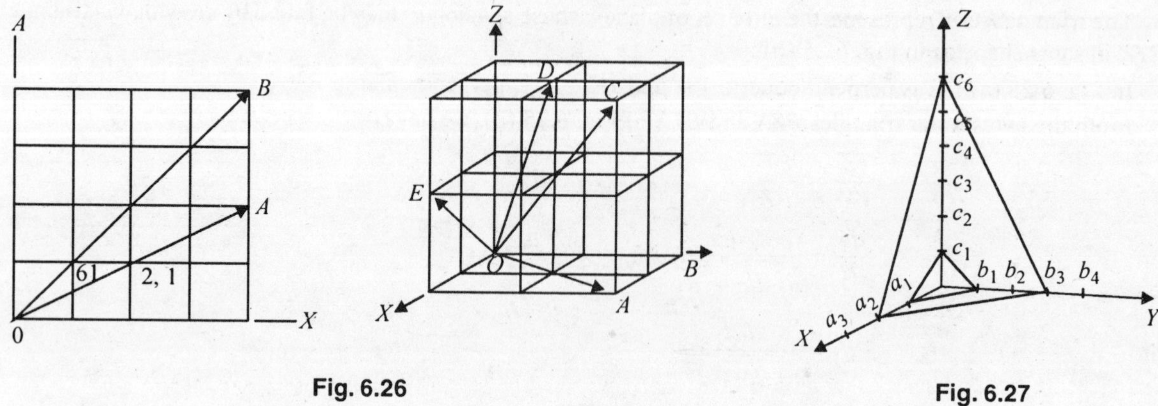

Fig. 6.26

Fig. 6.27

For example, + X-axis has indices of (100) whereas – X-axis has (100) indices. Similarly, + Y-axis has indices (010) and the – Y-axis (010). Positive Z-axis has indices (001) whereas – Z-axis has (001).

6.17 UNIT FACE AND AXIAL UNITS

Let OX, OY and OZ be taken as the reference axes as shown in Fig. 6.27. The faces $a_1\ b_1\ c_1$ and $a_2\ b_3\ c_6$ make certain intercepts or segments on these axes. The first face or plane intercepts segments of oa_1, ob_1 and oc_1 and the second – oa_2, ob_3 and oc_6. These segments represent the respective *linear parameters* of the two *faces*.

It is seen that the first face intercepts two, three and six spacings along the respective axes. The first face is called *unit face* which may be defined *as the face which makes one intercept on each axis.*

Obviously, the unit face produces three values which are related to each other as 1 : 1 : 1. The second face produces three values of $2a$, $3b$ and $6c$ which are related to each other as 2 : 3 : 6. These figures of 1, 2, 3, 6 represent the *numerical parameters* of the *faces* and must always be whole numbers.

Linear parameters intercepted by the unit *face* are called *axial units.* In the present case, these are a_1, b_1 and c_1 although they are represented generally by a, b and c along X, Y and Z-axes respectively.

6.18 MILLER INDICES

Orientations of *planes* in a crystal may also be described in terms of their intercepts on the three axes. For example, plane *PQR* of Fig. 6.28 (a) has intercepts of 2 axial units on x-axis, 2 axial units on y-axis and one axial unit on z-axis. In other words, the numerical parameters of the face are 2, 2 and 1. Hence, its orientation is (2, 2, 1). However, as suggested by Miller, it is more useful to describe the orientation of a plane by the *reciprocal* of its numerical parameters rather than by its linear parameters. These reciprocals when appropriately converted to whole numbers are called *Miller indices.* Hence, Miller indices of plane *PQR* of Fig. 6.28 (a) are (112). It is seen that for getting whole numbers, all the three reciprocals have been multiplied by 2. As is well known, multiplying all intercepts or reciprocals of intercepts by the same number does not change the orientation of a plane. The numbers for planes are written in parenthesis and not in square brackets. The general expression for indices of a plane is $(h\ k\ l)$. The symbol for a family of parallel planes is $< h\ k\ l >$.

Similarly, numerical parameters of face $P'Q'R'$ of Fig. 6.28 (b) are 4, 4 and 2. Hence, its Miller indices are (¼ : ¼ : ½) or (112) which shows that both faces belong to the same family of parallel planes.

Now, consider face *PQR* whose intercepts expressed in axial units are $a/2$, $b/2$, $3c/4$. Its numerical parameters are ½, ½ and ¾. Hence, its Miller indices which are nothing but reciprocal of its numerical parameters are $\left(\frac{1}{1/2} : \frac{1}{1/2}\ \frac{1}{3/4}\right)$ or $\left(2\ 2\ \frac{4}{3}\right)$ or (6 6 4) or (3 3 2).

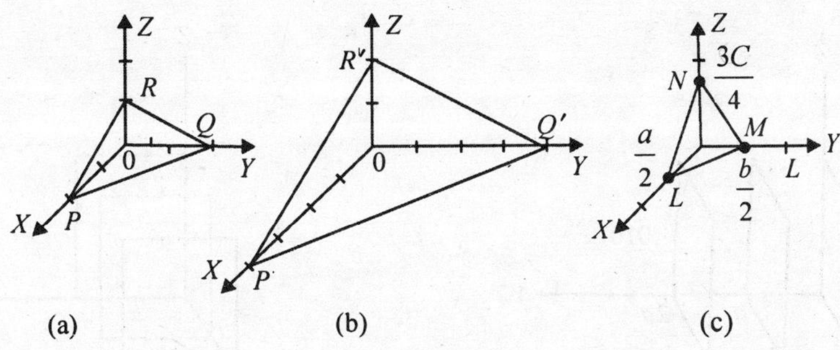

Fig. 6.28

Hence, Miller index of a crystal face or plane is given by three smallest figures which are inversely proportional to the numerical parameters of the face.

6.18.1 General Case

Let the intercepts of a crystal face on the three reference axes as expressed in terms of the axial units be *am*, *bn* and *co*. Obviously, its numerical parameters are *a*, *b* and *c*. Then Miller indices of the face are given by three numbers *h*, *k*, *l* such that

$$h : k : l :: \frac{1}{a} : \frac{1}{b} : \frac{1}{c}$$

with the condition that *h*, *k* and *l* are the smallest possible integers satisfying the above equation.

Note: If a plane has Miller indices of $(h\ k\ l)$, then its intercepts on the three axes are p/h, q/k and r/l respectively.

6.19 MILLER INDICES OF CUBIC CRYSTAL PLANES

The following are the points that should be kept in mind while finding Miller indices of a plane.

(*i*) When a plane is parallel to one of the coordinate axes, it is said to meet that axis at infinity. Since $1/\infty = 0$, the Miller index for that axis is zero.

(*ii*) When the intercept of a plane is on the negative part of any axis, the Miller index is distinguished by a minus sign.

Example: Let us consider the shaded planes of Fig. 6.24. The face plane *NRSO* of Fig. 6.24 (a) has Miller indices of (100). It is so because it has an intercept of unity on *x*-axis but is parallel to both *y* and *z*-axis. Since the shaded plane MRSP of Fig. 6.24 (b) intercepts *x* and *z*-axes but not the *y*-axis, it has indices of (101). The plane *PRT* of Fig. 6.24 (c) has equal intercepts on the three axes, hence it has indices of (111).

The set of parallel planes perpendicular to *y*-axis are as shown in Fig. 6.29 (a) has Miller indices (010). Similarly, planes normal to *x*-axis are designated as (100) and are shown in Fig. 6.29 (b).

Planes having equal intercepts on *x* and *y*-axis and parallel to *z*-axis have Miller indices (110) and are shown in Fig. 6.30 (a). The planes of Fig. 6.30 (b) which have equal intercepts on the three axes are (111) planes.

Figure 6.31 (a) shows a few planes which are designated as (112) and the Fig. 6.31 (b) shows a family of planes in which there is negative intercept on the *x*-axis. Their Miller indices are (110).

6.20 PROCEDURE FOR FINDING MILLER INDICES

Following are the steps to find out the Miller indices of the given plane.

(*i*) First, find the intercepts *x*, *y* and *z* of the plane on the three reference axes.

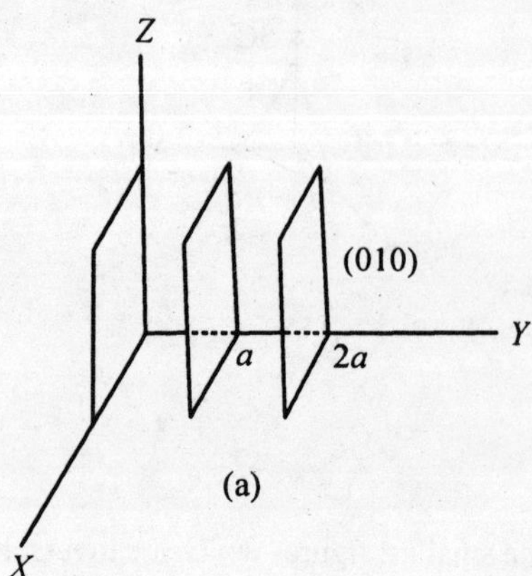

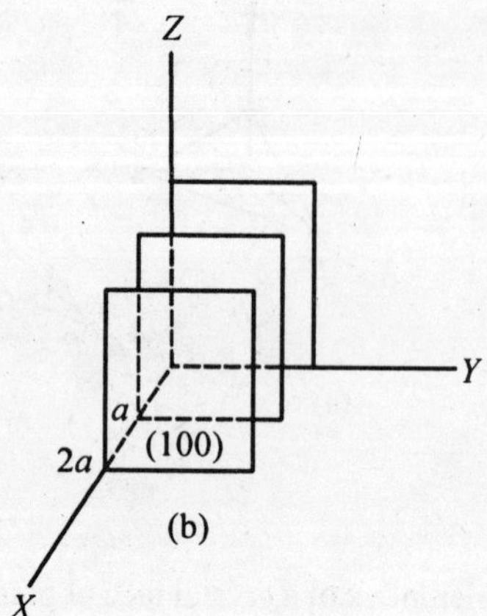

Fig. 6.29

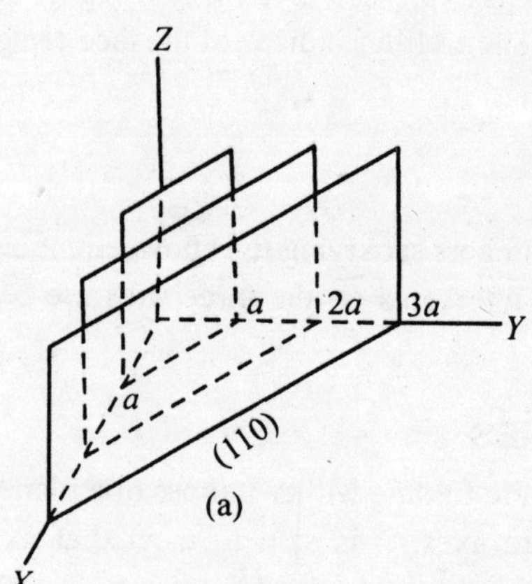

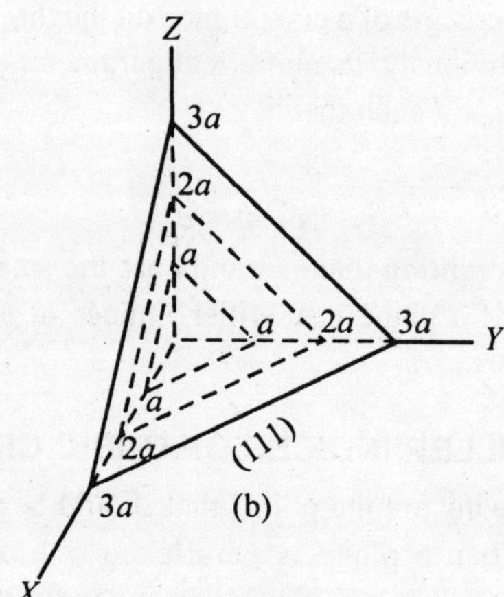

Fig. 6.30

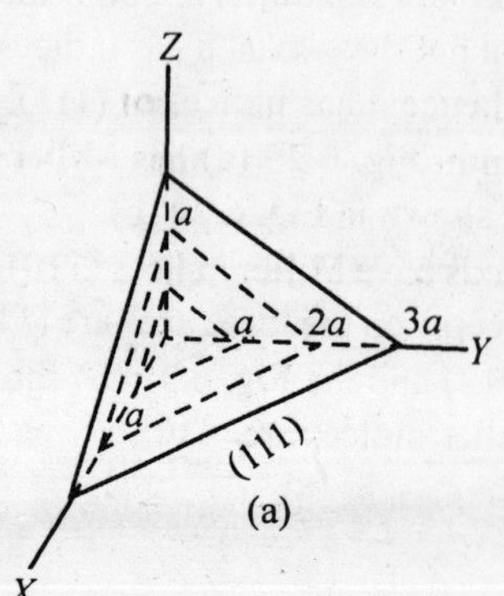

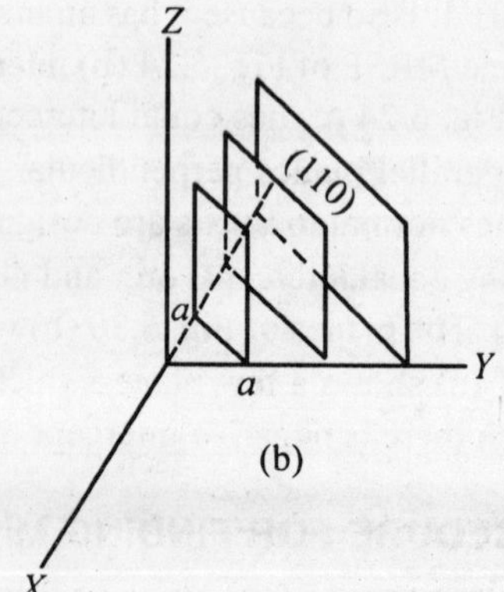

Fig. 6. 31

(*ii*) Then express them in terms of the axial units. Let them be $x = 4$ m, $y = 2n$ and $z = 0$ (or in general), $x = am$, $y = bn$ and $= co$).

(*iii*) Next, take the numerical parameters of the plane, *i.e.* 4 : 2 : 1 ($a : b : c$) and find the ratio of their reciprocals $\left(\frac{1}{4} : \frac{1}{2} : \frac{1}{1}\right)$ or $\left(\frac{1}{a} : \frac{1}{b} : \frac{1}{c}\right)$.

(*iv*) Convert these reciprocals into whole numbers by multiplying each with their LCM.
In the present case, LCM = 4. Hence, reciprocals when converted to whole numbers become (1 : 2 : 4).

(*v*) The Miller indices of the plane are (124).

6.21 RELATION BETWEEN INTERPLANAR SPACING *D* AND CUBE EDGE *A*

Suppose that the plane shown in Fig. 6.32 belongs to a family of planes whose Miller indices are <*hkl*>. The perpendicular *ON* from origin to the plane represents the interplanar spacing *d* of this family of planes. Let the direction cosines of *ON* be cos α, cos β and cos γ as shown. The intercepts of the plane on the three axes are

$$OP = a/h \,;\; OQ = a/k \text{ and } OR = a/l$$

where a is the length of the cube edge.

$$\cos \alpha' = \frac{d}{OP} = \frac{d}{a/h} = \frac{dh}{a}$$

$$\cos \beta' = \frac{d}{OQ} = \frac{d}{a/k} = \frac{dh}{a}$$

$$\cos \gamma' = \frac{d}{OR} = \frac{d}{a/l} = \frac{dh}{a}$$

where $d = ON$ is the perpendicular distance between adjacent members of the same family of planes.

Now, $\cos^2 \alpha' + \cos^2 \beta' + \cos^2 \gamma' = 1$

$$\therefore \quad \left(\frac{dh}{a}\right)^2 + \left(\frac{dk}{a}\right)^2 + \left(\frac{dl}{a}\right)^2 = 1$$

$$\therefore \quad \frac{d^2}{a^2}\left(h^2 + k^2 + l^2\right) = 1$$

$$d^2 = \frac{a^2}{h^2 + k^2 + l^2}$$

or

$$d = \frac{a^2}{\sqrt{h^2 + k^2 + l^2}}$$

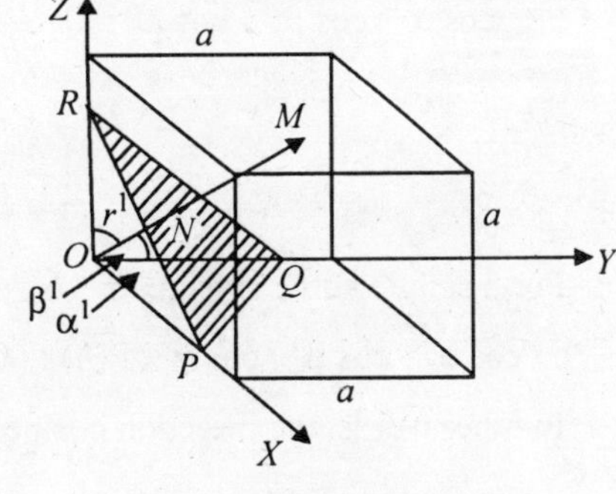

Fig. 6.32

This is the expression for the interplanes spacings *d* and the cube edge *a*.

6.22 IMPORTANT FEATURES OF MILLER INDICES

The following are the important features of the scheme of Miller indices particularly for the cubic system

(*i*) all equally-spaced parallel planes have the same index numbers ($h\ k\ l$).

(*ii*) a plane parallel to one of the coordinate axes has an intercept of infinity and index of zero for that axis.

(*iii*) Miller indices do not define a particular plane but a set of parallel planes.

(*iv*) it is only the ratio of the indices which is of importance, *i.e.* the (422) planes are the same as (211) planes.

(*v*) a plane passing through the origin is defined in terms of a parallel plane having non-zero intercepts.

The Miller index notation is especially useful for cubic system of crystals. Some of its desirable features are

(*a*) The angle θ between two direction (u_1, v_1, w_1) and (u_2, v_2, w_2) can be easily calculated from the expression.

$$\cos\theta = \frac{u_1u_2 + v_1v_2 + w_1w_2}{\left(u_1^2 + v_1^2 + w_1^2\right)^{1/2}\left(u_2^2 + v_2^2 + w_2^2\right)^{1/2}}$$

(*b*) The normal to the plane with index number $(h\,k\,l)$ is the direction $(h\,k\,l)$.

This fact can be utilized to find the angle between two planes by using the above relation.

(*c*) The distance d between adjacent planes of a set of parallel planes of the indices $(h\,k\,l)$ is given by

$$d = \frac{a}{\left(h^2 + k^2 + l^2\right)} \quad \text{where } a \text{ is the edge of the cube.}$$

In general, it may be noted that planes with low index numbers have wide interplanar spacing compared with those having high index numbers. Moreover, low index planes have a higher density of atoms per unit area than the high index planes. In fact, it is the low index planes which play an important role in determining the physical and chemical properties of solids.

6.23 CRYSTAL PLANE AND ITS NORMAL

In the Fig. 6.32, let the linear parameters of the shaded plane be pa, qb and rc respectively. The direction cosines of the normal ON are $\cos\alpha'$, $\cos\beta'$ and $\cos\gamma'$. The analytical relation between the axial intercepts of a plane and direction cosines of its normal is represented as

$$\cos\alpha' : \cos\beta' : \cos\lambda' :: \left(\frac{1}{pa}\right) : \left(\frac{1}{qb}\right) : \left(\frac{1}{rc}\right)$$

$$:: \left(\frac{h}{a}\right) : \left(\frac{k}{b}\right) : \left(\frac{l}{c}\right)$$

For a simple cubic crystal, $a = b = c$

$\therefore \cos\alpha' : \cos\beta' : \cos\gamma' :: (h) : (k) : (l)$

In other words, the direction cosines of a normal to a plane are the same as the Miller indices of the plane

UNSOLVED NUMERICAL PROBLEMS

1. Classify the following unit cells into proper system

(*i*) $a = 10.8$ Å, $b = 9.47$ Å, $c = 5.2$ Å

$\alpha = 41°$, $\beta = 83°$ and $\gamma = 93°$

(*ii*) $\alpha = b = 10.73$ Å, $c = 14.3$ Å

$\alpha = \beta = 90°$ and $\lambda = 120°$

Here, $a = b \neq c$ and $\alpha = \beta = 90°$ and $\gamma = 120°$. This unit cell belongs to *hexagonal* crystal system.

2. Calculate the lattice constant of a NaCl crystal (FCC structure) if its molecular weight and density are 58.46 and 2.167 g/cm^3 respectively. Also calculate the spacing between the 110 planes of NaCl crystal.

3. If density of NaCl is 2.163 g/cm^3 and its molecular weight 58.45, find the spacing betweenplanes parallel to the cubic lattie faces of the NaCl crystal. Given Avogadro's number is 6.02×10^{23} molecules/g-mol.
4. Calculate the density of copper. Given that copper has FCC structure, atomic weight = 632.54 and lattice constant 3.61 Å.
5. Obtains the Miller indices of a plane which intercepts at a, b | 2, $3c$ in a simple cubic unit cell. (361)
6. A certain orthorhombic crystal has axial units $a : b : c$ are 0.424 : 1 : 0.367. Find the Miller indices of crystal faces whose intercepts are :

   (*i*) 0.212 : 1 : 0.183

   (*ii*) 0.848 : 1 : 0.732

   (*iii*)0.424 : ∞ : 0.123
7. Sketch the (112) and (111) planes in a simple cubic cell.

$\frac{1}{1}, \frac{1}{1}, \frac{1}{1}$. This plane is shown in Fig. 6.33 (b).

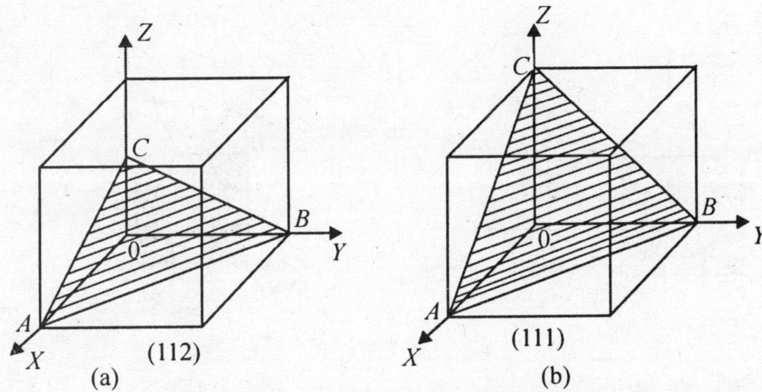

Fig. 6.33

8. Draw the different (111) planes in the unit cell of a simple cubic structure.
9. Lead is face-centred cubic with an atomic radius of r = 1.746 AU. Find the spacing of (*i*) (200) planes (*ii*) (220) planes and (*iii*) (111) planes.
10. In a unit cell of simple cubic structure, find the angle between the normals to pair of planes whose Miller indices are (*i*) (100) and (010) (*ii*) (121) and (111).
11. Sketch all the (110) planes which pass through an FCC unit cell.
12. How many atoms per mm^2 surface area are there in (*i*) (100) plane (*ii*) (110) plane and (*iii*) (111) plane for copper which has FCC structure and a lattice constant $a = 3.61 \times 10^{-10}$ m.
13. Draw the (110) and (111) planes and the (110) and (111) directions in a simple cubic crystal. What do you infer from these diagrams?
14. The atomic weight of silver is 108. It has a lattice constant of 4.077 Å. Calculate the density of silver if it has fcc lattice structure.
15. A KCl crystal which has FCC lattice structure has a density of 1.98×10^3 kg/m^3. Its molecular weight is 74.55. Find the distance between adjacent atoms.
16. Potassium chloride (KCl) is a face-centred cubic crystal having a density of 1.98 g/cm^3. If its molecular weight is 74.6, calculate (*i*) the distance from one atom to the next atom of the same kind and (*ii*) the distance between adjacent atoms.

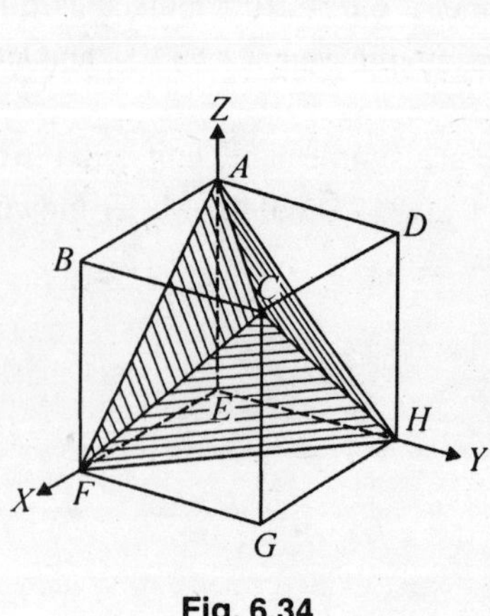

Fig. 6.34

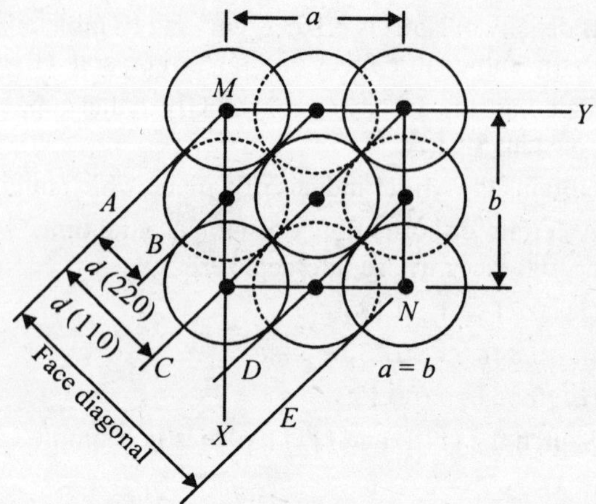

Fig. 6.35

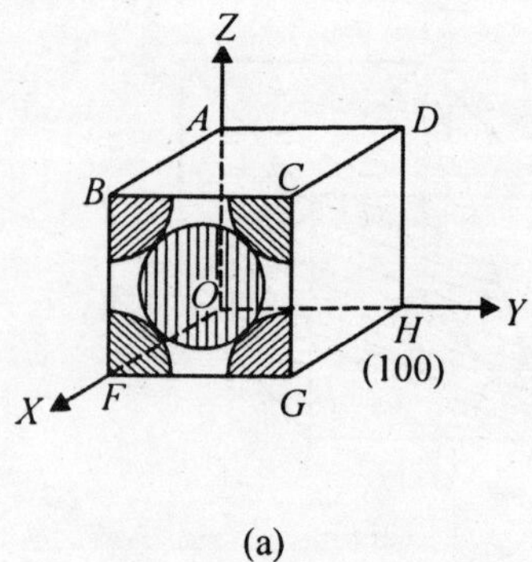

(a)

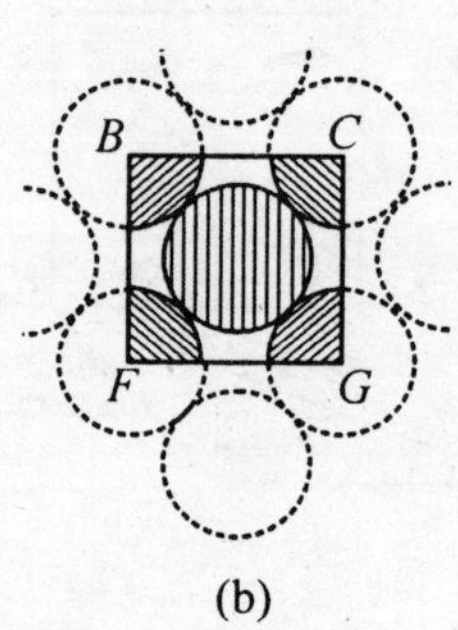

(b)

Fig. 6.36

(a)

(b)

Fig. 6.37

17. Nickel has face-centred cubic structure and its atomic radius is 1.243×10^{-10} m. Calculate the interplanar spacing for (*i*) (200) planes, (*ii*) (220) planes, and (*iii*) (111) planes.

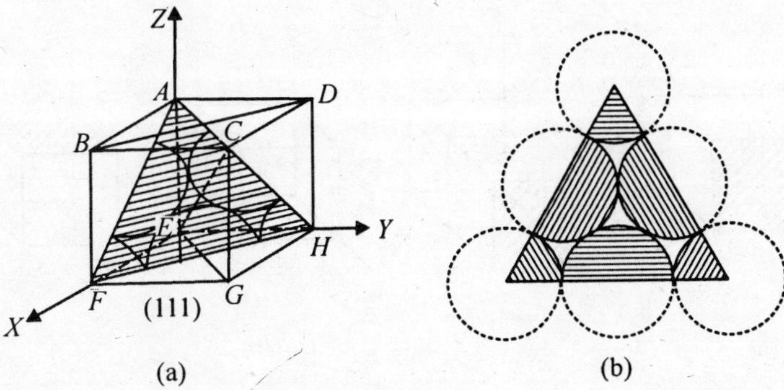

Fig. 6.38

18. The lattice constant of a unit cell of lead is 4.93 AU. Find the number of atoms/mm^2 of (*a*) (100) plane, (*b*) (111) plane if lead has FCC structure.
19. Draw the (111) planes of Cu which has FCC structure and a lattice constant of 3.61 AU. Calculate (*a*) the number of atoms/mm^2, and (*b*) inter-planar spacing.
20. Silver has FCC structure and its atomic radius is 1.441 Å. Find the spacing of the (220) planes.
21. Compute the Miller indices of the two planes as shown in Fig. 6.39.

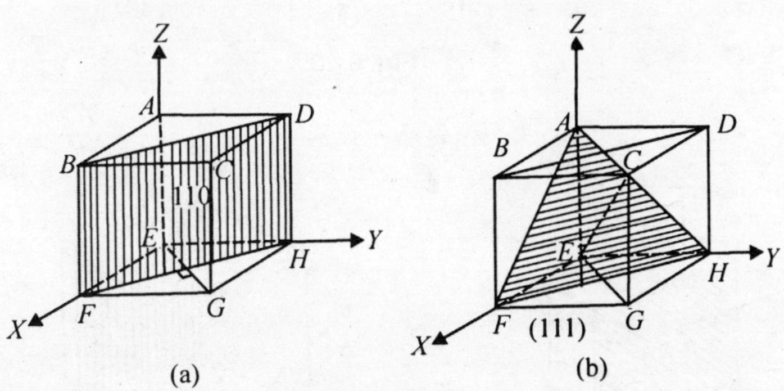

Fig. 6.39

22. Find the Miller indices of planes shown in Fig. 6.40.
[(*a*) (100), (*b*) (100), (*c*) (010), (*d*) (001), (*e*) (110), (*f*) (111), (*g*) (212)]
23. Prove that the volume V of the unit cell for the seven crystal systems is given by the following formulae:

(*i*) Triclinic system $V = abc\sqrt{1 - \cos^2\alpha - \cos^2\beta - \cos^2 + \gamma + 2\cos\alpha\cos\beta\cos\gamma}$

(*ii*) Monoclinic system $V = abc \sin\beta$

(*iii*) Orthorhombic system $V = abc$

(*iv*) Tetragonal system $V = a^2c$

(*v*) Cubic system $V = a^2$

(*vi*) Hexagonal system $V = \frac{\sqrt{3}}{2}a^2c$

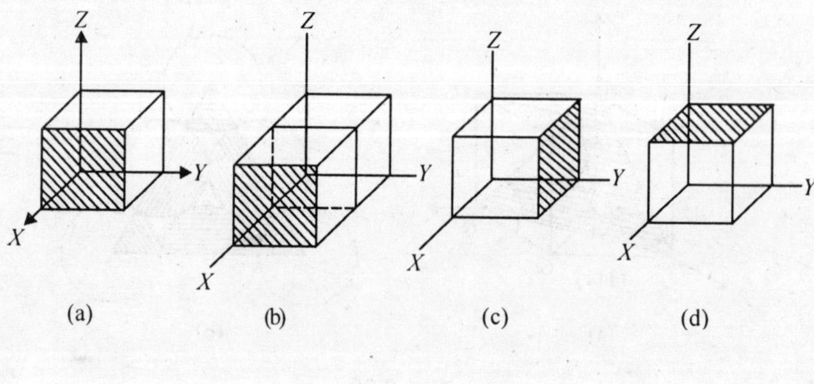

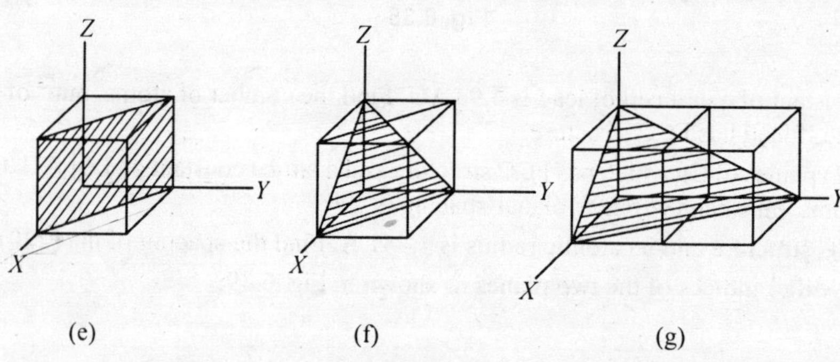

Fig. 6.40

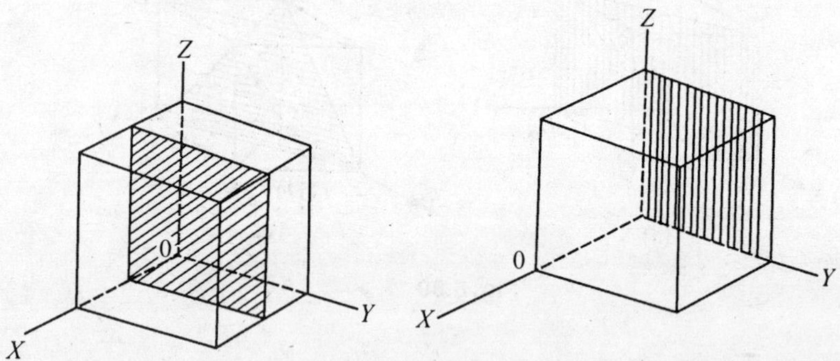

Fig 6.41

(*vii*) Rhombohedral system $V = a^3 \sqrt{1 - 3\cos^2 a + 2\cos^3 a}$

24. If the density of NaCl is 2.163 g/cm^2 and its molecular weight is 58.45, find the spacing between planes parallel to the cubic lattice faces of NaCl crystal. Given that Avogardro's number is 6.02×10^{23} molecules/g-mole.

25. Calculate the lattice constant of rocksalt (NaCl) crystal from the following data :

Molecular weight of NaCl = 58.45

Density of NaCl = 2170 kg/m^2

Avogadro number = 6×10^{26} per kg mole

26. Calculate the lattice constant for potassium bromide. Density of potassium bromide is 2700 kg/m^2 and belongs to FCC lattice. Molecular weight of potassium bromimde is 119.0 and Avogadro number is 6.02×10^{26} $(kg \cdot mole)^{-1}$
27. Obtain the Miller indices of a plane which intercepts at a, $b/2$, $3c$ in a simple cube unit cell.
28. A certain orthorhombic crystal has axial units $a : b : c$ of 0.424 : 1 : 0.367. Find the Miller indices of crystal faces whose intercepts are
    (*i*) 0.212 : 1 : 0.183
    (*ii*) 0.424 : ∞ : 0.123
29. For a simple cubic lattice of the lattice parameter 2.04 Å, calculate the spacing of lattice planes (212).
30. Deduce the spacing between (112) planes in a cubic crystal of lattice constant a.
31. Calculate the interplaner spacing for a (321) plane in a simple cubic lattice whose lattice constant is 4.2×10^{-8} cm.
32. In a simple cubic crystal, find the ratio of
    (*i*) spacing of the (110) and (111) planes.
    (*ii*) nearest neighbour distance to the next nearest neighbour distance.
33. In a crystal whose primitives are 1.2 Å, 1.8 Å and 2 Å, a plane (231) cuts an intercept 1.2 Å along the x-axis. Find the length of intercepts along y and z axes.

EXERCISE

1. Define the terms single crystal, unit cell, space lattice and Miller indices.
2. Derive the relation for the number of atoms per unit cell for cubic type of crystal and hence estimate the total number of atoms per unit cell of simple cubic, body-centred cubic and face-centred cubic.
3. Draw a diagram to indicate the structure of a sodium chloride crystal and with this explain the terms unit cell and lattice parameters of the unit cell.
4. Draw the different (111) planes in the unit cell of a simple cubiç structure.
5. Draw neat diagrams to indicate the important plane systems in a simple cubic crystal and find a relation between the spacings of the planes in the indicated systems.
6. What are Miller indices ? Indicate what should be the Miller indices for a typical plane belonging to each system.
7. Explain the following terms as applied to crystals: (*i*) space lattice, (*ii*) primitives, and (*iii*) coordination number.
8. Find the coordination numbers for *BCC* and the *FCC* structures.
9. Write a short note on Miller indices.
10. What is a primitive cell of a 2 dimensional lattice?
11. Draw the unit cell for a simple cubic structure and show the four (111) planes in it. How many atoms effectively contribute to this unit cell?
12. Draw the (110) and (111) planes and the (110) and (111) directions in a simple cubic crystal. What do you infer from these diagrams? Then explain the use of Miller indices notation in crystal studies.
13. What are Bravais' lattices and Miller indices in crystallographic notations? Show analytically that the (111) planes are perpendicular to the (111) direction in a simple cubic crystal.
14. With reference to crystals, state what you understand by the terms (*i*) unit cell, (*ii*) coordination number, (*iii*) packing factor.

15. Explain the term symmetry in crystal structure. Qualitatively or otherwise show that a five-fold symmetry is not possible in crystal structures.
16. Show that for BCC and FCC crystal structures, the lattice constanats are given by

$$a_{bcc} = 4r/\sqrt{3} \quad \text{and} \quad a_{fcc} = 4r/\sqrt{2} \text{ where } r \text{ is atomic radius.}$$

17. Define crystal lattice, basis and crystal structure.
18. What do you mean by a crystal lattice? With the help of an example distinguish between crystal lattice and crystal structure.
19. What do you understand by space lattice? Describe briefly the seven systems of crystals. Mention and explain with examples the types of lattices in cubic system.
20. What is coordination number? Calculate the coordination number for simple cubic, *BCC* and *FCC* lattices.
21. Derive the relation for the number of atoms per unit cell for cubic type of crystal and hence estimate the total number of atoms per unit cell of simple cubic, body centred cubic and face centred cubic.
22. What do you understand by packing density or packing factor? Show that the packing factor for simple lattice, body centred lattice and face centred lattice are $\pi/6$, $\sqrt{3\pi}/8$ and $\sqrt{2\pi}/6$ respectively.
23. Define atomic radius. Calculate the atomic radii in case of simple cubic, *FCC* and *BCC* lattices.
24. Explain the various types of bondings in crystals. Illustrate with examples.
25. Write a short note on ionic and covalent bonding.
26. Distinguish between covalent and ionic bonds in solids. Illustrate with examples.
27. What do you understand by Miller indices of a crystal plane ? Show that the spacing between consecutive planes defined by Miller indices (hkl) is given by,

$$d_{nkl} = \left(\frac{h^2}{a^2} + \frac{k^2}{b^2} + \frac{l^2}{c^2}\right)^{\frac{1}{2}}$$

where *a*, *b* and *c* are primitives along three mutually perpendicular axes.
28. For body centred cubic lattice, find the number of nearest neighbours, the nearest neighbours distance and the spacing of (110) planes.
29. What are Bravais lattice and Miller indice in crystallographic notations? Show analytically that the (111) planes are perpendicular to (111) direction in a simple cubic crystal.

7 Materials Science

7.1 IMAGING LITHIUM ATOMS

For the first time researchers have used a transmission electron microscope—the one angstrom microscope (OÅM) at the Department of Energy's National Center for Electron Microscopy (NCEM) at Lawrence Berkeley National Laboratory—to image lithium atoms. Only atoms of hydrogen and helium are smaller and lighter than those of lithium, which under ordinary conditions is not a gas but a soft, white metal.

Yang Shao-Horn of the Department of Mechanical Engineering at the Massachusetts Institute of Technology and Michael O'Keefe of Berkeley Lab's Materials Sciences Division used the OÅM to simultaneously resolve columns of lithium, cobalt, and oxygen atoms in the compound lithium cobalt oxide ($LiCoO_2$). They and their colleagues report their work in the July, 2003 issue of the journal nature materials, in a report currently accessible online.

$LiCoO_2$ is commonly used in the positive electrodes of lithium rechargeable batteries, whose operation is based on reversible insertion and removal of lithium ions to and from their positive and negative electrodes. Widely used in laptop computers, digital cameras, and many other devices, lithium ion batteries store more energy for their weight, operate at a higher voltage, and hold a charge much longer than other rechargeable batteries.

To improve their performance will require understanding how the atoms in the electrode materials—and the vacancies left by moving ions—are arranged in 3-D on the atomic scale. The structure of $LiCoO_2$ is known theoretically and has been confirmed with techniques like X-ray diffraction and neutron powder diffraction: layers of lithium atoms lie between slabs of cobalt and oxygen, which are arranged in octahedrons. But lithium ions have never been seen by these techniques, nor have they been seen in previous attempts to image $LiCoO_2$ with electron microscopy.

Thus, when Shao-Horn, who had been investigating lithium ion batteries with colleagues at the University of Bordeaux I, approached microscopist O'Keefe in the summer of 2001 to ask whether it would be possible to image lithium ions in the material with a high-resolution transmission electron microscope, his first answer was, 'I don't think so'.

7.2 HOW TRANSMISSION ELECTRON MICROSCOPY WORKS?

A transmission electron microscope (TEM) sends a beam of electrons through a thin sample of material; as the beam scatters from the electrical field of the atomic nuclei and their surrounding clouds of electrons, their unique arrangement affects the phase of the beam and to some extent its amplitude. When the altered beam exits the surface of a precisely oriented sample, it can be focused by an electromagnetic lens to project an image of the sample's columns of atoms.

Because atoms with small dimensions and very little mass barely affect the electron beam, they are difficult to resolve—a problem that worsens if there are heavier atoms nearby. Heavy cobalt, with atomic number 27 and atomic mass approaching 60, is relatively easy to image, but light oxygen, with atomic number 8 and atomic mass about 16, scatters electrons weakly. Lithium is smaller still, its atomic number only 3, its atomic mass only 7.

A TEM's ability to image these wispy particles depends on many factors including the microscope beam's energy, energy spread, and steadiness, and the distortion, or aberration, of the lens. All these combine to determine the smallest distance a microscope can distinguish between two adjacent objects, its native resolution. NCEM's one angstrom microscope is a medium-energy TEM with a native resolution of 1.6 angstroms (an angstrom, symbolized Å, is a ten-billionth of a meter), good enough to resolve cobalt atoms directly.

For the highest possible resolution it is necessary to go beyond native resolution to a microscope's information limit—the maximum amount of information about the sample that can be extracted from the scattered electron wave, even those portions of it that may be out of phase. One method of achieving this, called focal-series reconstruction, uses a computer to combine successive images, each made at a slightly different focus. In this way the one angstrom microscope has achieved a resolution as high as 0.78 angstrom. But there's a catch. Because the separate images in the series are focused differently, in-phase images of atoms in one image will be out of phase in another. How is the computer to know whether bright spots, dim spots, or no spots at all belong at particular locations in the composite image? How does the thickness of the sample affect the outcome?

7.3 THE RIGHT NUMBER OF BLOBS

To answer these questions, researchers use two kinds of computer programs. With the first, they can create a high-resolution simulation of what the microscope ought to see—'the right number of blobs for the right number of atoms', as O'Keefe puts it. The simulation program starts with a model of the material's crystal structure, then dials in atomic specifications, the thickness and orientation of the sample, such parameters as the energy of the microscope's electron beam, lens aberrations, beam misalignments, and other characteristics of both the sample and the instrument. Out comes a series of simulated images. When O'Keefe, who wrote the first such image simulation program in the late 1970 s, applied a recent version to the theoretical structure of $LiCoO_2$, the simulation provided such essential information as what resolution the microscope had to achieve to see anything at all, and at what resolution all the different kinds of atoms should be sharpest. His initial skepticism about whether lithium atoms could be imaged began to wane.

For the simulation showed that in a sample of the right thickness, lithium ions should become visible at 1 Å. And at 0.8 Å all three kinds of atoms should be clearly visible — a resolution that was, indeed, just possible to reach with the One Angstrom Microscope. O'Keefe says, 'The simulation showed that the columns of oxygen atoms should look bright and sharp-edged, the cobalts should be fuzzy, and the lithium should be small, weak, and look a little stretched'.

In 2002, working with NCEM's Chris Nelson to master the operation of the OÅM, Shao-Horn obtained many series of 20 differently focused images of individual crystals from a $LiCoO_2$ powder sample. The sample was prepared and well characterized by conventional X-ray diffraction in collaboration with her colleagues Laurence Croguennec and Claude Delmas, of the Institute de Chimie de la Matière Condensée de Bordeaux-CNRS and the Ecole National Supèrieure de Chimie et Physique de Bordeaux at the University of Bordeaux I. With a second computer program and their measurements of the microscope parameters, researchers take the experimental images and work backwards to produce a representation of the electron wave leaving the exit surface of the specimen. After confirming the focal value for each image in the series, Shao-Horn and her colleagues used the reconstruction program to produce an image from a small area of a thin edge of a $LiCoO_2$ crystal which matched the one predicted by the simulation program.

'Because material from which some lithium ions are removed is somewhat unstable under the electron beam, experimental imaging of the lithium ions and vacancies proved difficult in this study', says Shao-Horn. 'Nevertheless, the atomic resolution of lithium atoms is a novel and significant achievement, with implications for better understanding not only of lithium ion battery materials but of many other electroceramic materials as well. Says O'Keefe, 'We have shown that the range of the OÅM, and mid-voltage microscopes like it, can be extended all the way from heavy atoms down through oxygen, nitrogen, and carbon to the lightest metal — in fact, except for helium and hydrogen, the lightest atoms of all'.

'Atomic resolution of lithium ions in $LiCoO_2$', by Yang Shao-Horn, Laurence Croguennec, Claude Delmas, E. Chris Nelson, and Michael O'Keefe, appears in the July, 2003 issue of nature materials, and is available online at this web page. Berkeley Lab is a US department of energy national laboratory located in Berkeley, California. It conducts unclassified scientific research and is managed by the University of California.

7.4 WORLD'S FIRST TUNABLE 'PHOTON COPIER' ON A CHIP

A research team at the University of California at Santa Barbara (UCSB) has for the first time incorporated on a single chip both a widely tunable laser and an all-optical wavelength converter, thereby creating an integrated photonic circuit for transcribing data from one colour of light to another. Such a device is key to realizing an all-optical network. This research is being funded by a Defense Advanced Research Projects Agency (DARPA) Microsystems Technology Office (MTO) grant to push the boundary for photonic-circuit functional integration. Think about data transmission over the internet in terms of a telephone call between Los Angeles and New York. What enables two people to talk is the 'dedicated' line between them. On the Internet the long-distance transport of information between the two cities is via optical fibers or light pipes, which can move numerous colours of light over a single fiber at the same time with each colour representing a 'dedicated' line for the transmission of data (*i.e.* wavelength division multiplexing (WDM). Data moves between coasts through nodes of the internet located in cities like Phoenix or Houston, where the capability is needed to switch information arriving on one fiber as orange photons to continue the next leg of their journey on another fiber as red photons because the channel for orange on that fiber is in use. Today, this switching from one colour to another has to be done by converting photons to electrons, switching electronically, and converting electrons back to photons.

The new postage-stamp-size device is a tunable 'photon copier', which eliminates electronics as the middleman. It is as if information on orange paper were being copied onto red paper. The information stays the same, but the colour of the paper on which it is conveyed is different. By that analogy the tunable laser supplies the red paper, and the wavelength converter functions as the copy-machine transcribing the information of the orange original to red. The tunable laser is able to supply a wide range of colours and hues to copy onto, and the wavelength converter is able to maintain or improve the quality of the image, a process called 'regeneration'.

Past attempts to engineer photonic circuits with tunable lasers and wavelength converters have met with limited success, and the two components have heretofore lived on separate chips. Fabricating them on the same indium phosphide platform and thereby integrating them represents a technological breakthrough and greatly enhances performance and commercial feasibility. Integration provides many benefits, which include, notably, maintaining the quality of the signal by avoiding the taking of light on and off the chip many times. In electronics, one semiconductor component incorporating two capabilities typically costs much less than half of what two separate components, each supplying one of those capabilities, costs. Photonics requires breakthroughs to follow these same laws of integration.

Daniel Blumenthal, leader of the research group and a UCSB professor of electrical and computer engineering, figures this successful demonstration of two capabilities on one photonic chip means more to come. 'People are now accustomed', he said, to think of a silicon substrate with a seemingly innumerable number of transistors on it. But that sophisticated electronic technology of today began back in the 1960 s with a couple of transistors, and photonics is just beginning to enter that stage. We have been working long and hard at cracking this problem, and we are finally at the demonstration phase. UCSB graduate student Milan Masanovic presented the findings (in a paper entitled 'demonstration of monolithically-integrated InP Widely-Tunable Laser and SOA-MZI Wavelength Converter') at the 15th annual meeting of the Indium Phosphide and Related Materials Conference, held in May in Santa Barbara. That presentation earned him the conference Best Student Paper Award.

The paper states, 'The monolithic integration of tunable lasers and all-optical wavelength converters is a critical step towards realizing truly optical switches and networks'. These structures allow data to be imprinted from an input wavelength to a tunable output wavelength without passing the signal through electronics. Masanovic is a student jointly of Blumenthal and the paper's other faculty author and research team co-leader, Larry Coldren, who is the inventor of the tunable laser used in this device, which is called a Sampled-Grating Distributed-Bragg-Reflector laser (SGDBR). A 'tunable' laser enables the dialing up of different colours (*i.e.* a single miniscule device can be programmed to emit at one frequency and then reprogrammed to emit at another by electronically changing the effective optical length of its cavity and the shape of its internal colour filter).

The all-optical wavelength converter is an SOA-based Mach-Zehnder interferometer (SOA-MZI), which the paper's authors note 'also implements the significant feature of digital signal regeneration'. Digital regeneration is key to moving optical signals around a network with many nodes and long stretches of fiber in between and is a feature that electronics uses pervasively. The other three authors on the paper are Blumenthal's graduate student Vikrant Lal and Coldren's students Erik Skogen and Jonathon Barton.

The research is being conducted under the auspices of a four-year, $ 3.5 million DARPA grant through the microsystems technology office (MTO) Chip-Scale WDM program to investigate how to pipe digital and analog information through a photonic circuit on a single compound semiconductor chip; Dr. Jagdeep Shah is the program manager. Blumenthal, who serves as associate director of the UCSB multidisciplinary optical switching technology (MOST) center, is the principal investigator for the grant. The other four UCSB researchers are all professors of electrical and computer engineering: Coldren, director of the optoelectronics technology center (OTC); John Bowers, director of MOST; Evelyn Hu, acting director of the California Nano Systems Institute (CNSI); and Nandi Dagli.

7.5 ELECTRONICS/LIVING CELL TOXIC SENSOR

In experiments conducted at the University of California, Berkeley, researchers have found a way to tap into the telltale electrical signals that mark cell death, opening the door to the creation of a "canary on a chip" that can be used to sound the alarm of a biochemical attack or test drug toxicity on human tissue. In a study appearing in the June 15 issue of sensors and actuators, researchers used a microchip to electrically determine cell viability by detecting changes in the electrical resistance of a cell membrane within milliseconds after it is exposed to a toxic agent. They found that after a cell is exposed to a toxin, its electrical resistance experiences a quick spike before dropping dramatically when it dies. "The beauty of the device is that it detects the viability of a cell directly and instantaneously", said Boris Rubinsky, professor of mechanical engineering and bioengineering at UC Berkeley and co-author of the study, which is now available online. "This MEMS (micro-electromechanical) device will be invaluable in the detection of a biochemical attack because there you don't have the luxury of time and analysis. It's a new technology that will act like a canary on a chip".

The study is a continuation of the bionic chip research, part of UC Berkeley's Center for Information Technology Research in the Interest of Society, pioneered by Rubinsky and his former graduate student, Yong Huang, who received his PhD in mechanical engineering in 2001. Three years ago, Rubinsky and Huang invented a chip that merges a living biological cell with electronic circuitry. The bionic chip also exemplifies the type of advances made possible through the campus's Health Sciences Initiative, which combines research in diverse disciplines to launch innovations in health-related fields. The chip has since been patented by UC Berkeley and exclusively licensed to Excellin Life Sciences, a Milpitas-based biotech startup.

"This would benefit applications that rely upon the knowledge of whether a cell is alive or dead", said Huang, lead author of the current study and now the chief science officer of Excellin. "It has implications for use as a biochemical sensor that soldiers could wear on a nametag or that is stationed as a remote sensor in the field. Because this is a MEMS-based single-cell analysis technology, it's primed for miniaturization".

In this year's new design of the bionic chip, a cell essentially lives in a nutrient-rich culture between two electrodes on a silicon wafer. The electrodes continuously probe the cell membrane and track the amount of electrical current that is able to pass. An intact membrane will not allow ions to pass through, except in very constrained conditions.

"One of the properties of a dead cell is that the membrane becomes leaky", said Rubinsky. "It occurred to us that when the cell membrane breaks, it would allow various ions to move through the cell membrane and make a circuit that can be easily measured".

This change in membrane permeability forms the basis for traditional cell viability assays that use colourimetric or fluorescent dyes. The molecules of the dye can only pass through the membrane of dead cells. Drug researchers investigating a chemical's toxicity to cancer cells, for example, would look to see whether the dead cells appeared fluorescent through a microscope. The researchers used the traditional assays to help calibrate the bionic chip.

While the bionic chip has potential as a biochemical warning device, its possible use as a replacement for the traditional fluorescent dye assay may have more immediate impact, said Huang. Excellin is pursuing the use of the bionic chip as a tool for basic research on cell death and as a drug toxicity assay. Huang pointed out that the fluorescent dye assay only tells whether cells are alive or dead, shedding little light on what happens to the cell as it dies. The dyes also lack the sensitivity to measure what happens in a single cell. "With the bionic chip, the cell produces a dynamic electrical signature so you can monitor the cell death process in real time", said Huang. "This has not been done before, particularly on a single cell level". The researchers say the amount of cell membrane damage relates to the degree of ionic current, so the larger the damage, the greater the change in electrical resistance.

"*Not only can you tell when a cell dies or not, you know to what degree there is membrane impairment by comparing the electric signals*", *said Huang.* "*That's very difficult to do without our technology*".

Using electrical resistance to measure cell death is also more efficient than using fluorescent dyes, the researchers said. "The traditional protocol requires the addition of a dye that has been prepared and incubated, it requires that the dye gets into the cell, and it requires a measurement that is lengthy and complicated", said Rubinsky. "With the bionic chip, you don't need to add anything, and reactions happen very fast. You simply need two electrodes that measure the current as opposed to some expensive dyes and expensive measurement devices. It's orders of magnitude faster. It's a matter of milliseconds or seconds compared with half an hour or more". Other co-authors of the paper are Navdeep Sekhon and Ning Chen, former students at UC Berkeley's Department of Mechanical Engineering; and James Borninski, vice president of product development at Excellin.

7.6 CRACKING THE NANONEWTON FORCE BARRIER

How do you weigh a dust mite? Or determine the force required to pull a molecule apart? Such tasks require a device that measures nanonewtons—forces 1 billion times smaller than the force required to hold an apple against earth's gravity. Nanonewton forces are estimated with atomic force microscopes and instruments that measure the properties of ultrathin coatings like those used on computer hard drives or turbine blades. But the accuracy of such estimates is unknown because they haven't been calibrated with force standards based on the kilogram, the internationally accepted unit of mass. Luckily, there is hope on the horizon. In a paper presented June 4 at the annual conference of the Society of Experimental Mechanics in Charlotte, NC, national institute of standards and technology (NIST) engineers describe a prototype instrument that reliably measures forces as small as tens of nanonewtons and simultaneously ties those measurements to forces a thousand times larger based on the kilogram. The device works by connecting a well-calibrated spring-loaded scale with a set of electrodes that generates an electrostatic force. The instrument balances the downward force produced by a one-milligram mass artifact, by keeping the distance between the electrodes constant but varying the amount of voltage between them. The result is a force determination accurate to a few parts in 10,000 that is measured with voltage, electrical capacitance and distance (the location of the electrodes as measured in wavelengths of laser light). "It is much easier to make small changes in voltages, than to make small mass standards", explains NIST mechanical engineer Jon Pratt. The NIST researchers hope to extend the instrument's resolution to tens of piconewtons (trillionths of a newton).

In Professor Warren Chan's laboratory, nanosized quantum dots light up like multicoloured fireflies when exposed to ultraviolet light. But it's nothing compared with how Chan's face lights up when he talks about his goals for these microscopic particles—making it possible to detect, target and kill cancer cells. The 30-year-old assistant professor came to the Institute of Biomaterials and Biomedical Engineering three years ago. While completing his post-doctoral work in biomedical engineering at the University of California at San

Diego, Chan was working on a project that involved finding the best way to observe how a virus infects a cell. The theory behind it is similar to how biologists track whales—they tag them with some sort of transceiver and monitor the signal. But in the lab, trying to track viruses proved tricky. Whenever Chan put a tag—such as standard organic dyes—on the surface of a virus, he would lose the signal after three to five seconds. Along with his adviser, he started to examine quantum dots—nanoscale particles of semiconductors—as possible alternatives to the dyes. "Quantum dots were initially thought to be used for electronic circuitry", he says. "But what we saw in the literature was that these beautiful structures had never been applied to biology". Under a light source, the quantum dots glow like neon signs. "Under low-power or medium-power excitation, they can last for over 48 hours without much loss of light", Chan says. This meant that he could actually track a virus over the course of a biological process. The colour of light they emit can be changed by altering their size, with smaller dots emitting blue, green or yellow light and larger dots appearing orange, red or brown. "We can basically custom design the properties of the materials for whatever applications we need". Still, a major challenge remains. Quantum dots normally have a very oily surface and, like the old saying goes, oil and water don't mix: the dots wouldn't do well inside the water-based cellular environment. Chan is now looking for ways to modify the surface chemistry so that they interact with water-friendly molecules like proteins and DNA. Chan is now exploring whether these structures can be used to "light up" disease in animals, with future implications in humans. "When cells are diseased, they produce a unique set of proteins on their surface. If you find a matching molecule (to those proteins), you can take whatever you want to that site", he says. By attaching a quantum dot to a molecule that will target a specific type of cancer, for example, over time the dots will accumulate in the tumour. "As it accumulates, you see the animal light up in that particular region". Still, Chan cautions that there are limitations to light-based technology. "Light can only penetrate so far into the body. You can screen for surface cancers but the deeper the cancer, the harder it is to screen using quantum dots as a technique". For example, screening for skin, breast or prostate cancer might be possible but the technology might not work for lung or colon cancer. Chan also wants to determine where these nanostructures end up in the body and whether they are toxic. The quantum dots are made of potentially toxic heavy metals which, because of their size, can find their way into body structures that other materials cannot. "Nanotechnology has the potential to have many applications like contrast agents, biosensors and new diagnostic schemes", says Chan. "But you have to think about the other side of it—how does your body deal with these materials because we're going to be exposed to them on a more common basis. Right now, there is a need to handle these things carefully because we don't understand the after-effects".

Chan hopes that his current goal—to develop quantum dots that can target a disease site and light it up—can someday lead to an integrated system that will also use the quantum dots to bring drug therapies to the disease site. Beyond cancer, he is also exploring whether the dots could be used to detect pathogens such as malaria and HIV and he estimates that his quantum dots could be lighting up human disease within five to 10 years.

7.7 QUANTUM DOT

Imagine a piece of pure semiconductor. Electrons in this bulk are free to move around, in three dimensions. In solid state physics, these electrons are sometimes referred to as an electron gas. It is possible to constrain the movement of the electrons by disallowing one dimension—that is, by reducing one degree of freedom. The electrons can then move freely in only two dimensions. A system of electrons confined to a plane is called a two-dimensional electron gas, or 2 DEG. Quantum wires have only 1 degree of freedom, and quantum dots have 0 degrees of freedom. It is important to note that confinement is not perfect. A quantum dot is often defined as confinement to a very small volume of space, such as a sphere that has a diameter on the order of the material's Exciton Bohr Radius. At this size the energy bands of the semiconductor are no longer continuous, which results in the confinement. Quantum confinement gives rise to many interesting effects, and these effects can be exploited to develop new semiconductor devices. One such type of device are mesoscopic devices, such as single-electron transistors.

A group of scientists at the University of California, Berkeley, is giving new relevance to the term 'sharper image' by creating a superlens that can overcome a limitation in physics that has historically constrained the resolution of optical images. Using a thin film of silver as the lens and ultraviolet (UV) light, the researchers recorded the images of an array of nanowires and the word 'NANO' onto an organic polymer at a resolution of about 60 nanometers. In comparison, current optical microscopes can only make out details down to one-tenth the diameter of a red blood cell, or about 400 nanometers. The breakthrough, reported in the April 22 issue of the journal science, opens the door to dramatic technological advances in nanoengineering that could eventually lead to DVDs that store the entire contents of the Library of Congress, and computer processors that can quickly search through such a huge volume of data, the researchers said. 'The field of optics is involved in much of today's technology, including imaging and photolithography, which is used to make semiconductors and integrated circuits', said Xiang Zhang, UC Berkeley associate professor of mechanical engineering and principal investigator of the study. 'Our work has a far reaching impact on the development of detailed biomedical imaging, higher density electronic circuitry and ever-faster fiber optic communications. Nicholas Fang, one of Zhang's former Ph.D. students and lead author of the paper, said a nearer term application would be the development of medical imaging devices that could reveal never-before-seen details with optical microscopy. With current optical microscopes, scientists can only make out relatively large structures within a cell, such as its nucleus and mitochondria. With a superlens, optical microscopes could one day reveal the movements of individual proteins traveling along the microtubules that make up a cell's skeleton, the researchers said. Scanning electron and atomic force microscopes are now used to capture detail down to a few nanometers. However, such microscopes create images by scanning objects point by point, which means they are typically limited to non-living samples, and image capture times can take up to several minutes. 'Optical microscopes can capture an entire frame with a single snapshot in a fraction of a second', said Fang, who is now an assistant professor of mechanical engineering at the University of Illinois at Urbana-Champaign. 'That opens up nanoscale imaging to living materials, which can help biologists better understand cell structure and function in real time, and ultimately help in the development of new drugs to treat human diseases'. The study is the latest entry in a hotly debated topic among physicists and engineers surrounding the creation of a lens that can break the so-called diffraction limit of optics through negative refraction. Conventional lenses, whether manmade or natural, create images by capturing the propagating light waves all objects emit and then bending them. The angle of the bend is determined by the index of refraction and has always been positive. Yet objects also emit 'evanescent' waves that carry a great deal of detail but are far more elusive. Such evanescent waves decay exponentially and thus never make it to the image plane, an optics threshold known as the diffraction limit. Breaking this diffraction limit and capturing evanescent waves are critical to the creation of a 100 percent perfect representation of an object, considered the Holy Grail in optics. In 2000, British physicist John Pendry theorized that a material capable of a negative index of refraction could capture and 'refocus' evanescent waves into a perfect image. Pendry's proposed 'perfect lens' heory came more than 30 years after Russian physicist Victor Veselago first conceived of a negative refraction material that could reverse known optical phenomena.

7.8 NEW SUPERLENS OPENS DOOR TO NANOSCALE OPTICAL IMAGING

These theories are based on the fact that when electromagnetic waves of light reach the surface of a negative refraction lens, they excite a collective movement of surface waves, such as electron oscillations, also known as surface plasmons. That results in an enhancement of the evanescent waves and is different from the way light typically behaves when it reaches a conventional lens. Various negative refraction experiments were later conducted by researchers at UC San Diego, Boeing and Northeastern University, but they were limited to microwave beams. In 2003, Zhang's group was the first to confirm that optical evanescent waves are enhanced as they pass through a silver superlens in carefully designed conditions. But it wasn't until this latest experiment by Zhang's group that optical imaging with a superlens was demonstrated. Zhang and his research team used UV light at a 365-nanometer wavelength in the new experiments, so the image created actually has more detail than is possible with beams in the microwave range. The array of nanowires imaged measured 40 nanometers wide and the word NANO was about 60 nanometers wide. The objects, embedded onto a layer of chrome, were placed before the superlens, which was a layer of silver that was about 35 nanometers thick. The researchers recorded the image onto a photoresist, a polymer coating on the other side of the superlens that becomes insoluble when exposed to UV light.

'Our work provides a new imaging method that can beat the optical diffraction limit and that has tremendous potential to revolutionize a wide range of technologies', said Zhang. The key to the superlens is its ability to significantly enhance and recover the evanescent waves that carry information at very small scales. This enables imaging well below the diffraction limit. Notably, no lens is yet able to completely reconstitute all the evanescent waves emitted by an object, so the goal of a 100 percent perfect image is still out there. However, many scientists believe that a true perfect lens is not possible because there will always be some energy absorption loss as the waves pass through any known material. 'We did not create a perfect image in our experiment', said Fang. 'But it's clear that our image is dramatically better than the one created without the silver superlens. In the long run, this line of research could lead to even higher resolution imaging for distant objects, the researchers said. This includes more detailed views of other planets as well as of human movement through surveillance satellites. Other authors of the paper are Hyesog Lee, a graduate student in mechanical engineering, and Cheng Sun, a research scientist in Zhang's group. The research was supported by the Office of Naval Research, the Defense Advanced Research Projects Agency Multidisciplinary University Research Initiative, and the National Science Foundation Center for Nanoscale Science and Engineering.

The imaging or detection of their signals is easy for Hamamatsu's cameras. And we can make it easy for you, too!

7.9 EGANINTERE STING SETUP FOR PATCH CLAMPING

You have a thick sample with some fluorescence marke dcellsinit. You use an ORCA-camera to find those cells with fluorescence.Then with ju stone switch in the software, you can patch-clamp this cell in the NIR bright field image. With one camera, without loss of the focus! By the way, we have some nice digital camer as as Newvicon-successors available!

7.10 SCALING FRICTION DOWN TO THE NANO/MICRO REALM

An improved method for correcting nano- and micro-scale friction measurements has been developed by researchers at the National Institute of Standards and Technology (NIST). The new technique should help designers produce more durable micro- and nano-devices with moving parts, such as tiny motors, positioning devices or encoders.

Friction measurements made at the micro- and nano-scale can differ substantially due to changes in applied load. In a series of experiments described by nanotribologist Stephen Hsu at a technical meeting held May 17–20 in Toronto,* NIST scientists confirmed that many of the measured differences appear to be caused by unintended scratching of the surface by the sharp tips used in making the measurements themselves.

The NIST team used a specially designed friction tester developed jointly by NIST and Hysitron Inc. of Minneapolis. A carefully calibrated force was applied to diamond tips having a range of sizes. Friction forces were then measured as each tip was slid across a very smooth surface of silicon. Friction at the macroscopic scale is usually straightforward—doubling the force between two objects produces twice the friction. However, work at NIST and elsewhere has shown that friction at the microscale does not always obey this scaling rule. Forces greater than about 2 millinewton** produced substantially greater friction values than expected.

Images of the test surface made with an atomic force microscope confirmed that unintentional scratching produced the extra friction. To correct for this effect, NIST researchers developed a way to measure precisely the size, shape and orientation of the diamond tips so that friction forces caused by "plowing" can be subtracted to produce a more accurate final measurement.

7.11 DEVISING NANO VISION FOR AN OPTICAL MICROSCOPE

Contrary to conventional wisdom, technology's advance into the vanishingly small realm of molecules and atoms may not be out of sight for the venerable optical microscope, after all. In fact, research at the National Institute of Standards and Technology (NIST) suggests that a hybrid version of the optical microscope might be able to image and measure features smaller than 10 nanometers—a tiny fraction of the wavelength of visible light. In a preliminary test of the embryonic technique, NIST scientists used violet light with a wavelength of 436 nanometers to image features as small as 40 nanometers, about five times smaller than possible with a conventional optical microscope. Roughly speaking, such a feat is akin to picking up a solitary dime with a clumsy front-end loader. If successfully developed, the imaging technology could be readily incorporated into chip-making and other commercial-scale processes for making parts and products with nanometer-scale dimensions. The wavelengths of light in the visible part of the spectrum greatly exceed nanoscale dimensions. Consequently, the resolution of conventional light-based imaging methods is limited to about 200 nanometers—too large to resolve the details of nanotechnology, which, by definition, are no more than half that size.Resembling a 3-D checkerboard, this mirage-like pattern was formed by light waves after bouncing off a specimen with a peak-and-valley arrangement of etched lines. Such complicated wave patterns, the result of light scattering and interference effects, may be used to discern the dimensions of nanoscale features. However, a newly begun, five-year research effort at NIST suggests that a novel combination of illumination, detection and computing technologies can circumvent this limitation. Success would extend the technology's 400-year-long record as an indispensable imaging and measurement tool well into the expanding realm of nanotechnology. Called phase-sensitive, scatter-field optical imaging, the computer-intensive technique under development at NIST uses a set of dynamically engineered light waves optimized for particular properties (such as angular orientation and polarization). How this structured illumination field—engineered differently to highlight the particular geometry of each type of specimen—scatters after striking the target can reveal the tiniest of details.

"The scattering patterns are extremely sensitive to small changes in the shape and size of the scattering feature", explains Rick Silver, a physicist in NIST's Precision Engineering Division.

7.12 NEW MICROSCALE INCHWORM ACTUATOR

Creating a tool small enough to measure friction on a microelectromechanical systems (MEMS) device is not an easy task. The tool has to be about the width of a human hair. Yet, researchers at the at the National Nuclear Security Administration's Sandia National Laboratories have developed a new "inchworm" actuator instrument that provides detailed information about friction at the microscale. The main objective of the project was to study the validity of Amonton's Law at the microscale. This law, first stated 300 years ago, says friction force is proportional to normal force (normal means perpendicular to the surfaces). Although it remains a good description of friction today, there are interesting deviations from Amonton's Law, especially at low normal forces, where adhesion between the two surfaces is thought to contribute an extra force. Because of the large surface-to-volume ratio at the microscale, these adhesive forces could cause a strong deviation from Amonton's Law. "We must accomplish several different functions in one device to study the laws of friction at the microscale", said Sandia researcher and inchworm creator Maarten de Boer. "We need to build an actuator that controllably and accurately generates both very low and very high forces". These forces need to be applied both perpendicularly and tangentially to test the validity of Amonton's Law over a wide force range, he said. Another goal is to see if the friction measurements can describe the actual operation of the device.

Inchworm design

Parallel plates are easily fabricated in MEMS and can be used to obtain large normal forces. However, conventional MEMS tangential force actuators, known as comb drives, provide only 10 micronewtons of tangential force. De Boer wanted to achieve millinewtons of force and resorted to a mechanical amplification scheme. "We can bend plates to convert normal force to tangential force", he said. The inchworm has a force amplifying plate that spans two frictional clamps. While "inching" along, its step size is very small, about 40 nanometers, but its stepping cycle can be repeated over and over.

De Boer and Sandia postdoctoral researcher Alex Corwin developed a measurement methodology to confirm that the inchworm operates at up to 80,000 cycles a second, with a velocity of up to 3 millimeters per second. The force the inchworm develops is measured by attaching it to a load spring that has a special nonlinear characteristic — for each step the inchworm walks, the force increases more than it did in the previous step. A maximum tangential force of 2.5 millinewtons is achieved when the inchworm stalls out, about 250 times more force than a comb drive.

Investigating Amonton's law: So far this describes the inchworm's ability to move and provide forces. But de Boer asks the question: How does the inchworm give information on friction?

To determine the coefficient of static friction, the inchworm is walked out a long distance of about 20 micrometers against the nonlinear load spring. Then, a large voltage on one clamp holds it in place. This voltage is then gradually reduced, which decreases the frictional force it can sustain. At a certain point, the load spring overcomes that frictional force, and the inchworm slides forward a short distance (several micrometers) until the load spring force decreases enough that it stops again. This measurement is now repeated and the next measurement is at a much lower load due to the load spring's nonlinearity. Now, Amonton's law can be studied over a wide force range.

Working with Bob Ashurst, University of California, Berkeley, de Boer found that the friction force depends on the normal force from 1 millinewton down to normal loads as small as 50 micronewtons, in accordance with Amonton's law. Corwin further decreased the load and found that the coefficient of static friction began to increase. This was the deviation from Amonton's law due to adhesive forces, de Boer said. De Boer and Corwin also mapped out a way to measure dynamic friction, accounting for the friction forces in the presence of inertial forces, air damping forces, and the spring force. The result was that dynamic friction is about 80 percent of static friction. Gross slip before the static friction event is reached De Boer worked with David Luck, Sandia researcher, to determine whether the frictional measurements made so far described the behaviour of the device when it was operated. This test was different from the friction test and was conducted by loading the trailing clamp during the stepping cycle. They expected the average step size to decrease and it did. But when Luck tried to model the reduction in step size using the measured coefficients of friction, he could not get good agreement — the model predicted much larger step sizes than observed experimentally. Something other than the coefficient of friction had to be affecting the step size of the inchworm in operation. The researchers had assumed that the leading clamp was stationary in the stepping cycle, because the load applied to it was much larger than the load applied to the trailing clamp. De Boer reasoned that even though the leading clamp load was much larger, perhaps it was slipping back rather than remaining in position. Tests were devised to determine whether this slippage was occurring and the researchers inferred that it was. Corwin wanted to see if this sliding could be directly observed. By using high-resolution imaging, he looked for movement of the inchworm before the static friction event. He observed slip over distances as large as 200 nanometers.

"A slip this large was very surprising". De Boer said, "Theories predict about one or two nanometers of slip given the details of our frictional interfaces. But in our experiments, the slip is one hundred times larger than this".

This means there is some gross slip mechanism going on before the static friction limit is reached, he said.

"This gross sliding is clearly very important for MEMS, where you want to position an object with nanometer accuracy for optical applications", he said. "It is not something indicated by other friction instruments, but knowing about it will enable us to predict friction-related motion in MEMS with much better confidence".

Other Inchworm uses: Besides serving as a test structure useful for model friction studies, the inchworm actuator is attractive for use in actual MEMS applications. Marc Polosky, Sandia staff member, demonstrated its use in a MEMS system that performs mechanical logic functions.

"I had been looking for a good linear actuator", Polosky said. "The inchworm provides a very high output force and can be moved controllably both backwards and forwards. On the MEMS scale, stiction forces can degrade system performance".

MEMS actuators that have large output forces improve design margin and reliability. "The inchworm actuator enabled us to replace a number of components in our discrimination system and improved our production yield", Polosky said.

Dustin Carr, Sandia staff member, believes the inchworm actuator can be especially useful in applications where precise positioning and control of microoptical elements is necessary.

"The ability to do nanoscale positioning over many microns of range can enable the miniaturization of an optical lab benchtop, without using bulky or power-hungry piezoelectric devices", Carr said.

7.13 INCHWORM A MULTIDISCIPLINARY RESEARCH PROGRAM

The inchworm work is funded by Sandia's Laboratory Directed Research and Development (LDRD) project, titled "High Fidelity Friction Models for MEMS".

Dave Reedy, Sandia staff member, is serving as LDRD principal investigator.

"Modelers are excited by the inchworm test results, and now face the challenge of understanding and modeling newly observed phenomena such as gross slip prior to sliding", Reedy said. "Our goal is to develop a capability to perform finite element simulations of MEMS components that accurately predicts response in the presence of adhesion and friction".

Sandia is also working with Rob Carpick, a nanotribology expert and professor at the University of Wisconsin.

"Working with the inchworm team gives us a chance to connect the nano-scale measurements we do at Wisconsin with the microscale measurements that come from the inchworm", Carpick said. By looking at the same materials under the same conditions, we want to see if we can determine the fundamental underpinnings of Amonton's law. This could have implications in the study of friction that reach beyond the inchworm".

7.14 NANORULER A PROMISING LINE OF WORK

An MIT device that makes the world's most precise rulers—with "ticks" only a few hundred billionths of a meter apart—could impact fields from the manufacture of computer chips to space physics.

The nanoruler is 10 to 1,000 times faster and more precise than other methods for patterning parallel lines and spaces (known collectively as gratings) across large surfaces more than 12 inches in diameter. Such large surfaces are key to a number of applications involving gratings, such as larger wafers for the production of computer chips and higher-resolution space telescopes "Patterning gratings with precise control across large areas has bedeviled labs around the world for a long time, despite great efforts", said Mark L. Schattenburg (Ph.D. 1984), leader of the team and director of MIT's Space Nanotechnology Laboratory in the Center for Space Research (CSR). The nanoruler can pattern gratings of lines and spaces separated by only a few hundred nanometers, or billionths of a meter, across a surface 300 millimeters in diameter. It does so with a precision of less than one nanometer. "That is the equivalent of shooting a target the size of a nickel in Manhattan all the way from San Francisco", said Carl G. Chen (Ph.D. 2003), one of Schattenburg's colleagues. The researchers reported the results of initial trials of the nanoruler in the november-december issue of the journal of vacuum science and technology B. In addition to Schattenburg and Chen, they include Paul T. Konkola (Ph.D. 2003) and CSR research scientist Ralf K Heilmann. The team also received significant technical assistance from Robert Fleming of the CSR sponsored research technical staff. The nanoruler continues a line of research in advanced grating fabrication technology initiated at MIT in the late 1940 s. Gratings are of particular interest to scientists and engineers because, among other things, they allow the analysis of light. When the distance between one line and its neighbor, or period, is comparable to the wavelength of light, a phenomenon called diffraction occurs. Essentially the grating spreads the light into a spectrum, much as a prism spreads light into its colours. That spectrum, in turn, can be analyzed for information about the source.

For example, one such grating is key to NASA's Chandra X-ray observatory. The high energy transmission grating, also developed by Schattenburg's lab at MIT, spreads the X-rays from Chandra's mirrors into a spectrum that can then be "read" like a kind of cosmic bar code. From there, scientists can deduce the chemical composition and temperature of the source (such as the corona of a star).

7.15 BUILDING A BETTER RULER

Schattenburg began the nanoruler project because he wanted to create a better ruler for the semiconductor industry. "Today's advanced computer chips are packed with millions of transistors", he said. "Increasingly, however, it becomes a challenge to stuff more and more of these ever-shrinking features into an area no larger than a thumbnail".

What Schattenburg wanted, in essence, was an extremely well-made ruler whose ticks are spaced not millimeters but nanometers apart, and whose size was comparable to the largest commercial silicon wafers. "If such a ruler could be created, it would help chip makers do a much better job of laying down the Lilliputian circuitry", he said.

The nanoruler does the trick by combining two conventional methods to create gratings: mechanical ruling and interference lithography. Mechanical ruling essentially involves dragging a very sharp tool—almost always a diamond point—across the surface to form the sets of lines and spaces. Among other drawbacks, however, it is time-consuming. In interference lithography, two beams of light interfere with each other to produce interference "fringes", or parallel planes of high and low light intensity. These fringes can then be recorded in the surface as lines and spaces, using the same techniques common to the patterning of miniscule designs on computer chips. Interference lithography is much faster than mechanical ruling since all the grooves are formed simultaneously in a single exposure.The nanoruler essentially moves the surface to be patterned on a stage as a laser creates the interference fringes that in turn become the lines and spaces. "While other tools can make gratings with smaller periods, none is as fast or as accurate as the nanoruler", Schattenburg said. The project, which took five years, has had its share of trials and tribulations, Chen recalls. For example, toward the end he was stuck in Beijing for six months due to a visa delay. During that time he was able to work out some of the fundamental physics validating the nanoruler concept.

7.16 CELL PHONE STILL TOO BIG? MICRO-OSCILLATORS MAY HELP

A tiny, novel device for generating tunable microwave signals has been developed by researchers at the national institute of standards and technology (NIST). Described in the January 16 issue of physical review letters, the device measures just a few micro-meters square and is hundreds of times smaller than typical microwave signal generators in use today in cell phones, wireless internet devices, radar systems and other applications.

The device works by exploiting the fact that individual electrons in an electric current behave like minuscule magnets, each one with a "spin" that is either up or down, just as an ordinary magnet has a north and a south pole.

The NIST device consists of two magnetic films separated by a non-magnetic layer of copper. As an electric current passes through the first magnetic film, the electrons in the current align their spins to match the magnetic orientation in the film. But when the now aligned electrons flow through the second magnetic film, the process is reversed. This time the alignment of the electrons is transferred to the film. The result is that the magnetization of the film rapidly switches direction, or oscillates, generating a microwave signal. The microwave signal can be tuned from less than 5 gigahertz (5 billion oscillations a second) to greater than 40 GHz. The NIST experiments confirm predictions made by theorists at IBM Corp. and Carnegie Mellon University in 1996. NIST physicist, William Rippard says the new oscillators can be built into integrated circuits with the same technologies now used to make computer chips and that they may eventually replace bulkier technologies at a greatly reduced cost.

7.17 NANO CHEMISTS MAKE CURVED STRUCTURES

The natural world is full of curves and three dimensions, but the ability to deliberately and rationally construct such complex structures using nanoscale building blocks has eluded nanotechnologists who are eager to add curved structures to their toolbox. Now a team of northwestern university chemists report they have discovered ways to construct nanoscale building blocks that assemble into flat or curved structures with a high level of predictability, depending on the architecture and composition of the building blocks. The results are published in the Jan. 16 issue of the journal science. Using hybrid nanorods consisting of segments of gold and conducting polymers as their building blocks, the researchers created a number of unusual structures, including bundles, sheets and tubes of varying diameters. The extraordinary control that they were able to demonstrate over the process holds promise for building new and powerful drug delivery systems, electronic circuits, catalysts and light-harvesting materials. "We are trying to mimic life itself", said Chad A. Mirkin, director of Northwestern's Institute for Nanotechnology, who led the research team. "Much like proteins which must fold into complex structures in order to function properly, we have designed new materials that also form complex structures through the process of self-assembly". Mirkin and his team made the different structures by varying the diameter of the gold-polymer rods or adjusting the ratio of polymer segment to gold segment in the rods. Both methods should enable researchers to design structures with interesting electronic and optical properties. "We also discovered that the alumina template we used to build the rods initially is essential in guiding the assembly process", said Mirkin, also George B. Rathmann Professor of Chemistry. "Without the orientation the template provides, the rods do not form bundles, sheets or tubes". The nanorods were made by the sequential deposition of gold and conducting polymer into the pores of an aluminum template. After the gold-polymer rods were synthesized, the template was dissolved, leaving the rods parallel to one another, gold end to gold end and polymer end to polymer end. The strong interactions between the polymer ends built stress, causing curves to form. In a subsequent experiment, the researchers observed that self-assembly did not take place when the rods were randomly dispersed in solution. "The research clearly shows that some unnatural building blocks, such as the gold-polymer rods, need assistance in order to form higher-ordered structures", said Mirkin. "This means that when we work with building blocks that are larger than molecules but smaller than macroscopic objects, we should consider building materials in a completely new way—by using templates to help guide the assembly process and reduce the large number of assembly pathways potentially available to the building blocks". In addition to Mirkin, other authors on the science paper are Sungho Park (lead author), Jung-Hyurk Lim and Sung-Wook Chung, all from Northwestern University. The research was supported by the National Science Foundation and the US Air Force Office of Scientific Research.

7.18 RESEARCHERS CREATE FIRST EVER INTEGRATED SILICON CIRCUIT WITH NANOTUBE TRANSISTORS

The discovery of carbon nanotubes heralded a new era of scientific discovery that included the promise of ultra-sensitive bomb detectors and super-fast computer memory chips. But finding a way to incorporate nanomaterials into a working nanoelectronic system has been a frustratingly elusive achievement—until now. In an important milestone in the fields of nanosciences and nanoengineering, researchers at the University of California, Berkeley and Stanford University are announcing that they have created the first working, integrated silicon circuit that successfully incorporates carbon nanotubes in its design. "Until our work, no group has publicly reported success in directly integrating nanotubes onto silicon circuits," said Jeffrey Bokor, UC Berkeley professor of electrical engineering and computer sciences and principal investigator of the project. "It is a critical first step in building the most advanced nanoelectronic products, in which we would want to put carbon nanotubes on top of a powerful silicon integrated circuit so that they can interface with an underlying information processing system". Researchers say the development brings them a significant step closer to using carbon nanotubes for memory chips that can hold orders of magnitude more data than current silicon chips-10,000 times greater, according to some estimates—or for sensors sensitive enough to detect traces of explosives or biochemical agents at the molecular level. UC Berkeley, engineers teamed up with chemists at Stanford to develop an integrated circuit that can dramatically speed the analysis of thousands of synthesized carbon nanotubes. The description of this work appears in the January 2004 issue of nano letters, a publication of the American Chemical Society.

"These results represent a dream come true", said Hongjie Dai, associate professor of chemistry at Stanford and co-principal investigator of the project. "This achievement opens up a vast number of possible applications in nanotechnology". A carbon nanotube, which looks like rolled chicken wire when examined at the atomic level, is tens of thousands of times thinner than a human hair, yet remarkably strong. It has attractive electrical properties, which several research groups—including the one led by Dai at Stanford—have harnessed to create high performance transistors. The road to creating the first nano-silicon hybrid circuit began as a solution to a practical research problem: How to refine the process of growing nanotubes so that they are created with predictable qualities. Depending on the molecular structure specific to each carbon nanotube, it can either be metallic and capable of conducting electricity, or act like semiconductors, with conductivity that can be turned on and off. But the current synthesis process results in an unpredictable proportion of metallic and semiconducting nanotubes, leaving researchers uncertain as to how much of each type they'll get in any one batch. Analyzing whether a batch yielded metallic or semiconducting nanotubes involved a labor-intensive processing of manually checking the electrical conductivity of each carbon nanotube. To resolve this problem, the researchers set out to build a device that would automate the process of decoding thousands of carbon nanotubes on a silicon chip. Working with UC Berkeley's Microfabrication Laboratory, they created a chip with silicon metal oxide semiconductor (MOS) circuitry. The chip, dubbed the random access nanotube test chip, or RANT, contains a network of silicon wires and switches that form a circuit. Researchers then proceeded to grow carbon nanotubes directly onto "islands" on the circuit platform that contained the necessary catalyst for nanotube synthesis. The extreme heat required to grow nanotubes would typically melt the circuitry of traditional semiconductors, but the researchers got around that problem by interconnecting the silicon transistors with molybdenum, a refractory metal that can withstand very high temperatures.

"We first envisioned a patterned growth of carbon nanotubes on silicon wafers five years ago, but it wasn't clear at that time whether that approach would work as an integrated nanotube-silicon hybrid circuit", said Dai. "It was the combined expertise in chemistry, materials science and electrical engineering that made this a reality".

The resulting chip contained thousands of carbon nanotubes connected to the circuit on a 1-square-centimeter silicon chip. By turning certain switches on and off, researchers were able to isolate the path that leads to an individual nanotube. Not only could researchers pinpoint which nanotube was responding to electrical current passing through the system, they could tell whether the conductivity could be turned on or off. If they were able to change the conductivity of the nanotube, they knew that it was a semiconductor and not metallic.

"The circuit is interconnected in such a way that only 22 control signals are needed in testing more than 2,000 nanotubes", said Yu-Chih Tseng, a UC Berkeley graduate student in electrical engineering and computer sciences and lead author of the paper. "The key is that this can all be done by a machine and computer. We succeeded in making a tool for nanotechnology researchers, and in the process, we demonstrated the broader proof of principle that nanotubes can be successfully integrated in a complex circuit". Research such as this is an important component of the UC Berkeley-based Center for Information Technology Research in the Interest of Society, or CITRIS. The center includes a major emphasis on nanosciences and nanoengineering, and is funding the construction of a new nanofabrication laboratory on the UC Berkeley campus that would significantly enhance researchers' ability to conduct such fundamental and innovative work, said Ruzena Bajcsy, director of CITRIS. Bokor cautions that the integrated circuit they have built is not a likely candidate for commercialization just yet. For one, the molybdenum they used to protect the circuit from heat damage is not a typical material used in the semiconductor industry because it is a high-resistance metal. Nevertheless, the achievement opens the door for other promising research on nanotechnology devices, including those made of silicon nanowires and organic polymers, researchers said.

"Carbon nanotubes have fascinated many scientists and those interested in science ever since they were discovered", said Ali Javey, a graduate student in chemistry at Stanford and co-author of the paper. "This work takes us an important step forward by proving the compatibility of the nanotube synthesis process with modified silicon technology and leading the way to future nanotube-based commercial applications". The research was conducted as part of the materials, structures and devices Center, one of five multi-university focus centers funded by the Microelectronics Advanced Research Corporation (MARCO), a subsidiary of the semiconductor research corporation. The project was also supported by the defense advanced research projects agency (DARPA). Other co-authors of the study include Peiqi Xuan and Ryan Malloy at UC Berkeley, and Qian Wang at Stanford.

7.19 SUPERSONIC CRACKS

Glass breaks, rubber bursts—there are numerous ways in which materials can fail under extreme conditions. Many of the atomic mechanisms of materials failure however still remain a mystery. Some materials harden when they are stretched, others soften under large deformation. This phenomenon is referred to as hyperelasticity. Scientists from the Max Planck Institute for Metals Research in Stuttgart and IBM Almaden Research Center in San Jose, California study the dynamics of cracks using the world's most powerful computers, whereby the motion of every single atom in the material is calculated according to Newton's laws of motion. Combining theoretical considerations and large scale molecular dynamics simulations, the scientists derived the conditions under which hyperelasticity governs dynamic fracture. The researchers discovered that cracks can propagate supersonically when hyperelasticity, the elasticity at large strains, becomes dominant within a zone of high energy transport near the crack tip. This is important in understanding the dynamics of earthquakes or nucleation and propagation of cracks in aircrafts and space shuttles (Nature, November 13, 2003). The results are in clear contrast to classical theories in which the speed of elastic waves was considered the limiting speed of fracture, analogous to the speed of light in the theory of relativity. Historically, the classical physics of continuum has been the basis for most theoretical and computational tools in engineering, and theories relying on numerous phenomenological assumptions have been used. Over the last decades, there has been a new realization that atomic scale behaviour is crucial to achieving a more fundamental understanding of how materials fail. There is also the challenge created by the increasing trend to miniaturization in technology as relevant length scales of materials approach the nanoscale. At nanoscales which the continuum description of materials becomes questionable. Scientists use atomic simulations to study how materials behave under extreme conditions. The simulations yield fundamental "ab-initio" information about the mechanisms of deformation and failure at length and time scales unattainable by experimental measurement and unpredictable by continuum theories. With today's supercomputers, systems up to billions of atoms can be simulated, approaching materials size on the order of micrometers! Such length scales are of immediate interest to engineers. It is estimated that the billion-atom simulations will become routine in a few years, underlining the enormous potential of atomistic methods for technological innovations in materials science. In a recent article in nature, a team of scientists from Max Planck Institute for Metals Research in Stuttgart and IBM Almaden Research Center in San Jose, California study the dynamics of brittle crack propagation with large-scale atomic simulations. In this article the scientists report the discovery of an important missing feature in the existing theories of dynamic fracture: The elasticity of solids depend on their state of deformation. Metals will weaken, or soften, and polymers may stiffen as the strain approaches the state of materials failure. "It is only for infinitesimal deformation that the elastic moduli can be considered constant and the elasticity of the solid linear", says Prof Dr Huajian Gao, Director at the Max Planck Institute for Metals Research in Stuttgart, Germany. "However, many existing theories model fracture using linear elasticity, and neglect the considerable difference in material behaviour at small and large strains. Certainly, this can be considered questionable since materials fail at the tip of a dynamic crack because of the extreme deformation". The scientists hence postulated that hyperelasticity, the elasticity at large strains, can play a governing role in dynamic fracture. In their studies, the scientists show that hyperelasticity, the elasticity at large strains, can dominate the dynamics of fracture. Cracks moving in solids absorb and dissipate energy from the surrounding material. "We discovered a new length scale characterizing the zone near the crack tip from which the crack draws energy to sustain its motion", says Gao. "When materials are under extreme stress, this length scale extends only a few dozens nanometers".

This characteristic length scale is found to be proportional to the fracture surface energy and elastic modulus, and inversely proportional to the square of the applied stress. Contrary to the common belief, the crack does not need to transport energy from regions far away from its tip, rather only from a small local region described by the characteristic length scale.

If the region around the crack tip becomes stiff due to hyperelasticity, more energy can flow to the crack tip in shorter time. On the other side, energy transport gets slower when there is a local softening zone around the crack tip. Therefore, the scientists conclude that hyperelasticity is crucial for understanding and predicting the dynamics of brittle fracture. When hyperelasticity dominates, cracks can move faster than all elastic waves. This is in clear contrast to the classical theories in which it is believed that the longitudinal elastic wave speed is an impenetrable upper limit of crack speed. Figure shows snapshots of a computer simulation where the crack breaks through the sound barrier and propagates at a supersonic velocity (in glass, for instance with a speed higher than 5.300 m/sec, which is greater than 19,000 km/h). Such phenomenon can only be understood through hyperelasticity. Hyperelasticity dominates fracture energy transport when the size of the hyperelastic zone approaches the energy characteristic length. Under normal experimental conditions, the magnitude of stress may be one or two orders of magnitude smaller than that under atomic simulations. In such cases, the characteristic length is relatively large and the effect of hyperelasticity on effective velocity of energy transport is relatively small. At about one percent of elastic strain, the energy characteristic length is on the order of a few hundred atomic spacing and significant hyperelastic effects are observed. It seems that hyperelasticity can play the governing role especially in nanostructured materials such as thin films, or under high-impact conditions where huge stresses occur, so that the region from which the crack needs to draw energy is small. The computer simulations reported in the nature article to prove the concepts of hyperelasticity and energy characteristic length need to attain sufficiently large system sizes, and are hence very resource demanding. These simulations have been performed at the Max Planck Society's Supercomputer Center in Garching, which is among the most prominent scientific computer centers in the world.

"Supersonic crack motion. The plot shows the snapshots of a shear dominated crack (similar to an earthquake in California) propagating at a velocity faster than the longitudinal wave speed of the solid. The two Mach cones or shock fronts (analogous to the shock front or sonic boom around a supersonic jet airplane) indicate supersonic crack motion. This phenomenon can not be understood using the classical theories of fracture and is now explained by hyperelasticity models."

7.20 NEW SANDIA UV LEDS EMIT SHORT-WAVELENGTH, HIGH-POWER OUTPUT

Researchers at Sandia National Laboratories developing ultraviolet (UV) light-emitting diodes (LEDs) recently demonstrated two deep UV semiconductor optical devices that set records for wavelength/power output. One emits at a wavelength of 290 nanometers (nm) and produces 1.3 milliwatts of output power, and the other emits at a wavelength of 275 nm and produces 0.4 milliwatts of power. "Emission wavelengths of 275 to 290 nm with such high power outputs is a major breakthrough in UV LEDs development," says Bob Biefeld, manager of Sandia's Chemical Processing Science Department. "Only a handful of research groups around the world have come anywhere close". Operating at the shorter UV wavelengths makes it possible to build miniaturized devices that can detect biological agents, perform non-line-of-sight (NLOS) covert communications, purify water, cure polymers and other chemicals, and decontaminate equipment. The Sandia team is part of the Defense Advanced Research Projects Agency (DARPA) project SUVOS (for semiconductor ultraviolet optical source), which funds various research groups around the country to develop deep UV compact semiconductor optical sources. (Visible light can be seen in the 400 to 700 nm wavelength range. UV is shorter than about 400 nm. Deep UV is less than around 300 nm.) The SUVOS program has also been developing biological agent detection system prototypes such as UV fluorescence-based LIDAR. The team reached the 275–290 nm breakthrough a couple of months ago after working on similar research for more than three years — one-and-a-half of those years with DARPA funding. "When we started the project, one group had been clearly leading the field for several years without any other group making much progress", says Andy Allerman, the Sandia scientist who leads the growth of the LED semiconductor material. "Our LED performance increased rapidly the past year and recently we reached record power at 290 and 275 nm".

The device has a sapphire substrate with conductive layers of aluminum gallium nitride. It is well known that the more aluminum added to the semiconductor material, the shorter the output wavelength. But with increasing aluminum content the material becomes much harder to grow and harder to flow electrical current through it. The mix that reached the 275 nm is around 50 percent aluminum. A key step in achieving the high powers was getting high-quality material growth at these high aluminum percentages, considered to be very difficult. Also contributing to the advance is a smart packaging technology that has a flip-chip geometry. Instead of the standard top-emitting LED, the LED die is flipped upside down and bonded onto a thermally conducting submount. The finished LED is a bottom-emitting device that uses a transparent buffer layer and substrate. Having the device emit light from the bottom serves two purposes, says Kate Bogart who together with Art Fischer developed the advanced packaging at Sandia. "First, the light is two times brighter when the LEDs are in the flip-chip geometry, primarily because the light is not physically blocked by the opaque metal contacts on the top of the LED", Bogart says. "In addition, the flip-chip submount pulls heat away from the device because we make it out of materials with high thermal conductivity. This improves efficiency levels with less energy getting converted to heat and more to light". The result is a device that is low-weight, small, and resistant to vibrations and shock. Conventional UV sources are mercury vapour and other types of discharge lamps which are bulky, heavy, and power hungry — completely different from the new LEDs developed at Sandia that are no bigger than one square millimeter.

Biefeld says, that another aspect of the device that makes it unique is that the high power output of 1.3 milliwatts at 290 nm is obtained in a continuous wave (CW) mode.

"That was a continuous wave power measurement under direct current (DC) operation, not a pulsed current measurement like other UV LED groups have reported", Biefeld says. "We were able to control the heat issue to reach these powers in CW mode".

In October Sandia researchers Mary Crawford and Allerman attended a LED conference in Virginia where Mary gave a presentation on the advances of their devices.

"People were very interested", Crawford says. "One of our long-term DARPA program goals was to reach 1 milliwatt powers. And now our team has gone beyond that. Everyone was very excited about the power levels". While Crawford and Fischer continue to characterize the new UV LEDs and determine exactly how they can be used, LED devices are already being supplied to DARPA program participants making both non-line-of-sight communications and bio-sensor test beds.

7.21 'DARK MATTER' FORMS DENSE CLUMPS IN GHOST UNIVERSE

The "dark matter" that comprises a still-undetected one-quarter of the universe is not a uniform cosmic fog, says a University of California, Berkeley, astrophysicist, but instead forms dense clumps that move about like dust motes dancing in a shaft of light.

In a paper submitted this week to Physical Review D, Chung-Pei Ma, an associate professor of astronomy at UC Berkeley, and Edmund Bertschinger of the Massachusetts Institute of Technology (MIT), prove that the motion of dark matter clumps can be modeled in a way similar to the Brownian motion of air-borne dust or pollen.

Their findings should provide astrophysicists with a new way to calculate the evolution of this ghost universe of dark matter and reconcile it with the observable universe, Ma said.

Dark matter has been a nagging problem for astronomy for more than 30 years. Stars within galaxies and galaxies within clusters move in a way that indicates there is more matter there than we can see. This unseen matter seems to be in a spherical halo that extends probably 10 times farther than the visible stellar halo around galaxies. Early proposals that the invisible matter is comprised of burnt-out stars or heavy neutrinos have not panned out, and the current favorite candidates are exotic particles variously called neutrilinos, axions or other hypothetical supersymmetric particles. Because, these exotic particles interact with ordinary matter through gravity only, not via electromagnetic waves, they emit no light.

"We're only seeing half of all particles", Ma said. "They're too heavy to produce now in accelerators, so half of the world we don't know about".

The picture only got worse four years ago when "dark energy" was found to be even more prevalent than dark matter. The cosmic account now pegs dark energy at about 69 percent of the universe, exotic dark matter at 27 percent, mundane dark matter-dim, unseen stars-at 3 percent, and what we actually see at a mere 1 percent. Based on computer models of how dark matter would move under the force of gravity, Ma said that dark matter is not a uniform mist enveloping clusters of galaxies. Instead, dark matter forms smaller clumps that look superficially like the galaxies and globular clusters we see in our luminous universe. The dark matter has a dynamic life independent of luminous matter, she said. "The cosmic microwave background shows the early effects of dark matter clumping, and these clumps grow under gravitational attraction", she said. "But each of these clumps, the halo around galaxy clusters, was thought to be smooth. People were intrigued to find that high-resolution simulations show they are not smooth, but instead have intricate substructures. The dark world has a dynamic life of its own". Ma, Bertschinger and UC Berkeley graduate student Michael Boylan-Kolchin performed some of these simulations themselves. Several other groups over the past two years have also showed similar clumping.

The ghost universe of dark matter is a template for the visible universe, she said. Dark matter is 25 times more abundant than mere visible matter, so visible matter should cluster wherever dark matter clusters.

Therein lies the problem, Ma said. Computer simulations of the evolution of dark matter predict far more clumps of dark matter in a region than there are clumps of luminous matter we can see. If luminous matter follows dark matter, there should be nearly equivalent numbers of each. "Our galaxy, the Milky Way, has about a dozen satellites, but in simulations we see thousands of satellites of dark matter", she said. "Dark matter in the Milky Way is a dynamic, lively environment in which thousands of smaller satellites of dark matter clumps are swarming around a big parent dark matter halo, constantly interacting and disturbing each other". In addition, astrophysicists modeling the motion of dark matter were puzzled to see that each clump had a density that peaked in the center and fell off toward the edges in the exact same way, independent of its size. This universal density profile, however, appears to be in conflict with observations of some dwarf galaxies made by Ma's colleague, UC Berkeley professor of astronomy Leo Blitz, and his research group, among others. Ma hopes that a new way of looking at the motion of dark matter will resolve these problems and square theory with observation. In her physical review article, discussed at a meeting earlier this year of the American Physical Society, she proved that the motion of dark matter can be modeled much like the Brownian motion that botanist Robert Brown, described in 1828 and Albert Einstein explained in a seminal 1905 paper that helped garner him the 1921 Nobel Prize in Physics. Brownian motion was first described as the zigzag path traveled by a grain of pollen floating in water, pushed about by water molecules colliding with it. The phenomenon refers equally to the motion of dust in air and dense clumps of dark matter in the dark matter universe, said Ma.

This insight "let's us use a different language, a different point of view than the standard view", to investigate the movement and evolution of dark matter, she said. Other astronomers, such as UC Berkeley emeritus professor of astronomy Ivan King, have used the theory of Brownian motion to model the movement of hundreds of thousands of stars within star clusters, but this, Ma said, "is the first time it has been applied rigorously to large cosmological scales. The idea is that we don't care exactly where the clumps are, but rather, how clumps behave statistically in the system, how they scatter gravitationally".

Ma noted that the Brownian motion of clumps is governed by an equation, the Fokker-Planck equation, that is used to model many stochastic or random processes, including the stock market. Ma and collaborators are currently working on solving this equation for cosmological dark matter.

"It is surprising and delightful that the evolution of dark matter, the evolution of clumps, obeys a simple, 90-year-old equation", she said.

The work was supported by the National Aeronautics and Space Administration.

7.22 QUANTUM THEORY OF GRAVITY

When it was discovered in the early twentieth century that Newtonian physics, although it had stood unchallenged for hundreds of years, failed to answer basic questions about time and space, such as 'Is the universe infinite'? or 'Is time eternal'?, a new basis for physics was needed. This lead to the development of quantum theory by Bohr, Schrödinger and Heisenberg and relativity theory by Einstein. This was the first step in the development of a new basis for physics. Both theories, however are incomplete, and are limited in their abilities to answer many questions. Quantum physics deals with the behaviour of very small objects, such as atoms, why they do not disintegrate as Newtonian physics wanted. The theory of relativity, on the other hand deals with much large scales, celestial bodies and others. Both theories fail when confronted to the other's 'domain', and are therefore limited in their ability to describe the universe. One must unify these theories, make them compatible with one another. The resulting theory would be able to describe the behaviour of the universe, from quarks and atoms to entire galaxies. This is the quantum theory of gravity.

There are two fundamental areas of modern physics, each describes the universe on different scales. First we have quantum mechanics which talks about atoms, molecules and fundamental particles. Then we have general relativity which tells us that gravity is the bending and warping of space-time. There has been much work on finding a theory that combines these two pillars of physics.

There are three main aproches to quantum gravity all have there problems.

1. Loop quantum gravity.
2. String theory.
3. Others; Penrose spin networks, Connes non-commutative geometry, etc.

1. Loop quantum gravity is a way to quantise space time while keeping what general relativity taught us. It is independent of a background gravitational field or metric. So it should be if we are dealing with gravity. Also, it is formulated in 4 dimensions. The main problem is that the other forces in nature, electromagnetic, strong and weak cannot be included in the formulation. Nor it is clear how loop quantum gravity is related to general relativity.
2. Then we have string theory. String theory is a quantum theory where the fundamental objects are one dimensional strings and not point like particles. String theory is "large enough" to include the standard model and includes gravity as a must. The problems are three fold, first the theory is background dependant. The theory is formulated with a background metric. Secondly no-one knows what the physical vacuum in string theory is, so it has no predictive powers. String theory must be formulated in 11 dimensions, what happened to the other 7 we cannot see? (Also string theory is supersymmetric and predicts a load of new particles).
3. Then we have other approches, such as non-commutative geometry. This assumes that our space-time coordinates no longer commute, *i.e.* xy-yx is not zero. This formulation relies heavily on operator algebras.

All the theories have several things in common which are accepted as being part of quantum gravity at about Planck scale.

(*i*) Space-time is discrete and non-commutative.

(*ii*) Holography and the Bekenstin bound.

(*iii*) This is "simply" applying quantum mechanics to space-time. In quantum mechanics all the physical observables are discrete.

(*iii*) The holographic principle was first realised by Hawking. He realised that the entropy of a black hole was proportional to the surface area of the horizon and not the volume. That is all the information about a black hole is on the surface of the horizon. It is like a holograph, you only need to look at the 2-*d* surface to know everything you can about the black hole.

Bekenstin showed that there is a maximum amount of information that can pass through a surface. It is quantised in Planck units.

Is it possible for electrons to quantum tunnel through charged aluminum foil?

Yes. Quantum tunneling shouldn't be thought of as movement "through" something, exactly, as the name might seem to imply.

Various dynamic characteristics of electrons (really, all matter) are governed by statistics. Quantum mechanics then uses mechanisms like wave possibility functions to model these statistics. The convenience of this is that these wave functions can interact constructively or destructively with other wave functions; in these interactions, probability (the square of possibility) is conserved much like how energy is conserved when acoustic or electrical waves interact. That being said, in any system, an electron's position can be predicted with a statistical distribution of possible positions. The position of an electron does not exactly exist (using the words of one version of quantum reality) until being measured. When measured, we discover exactly where the electron exists. The next measurement may show an electron somewhere else, and that exact position will be unknown until measurement, but we know that it will fall somewhere on the probability curve that we've found analytically. So now imagine an electron in a box or a hole. The depth of this box is infinite. That means that regardless of how much energy the electron has, it could never hope to be able to exit the box. If the electron was a ball, regardless of how much energy it had to roll up the side of the hill, it would never reach the top of the hill and would thus never find its way out of the valley. Now make that box a little more realistic and simply make its walls pretty high instead of infinitely high. Here, it is possible that some electron may one day have enough energy to find itself outside of the box. However, there's no way to tell when or how it got there. We simply know there is some possibility (positive probability) that the electron will be outside of the box.

Over enough measurements, that means that eventually an electron will be found outside of the box. We say that the electron has "tunneled" out of the box, but it never really existed inside the box. It only had a probability that it could be inside the box, and this particular electron, when measured, was one of the few that were outside the box. So in your example, the aluminum foil would need to be charged with INFINITE potential in order to keep electrons from tunneling through it. Otherwise, if an electron could possibly have enough energy one day to find itself outside the aluminum foil, then one day electrons WILL be found on the other side of the aluminum foil. As you increase the charge, fewer will be found there as often... but still some will be found there. The ones that are found there shouldn't be thought of having once existed on the other side of the foil and happened to move through the foil... They should be thought of as perhaps always existing outside of the foil or possibly existing on either side of the foil. They just collapse to one possibility when you measure outside the foil. I recommend a book by Richard P. Feynman (QED: The Strange Theory of Light and Matter), who uses a different quantum mechanical model to explain the same phenomena. In his model, he uses a sort of "all possible paths" approach to modeling quantum probability. It produces the same answers, but perhaps motivates a different sort of intuition. It's a short collection of lectures. I recommend reading it.

7.23 NEW HYBRID MATERIAL HAS POTENTIAL USE IN MICROELECTRONICS

U of T scientists have developed a new class of hybrid materials combining organic and inorganic elements that could lead to improved computer chips, among other applications.

The computer industry is faced with a conundrum: as chip components become smaller and faster, the increased electrical resistance and capacitance they generate ultimately slows performance. The silica that insulates individual components becomes less effective as chip components shrink in size. A new material developed in the lab of University Professor Geoffrey Ozin of U of T's chemistry department may help address this problem, pushing computers to even faster performance. The material is a porous solid that assembles itself at the molecular level and displays superior insulating properties to silica; it is categorized as a nanocomposite because the nanoscale pore size of its honeycomb-like structure is so tiny, measuring in the billionths of a metre in diameter and organic and inorganic parts are integrated into a composite structure.

"I'd like to make thin films of this material, ultimately for use in microelectronics, for example, as packaging material", says Ozin, a Canada Research Chair in Materials Chemistry. Preliminary tests have shown the new material has promising insulating ability; researchers are now testing it on actual computer chips. A paper in the October 10, issue of the journal science describes how the research team developed the material by chemically combining silicon, an inorganic element, with an organic, methylene, in a one-to-one ratio. The resulting hybrid material—called a three-ring periodic mesoporous organosilica (PMO)—incorporates an unprecedented level of organic components in its structure compared to earlier nanocomposites, which reached an inorganic-organic ratio of only three to one. Boosting the organic component increases scientists' ability to exploit the function - for example, insulating ability of that organic, says Kai Landskron, a post-doctoral fellow in the Ozin lab and first author on the Science paper. The one-to-one ratio of organic to inorganic components is the first time this structural feature has been observed in a porous organic-inorganic hybrid material that has a regular array of single size pores, he says.

"As the organic function is the carrier of the desired material's properties, the technological impact of this discovery is expected to be profoundly important for scientists in various disciplines", says Landskron.

Boosting the organic component also gives scientists more options when it comes to manipulating the hybrid material's mechanical features, such as the ability to control how the material responds under stress and strain and how stiff or flexible it is, says Professor Doug Perovic, co-author and chair of U of T's department of Materials Science and Engineering. The inorganic component gives the necessary rigidity to the molecular structure that makes the material useable. This research shows that the three-ring PMO structure has enough resilience, even with the increased organic content (typically softer than inorganics), to maintain the stability of the molecular assembly, says Ozin, senior author on the paper. He believes the potential applications of this new class of self-assembling nanocomposites could impact a broad range of problems because simply changing the organic group from methylene to another organic with a different function could result in a material with an entirely different use. "Here's a new class of materials that has three silicons and three methylene groups as a basic building block", he says. "There's a whole family of potential new materials that could grow from that, just by using chemistry to change the organic group".

7.24 RESEARCHERS FIND THAT SUPERMAN'S TEETH CAN SUPERCONDUCT

Researchers at the University of Warwick have found that phosphorus, an element commonly found in teeth, can act as a "superconductor"—but you would have to have the strength of Superman to clench your teeth hard enough for it to work—as it happens at a pressure of around 2.5 megabars—some 30,000 times harder than an ordinary human can clench their teeth. Physicists were aware that lower pressures of around 0.1 megabars could convert the electrically insulating phosphorus to a form which can conduct and which allows limited superconductivity at 10 degrees kelvin and under. Recently experimentalists have found that another form of phosphorus occurs when 2.5 megabars of pressure is applied which causes it to form a "body centred cubic" or bcc crystal structure, comprised stacks of interpenetrating cubes of phosphorus atoms. Common metals such as iron and chromium have this structure at normal pressures. However, it was not known until now if this form of phosphorus would superconduct. Now University of Warwick physicists Sergey Ostanin and Professor Julie Staunton have used a number of theoretical physics techniques to describe the movement of the electrons and ionic vibrations, which proves that this version of phosphorus is an even better superconductor than the phosphorus held under 0.1 megabars of pressure. The University of Warwick researchers predict that the BCC structure of phosphorus will in fact superconduct at temperatures of around 14–22 Kelvin. In their research paper in physical review letters the University of Warwick researchers further suggest a clever means of preserving and using the bcc structure without having to resort to the high pressures. They suggest that the structure lends itself to being grown by depositing the atoms on a substrate of iron which itself organizes into a BCC structure. Anchoring the bcc phosphorus in this way would create and maintain its useful superconducting structure. Furthermore if the phosphorus BCC layer were positioned between a pair ferromagnets one could create a "superconducting switch" in which the phosphorus could be switched from superconductor to regular conductor and back again.

In Professor Warren Chan's laboratory, nanosized quantum dots light up like multicoloured fireflies when exposed to ultraviolet light. But it's nothing compared with how Chan's face lights up when he talks about his goals for these microscopic particles—making it possible to detect, target and kill cancer cells.

The 30-year-old assistant professor came to the Institute of Biomaterials and Biomedical Engineering three years ago. While completing his post-doctoral work in biomedical engineering at the University of California at San Diego, Chan was working on a project that involved finding the best way to observe how a virus infects a cell. The theory behind it is similar to how biologists track whales—they tag them with some sort of transceiver and monitor the signal.

But in the lab, trying to track viruses proved tricky. Whenever Chan put a tag — such as standard organic dyes—on the surface of a virus, he would lose the signal after three to five seconds. Along with his adviser, he started to examine quantum dots—nanoscale particles of semiconductors—as possible alternatives to the dyes. "Quantum dots were initially thought to be used for electronic circuitry", he says. "But what we saw in the literature was that these beautiful structures had never been applied to biology".

Under a light source, the quantum dots glow like neon signs. "Under low-power or medium-power excitation, they can last for over 48 hours without much loss of light", Chan says. This meant that he could actually track a virus over the course of a biological process. The colour of light they emit can be changed by altering their size, with smaller dots emitting blue, green or yellow light and larger dots appearing orange, red or brown. "We can basically custom design the properties of the materials for whatever applications we need".

Still, a major challenge remains. Quantum dots normally have a very oily surface and, like the old saying goes, oil and water don't mix: the dots wouldn't do well inside the water-based cellular environment. Chan is now looking for ways to modify the surface chemistry so that they interact with water-friendly molecules like proteins and DNA.

Chan is now exploring whether these structures can be used to "light up" disease in animals, with future implications in humans. "When cells are diseased, they produce a unique set of proteins on their surface. If you find a matching molecule [to those proteins], you can take whatever you want to that site", he says. By attaching a quantum dot to a molecule that will target a specific type of cancer, for example, over time the dots will accumulate in the tumor. "As it accumulates, you see the animal light up in that particular region".

Still, Chan cautions that there are limitations to light-based technology. "Light can only penetrate so far into the body. You can screen for surface cancers but the deeper the cancer, the harder it is to screen using quantum dots as a technique". For example, screening for skin, breast or prostate cancer might be possible but the technology might not work for lung or colon cancer.

Chan also wants to determine where these nanostructures end up in the body and whether they are toxic. The quantum dots are made of potentially toxic heavy metals which, because of their size, can find their way into body structures that other materials can't. "Nanotechnology has the potential to have many applications like contrast agents, biosensors and new diagnostic schemes", says Chan. "But you have to think about the other side of it — how does your body deal with these materials because we're going to be exposed to them on a more common basis. Right now, there is a need to handle these things carefully because we don't understand the after-effects".

Chan hopes that his current goal — to develop quantum dots that can target a disease site and light it up — can someday lead to an integrated system that will also use the quantum dots to bring drug therapies to the disease site. Beyond cancer, he is also exploring whether the dots could be used to detect pathogens such as malaria and HIV and he estimates that his quantum dots could be lighting up human disease within five to 10 years.

7.25 3-D MULTILAYERED OPTICAL STRUCTURES MADE WITHOUT A CLEAN ROOM

Achieving a first in the world of novel optical materials, researchers at the U.S. Department of Energy's Ames Laboratory are making 3-D photonic band gap crystals four millimeters square (approximately one-eighth of an inch square) and 12 layers high without benefit of a "clean room" environment or the multimillion dollar equipment traditionally required to create such structures. The fundamental research, supported by the Basic Energy Sciences Office of the DOE's Office of Science, holds potential for significantly reducing the costs associated with fabricating PBG crystals, devices that make it possible to route, manipulate and modify the properties of light.

PBG crystals can permit or block the transmission of light of certain frequencies in all directions. This characteristic makes them especially promising for applications in the field of optical communications, where the push is on to create a photonic crystal within a single computer chip.

The research path to that goal is an unbelievably expensive one. But Kai-Ming Ho, an Ames Laboratory senior physicist, and Kristen Constant, an Iowa State University associate professor of materials science and engineering, and their co-workers are easing the way by fabricating PBG crystal microstructures in the open air, something that has never been done before.

The project is based on Ho's original 1990 research that theoretically demonstrated the existence of the first PBG crystal through his diamond lattice structure design.

That unique design is key to the multilayered PBGs that are being fabricated by members of Ho's and Constant's research groups. They have adapted a technique called micro transfer molding to make templates for the fabrication of multilayered photonic band gap crystals.

"The micro transfer mold technique is not new", said Ho, who is also an ISU distinguished professor of physics and astronomy. "Modifying it to create multilevel lattice structures at micron- and submicron-length scales – that is the new advance".

The modified technique involves meticulous work at the micron-scale level. (For size reference, the period at the end of this sentence equals approximately 615 microns.) First, an elastomer mold is created with more than 1,000 micro channels on its surface. The channels are filled by hand with liquid polymer filler. The filler is then solidified by ultraviolet light. Next, the solidified polymer rods in the channels are coated with a second polymer that acts as glue, bonding the filler to a silicon wafer substrate. Once hardened, the elastomer mold is peeled off, leaving a set of parallel polymer rods on the substrate—one layer of the polymer template. By repeating the procedure, in principle, any number of multilayer structures is achievable. To convert the template to a ceramic photonic crystal, the template is over-infiltrated with Titania slurry. The structure is fired to 550 degrees Celsius (1022 F) to remove the template and sinter the Titania structure

Ho and Constant credit many of the fabrication advances to the unique skills of the young scientists they mentor: postdoctoral fellow Chang-Hwan Kim; current graduate students Jae-Hwang Lee, Yong-Sung Kim, and Ping Kuang; and former graduate student Henry Kang, now at Hewlett Packard in Oregon. They are conquering what is perhaps the biggest challenge—aligning the multiple layers that make up the PBG crystals. The 1,000 plus rods per layer in a four-millimeter-square PBG crystal are only 2.5 microns apart. "The placement of each rod is so precise", said Constant. "It's hard to imagine that we can put something down within a micron or half a micron".

Ho added, "If you make a mistake in one layer, it will disrupt the next one and spoil the rest of the sample. In order to build multilayer, you need to get things right successively".

Lee knows the kind of concentration that requires. He has constructed a 12-layer template for a PBG crystal and modestly admitted, "I can stack more than this; however, it will task my patience!" To improve the alignment,

Lee and Chang-Hwan Kim came up with an ingenious method based on diffracted moiré fringes that has proven indispensable. Ho explained, "Photonic crystals are periodic structures, so any shifts in periodicity will show up over a much larger area. Those shifts are called fringes", he said. The better the alignment, the farther apart those fringes are spaced, so the fringe pattern tells you how good the alignment is.

Constant praised the project's blended research team of physicists and materials scientists. "We've established an expertise with micro transfer molding. When people hear that we're doing this in open air, it really amazes them. It amazes me, too", she admitted, "especially when you realize that a speck of dust can disrupt the whole structure".

Ho noted that the care and expertise of the project's team members was overcoming the open-air obstacles. "It's a high-quality, low-cost process – that's the key—and it's achieved by a lot of engineering ingenuity", he said.

7.26 USING NANOTECHNOLOGY TO TARGET SICK CELLS

In Professor Warren Chan's laboratory, nanosized quantum dots light up like multicoloured fireflies when exposed to ultraviolet light. But it's nothing compared with how Chan's face lights up when he talks about his goals for these microscopic particles—making it possible to detect, target and kill cancer cells.

The 30-year-old assistant professor came to the Institute of Biomaterials and Biomedical Engineering three years ago. While completing his post-doctoral work in biomedical engineering at the University of California at San Diego, Chan was working on a project that involved finding the best way to observe how a virus infects a cell. The theory behind it is similar to how biologists track whales—they tag them with some sort of transceiver and monitor the signal.

But in the lab, trying to track viruses proved tricky. Whenever Chan put a tag—such as standard organic dyes—on the surface of a virus, he would lose the signal after three to five seconds. Along with his adviser, he started to examine quantum dots—nanoscale particles of semiconductors—as possible alternatives to the dyes. "Quantum dots were initially thought to be used for electronic circuitry", he says. "But what we saw in the literature was that these beautiful structures had never been applied to biology".

Under a light source, the quantum dots glow like neon signs. "Under low-power or medium-power excitation, they can last for over 48 hours without much loss of light", Chan says. This meant that he could actually track a virus over the course of a biological process. The colour of light they emit can be changed by altering their size, with smaller dots emitting blue, green or yellow light and larger dots appearing orange, red or brown. "We can basically custom design the properties of the materials for whatever applications we need".

Still, a major challenge remains. Quantum dots normally have a very oily surface and, like the old saying goes, oil and water don't mix: the dots wouldn't do well inside the water-based cellular environment. Chan is now looking for ways to modify the surface chemistry so that they interact with water-friendly molecules like proteins and DNA.

Chan is now exploring whether these structures can be used to "light up" disease in animals, with future implications in humans. "When cells are diseased, they produce a unique set of proteins on their surface. If you find a matching molecule (to those proteins), you can take whatever you want to that site", he says. By attaching a quantum dot to a molecule that will target a specific type of cancer, for example, over time the dots will accumulate in the tumor. "As it accumulates, you see the animal light up in that particular region".

Still, Chan cautions that there are limitations to light-based technology. "Light can only penetrate so far into the body. You can screen for surface cancers but the deeper the cancer, the harder it is to screen using quantum dots as a technique". For example, screening for skin, breast or prostate cancer might be possible but the technology might not work for lung or colon cancer.

Chan also wants to determine where these nanostructures end up in the body and whether they are toxic. The quantum dots are made of potentially toxic heavy metals which, because of their size, can find their way into body structures that other materials can't. "Nanotechnology has the potential to have many applications like contrast agents, biosensors and new diagnostic schemes," says Chan. "But you have to think about the other side of it — how does your body deal with these materials because we're going to be exposed to them on a more common basis. Right now, there is a need to handle these things carefully because we don't understand the after-effects".

Chan hopes that his current goal — to develop quantum dots that can target a disease site and light it up — can someday lead to an integrated system that will also use the quantum dots to bring drug therapies to the disease site. Beyond cancer, he is also exploring whether the dots could be used to detect pathogens such as malaria and HIV and he estimates that his quantum dots could be lighting up human disease within five to 10 years.

7.27 APPLICATION OF SEMICONDUCTOR NANOCRYSTALS TO PHOTOVOLTAIC ENERGY CONVERSION DEVICES

Solar radiation is a plentiful and clean source of power. However, due to the high cost of electrical conversion on a per watt basis, conventional solar cells have not been exploited to their full potential. The cost per watt can be reduced by increasing the efficiency of solar cells and by reducing their cost to manufacture. The high price for solar cells is largely due to the use of expensive substrate materials and costly micro fabrication processing. CdSe and PbSe semiconductor nanocrystals (EviDots) are ideally suited as solar cell material due to the ability to tune the electronic band gap and, hence, optimize a solar cell for maximum efficiency. Furthermore, EviDots are produced in a manner that is conducive to low temperature; liquid phase processing that eliminates the need for expensive substrates and micro fabrication. Evident Technologies, Inc., the world's leading manufacturer of semiconductor nanocrystals, has technology that could enable wide spread use of solar energy conversion-reducing our dependence on fossil fuels.

7.28 CORE SHELL EVIDOTS: HIGH YIELD, NARROW EMISSION, PHOTOSTABLE FLUORESCENT LABELS

EviDots are very small crystals of a semiconductor material, in the nanometer size range (billionths of a meter). Their unique optical, magnetic and electrical properties are dependent on the size and composition, both of which are controlled during synthesis, allowing these properties to be "tuned" for specific applications. EviDots fluoresce brightly under ultraviolet irradiation and this fluorescence is both enhanced and stabilized by surrounding the core EviDots with a "shell" of another semiconductor material. These "Core Shell EviDots" are premium material for a wide variety of fluorescence applications, particularly those which call for simultaneous emission of multiple colours.

7.29 OPTICAL SWITCHING SPECTROSCOPY OF PbS QUANTUM DOTS WITH DUAL-WAVELENGTH PUMP PROBE

Optical transmission modulation in PbS quantum dots is investigated using a dual-wavelength pump-probe technique. High-energy excitation at 800 nm efficiently alters the transmission of a probe beam 1100 – 1600 nm spectral range, with a clear enhancement observed at longer wavelengths towards the excitation peak (2100 nm). A decay time of less than 2.5 ps is measured for pump intensities of up to 3.8 GW/cm^2. The maximum modulation is delayed by approximately 1 ps relative to the perfect overlap of the pump and probe pulses, the time required for the electrons to reach lower energy states. Long term exposure shows photo-degradation by the process is reversible and recovers on its own after several hours.

7.30 SEMICONDUCTOR NANOCRYSTAL BASED SATURABLE ABSORBERS FOR OPTICAL SWITCHING APPLICATIONS

This technical report presents experimental and theoretical results of an effort to develop semiconductor nanocrystal (also known as quantum dots) doped thin films for nanotechnology applications in optical switching, optical transistors, optical computing or telecommunications packet switching. Films doped with high quality

PbS and PbSe nanocrystals (quantum dots) have been fabricated. Measurements of the absorption spectra are compared with theoretical predictions. The general theoretical framework for treating the optical properties of multi-layer dielectric structures containing thin nanocrystal doped layers is presented. A simple model (the particle-in-a-sphere) is used as a starting point; however, more sophisticated models (such as those based on kp theory, or fits to experiment) can easily be incorporated. A prototypical micron-scale Fabry-Perot optical switching cavity has been constructed and is discussed. Theoretical predictions for the optical switching behaviour of such a structure, upon introduction of the semiconductor nanocrystal material into the optical cavity, are presented.

7.31 OPTICAL SWITCHING APPLICATIONS USING PbSe IR-EVIDOT NANOCRYSTALS

We live in the information age where computing and data transport is based upon aging silicon technology. Scientists are starting to unlock the fundamental secrets of our DNA, requiring unprecedented amounts of computer processing to make sense of the abundance of information. Collaborators no longer need to work at the same facility but can be separated by miles with the internet. The public now demands increased data flow to their homes including such broadband applications as digital cable, video-on-demand services and real-time video phoning. With this explosion of information, comes an increased need to share data and transport information from across the room or across the company. Hence, there is a continuing need to increase the speed of computing and for ultra-fast, all-optical switching (*i.e.* switching light with light) for both the telecommunication and computer industries.

7.32 ENGINEERING SEMICONDUCTOR PROPERTIES WITHOUT COSTLY EPITAXIAL GROWTH

In general, semiconductors have been traditionally defined as "any of a class of solids whose electrical conductivity is between that of a conductor and that of an insulator". Ever since Thomas Seebeck, discovered semiconductor properties of PbS (Lead Sulfied) in 1821 scientists have studied the properties with an eye towards trying to improve proformance. The properties of a specific semiconductor material are fixed and determined by the electronic structure as given by nature. The challenge for semiconductor researchers is to make a semiconductor with an engineered electronic structure resulting in a more effective material. The only tools at their disposal to achieve this have been via epitaxial methods.

7.33 FIRST WHITE LED USING QUANTUM DOTS CREATED

In a different approach to creating white light several researchers at the department of energy's (DOE) Sandia National Laboratories have developed the first solid-state white light-emitting device using quantum dots. In the future, the use of quantum dots as light-emitting phosphors may represent a major application of nanotechnology.

"Understanding the physics of luminescence at the nanoscale and applying this knowledge to develop quantum dot-based light sources is the focus of this work", says Lauren Rohwer, principal investigator. "Highly efficient, low-cost quantum dot-based lighting would represent a revolution in lighting technology through nanoscience".

The approach is based on encapsulating semiconductor quantum dots—nanoparticles approximately one billionth of a meter in size—and engineering their surfaces so they efficiently emit visible light when excited by near-ultraviolet (UV) light-emitting diodes (LEDs). The quantum dots strongly absorb light in the near UV range and re-emit visible light that has its colour determined by both their size and surface chemistry.

This nanophosphor-based device is quite different from an alternative approach based upon growth of blue, green, and red emitting semiconductor materials that requires careful mixing of the those primary colours to produce white illumination. Efficiently extracting all three colours in such a device requires costly chip designs, which likely cannot compete with conventional fluorescent lighting but can be attractive for more specialized lighting applications.

Rohwer and the quantum dot team—Jess Wilcoxon, Stephen Woessner, Billie Abrams, Steven Thoma, and Arturo Sanchez—started on the project two-and-a-half years ago. Subsequently, their research has advanced significantly, including recently reaching a major milestone of creating white and blue lighting devices using encapsulated quantum dots.

"This accomplishment brings quantum dot technology from the laboratory demonstration phase to a packaged component", Rohwer says.

LEDs for solid-state lighting typically emit in the near UV to the blue part of the spectrum, around 380–420 nanometers. Conventional phosphors used in fluorescent lighting are not ideal for solid state lighting because they have poor absorption for these energies. So researchers worldwide have been investigating other chemical compounds for their suitability as phosphours for solid state lighting.

Quantum dots represent a new approach. The nanometer-size quantum dots are synthesized in a solvent containing soap-like molecules called surfactants as stabilizers. The small size of the quantum dots — much smaller than the wavelength of visible light — eliminates all light scattering and the associated optical losses. Optical backscattering losses using larger conventional phosphours reduce the package efficiency by as much as 50 percent.

Nanophosphours based upon quantum dots have two significant advantages over the use of conventional bulk phosphor powders. First, while the optical properties of conventional bulk phosphor powders are determined solely by the phosphor's chemical composition, in quantum dots the optical properties such as light absorbance are determined by the size of the dot. Changing the size produces dramatic changes in colour. The small dot size also means that, typically, over 70 percent of the atoms are at surface sites so that chemical changes at these sites allow tuning of the light-emitting properties of the dots, permitting the emission of multiple colours from a single size dot.

"This provides two additional ways to tune the optical properties in addition to chemical composition of the quantum dot material itself", Wilcoxon says.

For the quantum dots to be used for lighting, they need to be encapsulated, usually in epoxy or silicone.

"Doing this we had to take care not to alter the surface chemistry of the quantum dots in transition from solvent to encapsulant", says Thoma, who worked on the encapsulation portion of the project.

Quantum dot phosphors are integrated with a commercial LED chip that emits in the near ultraviolet at 400 nanometers by encapsulating the chip with a dot-filled epoxy, creating a dome. The quantum dots in the dome absorb the invisible 400 nanometer light from the LED and reemit it in the visible region — a principle similar to that used in fluorescent lighting.

However, a key technical issue in the encapsulation process had to be solved first. When altering the environment of the dots from a solvent to an encapsulant, the quantum dots would "clump up" or agglomerate, causing them to lose their light-emitting properties. By attaching the quantum dots to the "backbone" of the encapsulating polymer they are close, but not touching. This allows for an increase in efficiency from 10–20 percent to an amazing 60 percent, Thoma says.

The team notes that other people working in the field of quantum dots have reported conversion efficiencies of nearly 50 percent in dilute solutions. However, to their knowledge, Sandia's team is the first to make an encapsulated quantum dot device with such high efficiencies.

To date, the Sandia's quantum dot devices have largely been composed of the semiconductor material cadmium sulfide. Cadmium is a toxic heavy metal similar to lead so alternative nanophosphor materials is desired. Fortunately, quantum dot phosphors can also be made from other types of materials, including nontoxic nanosize silicon or germanium semiconductors with light-emitting ions like manganese on the quantum dot surface.

"Silicon, which is abundant, cheap, and non-toxic, would be an ideal material", says Woessner. "The scientific insights gained through the team's success with cadmium sulfide quantum dots will enable this next step in nanophosphor development".

In the next year the researchers will increase the concentration of the quantum dots in the encapsulant to obtain further increases in light output while extending the understanding of quantum dot electronic interactions at high concentrations.

While the researchers investigate the use of quantum dots as phosphors as part of an internally funded research project, they also have a grant from the DOE office of building technologies for a collaborative project with Lumileds Lighting, a joint venture between Agilent Technologies and Philips Lighting. In this project they are helping Lumileds measure quantum efficiency of light emission from various types of dots.

Jerry Simmons, who with James Gee, heads up the Sandia's Solid State Lighting grand challenge, says the quantum dot research is an integral part of the work at Sandia.

"We are very proud of these accomplishments", Simmons says. "The team has come a long way in a short time".

7.34 NEW SUPER LENS OPENS DOOR TO NANOSCALE OPTICAL IMAGING

A group of scientists at the University of California, Berkeley, is giving new relevance to the term 'sharper image' by creating a super lens that can overcome a limitation in physics that has historically constrained the resolution of optical images.

Using a thin film of silver as the lens and ultraviolet (UV) light, the researchers recorded the images of an array of nanowires and the word 'NANO' onto an organic polymer at a resolution of about 60 nanometers. In comparison, current optical microscopes can only make out details down to one-tenth the diameter of a red blood cell, or about 400 nanometers.

The breakthrough, reported in the April 22 issue of the journal science, opens the door to dramatic technological advances in nanoengineering that could eventually lead to DVDs that store the entire contents of the Library of Congress, and computer processors that can quickly search through such a huge volume of data, the researchers said.

'The field of optics is involved in much of today's technology, including imaging and photolithography, which is used to make semiconductors and integrated circuits,' said Xiang Zhang, UC Berkeley associate professor of mechanical engineering and principal investigator of the study. 'Our work has a far reaching impact on the development of detailed biomedical imaging, higher density electronic circuitry and ever-faster fiber optic communications'.

Nicholas Fang, one of Zhang's former Ph.D. students and lead author of the paper, said a nearer term application would be the development of medical imaging devices that could reveal never-before-seen details with optical microscopy.

With current optical microscopes, scientists can only make out relatively large structures within a cell, such as its nucleus and mitochondria. With a super lens, optical microscopes could one day reveal the movements of individual proteins traveling along the microtubules that make up a cell's skeleton, the researchers said.

Scanning electron and atomic force microscopes are now used to capture detail down to a few nanometers. However, such microscopes create images by scanning objects point by point, which means they are typically limited to non-living samples, and image capture times can take up to several minutes.

'Optical microscopes can capture an entire frame with a single snapshot in a fraction of a second,' said Fang, who is now an assistant professor of mechanical engineering at the University of Illinois at Urbana-Champaign. 'That opens up nanoscale imaging to living materials, which can help biologists better understand cell structure and function in real time, and ultimately help in the development of new drugs to treat human diseases'.

The study is the latest entry in a hotly debated topic among physicists and engineers surrounding the creation of a lens that can break the so-called diffraction limit of optics through negative refraction.

Conventional lenses, whether manmade or natural, create images by capturing the propagating light waves all objects emit and then bending them. The angle of the bend is determined by the index of refraction and has always been positive.

Yet objects also emit 'evanescent' waves that carry a great deal of detail but are far more elusive. Such evanescent waves decay exponentially and thus never make it to the image plane, an optics threshold known as the diffraction limit. Breaking this diffraction limit and capturing evanescent waves are critical to the creation of a 100 percent perfect representation of an object, considered the Holy Grail in optics.

In 2000, British physicist, John Pendry theorized that a material capable of a negative index of refraction could capture and 'refocus' evanescent waves into a perfect image. Pendry's proposed 'perfect lens' theory came more than 30 years after Russian physicist, Victor Veselago first conceived of a negative refraction material that could reverse known optical phenomena.

These theories are based on the fact that when electromagnetic waves of light reach the surface of a negative refraction lens, they excite a collective movement of surface waves, such as electron oscillations, also known as surface plasmons. That results in an enhancement of the evanescent waves and is different from the way light typically behaves when it reaches a conventional lens.

Various negative refraction experiments were later conducted by researchers at UC San Diego, Boeing and Northeastern University, but they were limited to microwave beams.

In 2003, Zhang's group was the first to confirm that optical evanescent waves are enhanced as they pass through a silver super lens in carefully designed conditions.

But it wasn't until this latest experiment by Zhang's group that optical imaging with a super lens was demonstrated. Zhang and his research team used UV light at a 365-nanometer wavelength in the new experiments, so the image created actually has more detail than is possible with beams in the microwave range.

The array of nanowires imaged measured 40 nanometers wide and the word NANO was about 60 nanometers wide. The objects, embedded onto a layer of chrome, were placed before the super lens, which was a layer of silver that was about 35 nanometers thick. The researchers recorded the image onto a photo resist, a polymer coating on the other side of the super lens that becomes insoluble when exposed to UV light.

'Our work provides a new imaging method that can beat the optical diffraction limit and that has tremendous potential to revolutionize a wide range of technologies', said Zhang. 'The key to the super lens is its ability to significantly enhance and recover the evanescent waves that carry information at very small scales. This enables imaging well below the diffraction limit'.

Notably, no lens is yet able to completely reconstitute all the evanescent waves emitted by an object, so the goal of a 100 percent perfect image is still out there. However, many scientists believe that a true perfect lens is not possible because there will always be some energy absorption loss as the waves pass through any known material.

'We did not create a perfect image in our experiment', said Fang. 'But it's clear that our image is dramatically better than the one created without the silver super lens'.

In the long run, this line of research could lead to even higher resolution imaging for distant objects, the researchers said. This includes more detailed views of other planets as well as of human movement through surveillance satellites.

Other authors of the paper are Hyesog Lee, a graduate student in mechanical engineering, and Cheng Sun, a research scientist in Zhang's group.

The research was supported by the Office of Naval Research, the Defense Advanced Research Projects Agency Multidisciplinary University Research Initiative, and the National Science Foundation Center for Nanoscale Science and Engineering.

7.35 DEVISING NANO VISION FOR AN OPTICAL MICROSCOPE

Contrary to conventional wisdom, technology's advance into the vanishingly small realm of molecules and atoms may not be out of sight for the venerable optical microscope, after all. In fact, research at the National Institute of Standards and Technology (NIST) suggests that a hybrid version of the optical microscope might be able to image and measure features smaller than 10 nanometers—a tiny fraction of the wavelength of visible light.

In a preliminary test of the embryonic technique, NIST scientists used violet light with a wavelength of 436 nanometers to image features as small as 40 nanometers, about five times smaller than possible with a conventional optical microscope.

Roughly speaking, such a feat is akin to picking up a solitary dime with a clumsy front-end loader. If successfully developed, the imaging technology could be readily incorporated into chip-making and other commercial-scale processes for making parts and products with nanometer-scale dimensions.

The wavelengths of light in the visible part of the spectrum greatly exceed nanoscale dimensions. Consequently, the resolution of conventional light-based imaging methods is limited to about 200 nanometers—too large to resolve the details of nanotechnology, which, by definition, is no more than half that size.

Resembling a 3-D checkerboard, this mirage-like pattern was formed by light waves after bouncing off a specimen with a peak-and-valley arrangement of etched lines. Such complicated wave patterns, the result of light scattering and interference effects, may be used to discern the dimensions of nanoscale features.

However, a newly begun, five-year research effort at NIST suggests that a novel combination of illumination, detection and computing technologies can circumvent this limitation. Success would extend the technology's 400-year-long record as an indispensable imaging and measurement tool well into the expanding realm of nanotechnology.

Called phase-sensitive, scatter-field optical imaging, the computer-intensive technique under development at NIST uses a set of dynamically engineered light waves optimized for particular properties (such as angular orientation and polarization). How this structured illumination field—engineered differently to highlight the particular geometry of each type of specimen—scatters after striking the target can reveal the tiniest of details.

"The scattering patterns are extremely sensitive to small changes in the shape and size of the scattering feature", explains Rick Silver, a physicist in NIST's Precision Engineering Division.

7.36 SCIENTISTS FIND ATOMIC CLUES TO TOUGHER CERAMICS

Advanced ceramics are wonderful materials – they withstand temperatures that would melt steel and resist most corrosive chemicals. If only they weren't so brittle. Poor resistance to fracture damage has been the major drawback to the widespread use of advanced ceramics as structural materials. Help, however, may be on the way.

A collaboration of scientists led by researchers with the US Department of Energy's Lawrence Berkeley National Laboratory (Berkeley Lab) has uncovered clues at the atomic level that could lead to a new generation of much tougher advanced ceramics to be used in applications like gas turbine engines.

Working with the unique facilities at Berkeley Lab's National Center for Electron Microscopy (NCEM), the collaboration has produced atomic-resolution images of silicon nitride ceramics that were sintered with oxides of rare earth elements to toughen them up and prevent cracks from spreading. These images revealed, for the first time, the exact location of each rare-earth atom in the final material and how their presence affected its toughness.

'Our findings are a prime factor in understanding the origin of the mechanical properties in advanced ceramics and should make it possible to do the precise tailoring in the future that will critically improve the performances of these materials over a wide range of applications,' says Robert Ritchie, a materials scientist who holds a joint appointment with Berkeley Lab's Materials Sciences Division and the University of California at Berkeley's Department of Materials Science and Engineering.

Ritchie and Alexander Ziegler, a member of Ritchie's research group, were the principal authors of a paper by the collaboration which appears in the December 3 issue of the journal science. The other co-authors were Christian Kisielowski and Nigel Browning of Berkeley Lab, Juan Idrobo of UC Davis, and Michael Cinibulk of the Air Force Research Laboratory in Ohio.

Ceramics are probably the oldest construction materials known, their use dating back thousands of years, when they were made from wet clay and baked at high temperatures until hard. Today's advanced ceramics are made from powders of complex chemical compounds and their production requires careful control at every stage of the process.

Much attention is currently being focused on silicon nitride advanced ceramics, which are considered to be leading candidates as structural materials for future gas turbine engines. These engines, which are projected for use in electrical power plants, among other applications, will burn fuel at temperatures of around 1,200 degrees Celsius (2,192 degrees Fahrenheit), well beyond the tolerance of metals, even nickel-base super-alloys. Running at such high temperatures, the advanced gas turbines are expected to achieve a much higher thermal-to-electricity efficiency than today's steam-driven electrical power plants, while emitting far less greenhouse gases. For this to happen, however, the brittleness problem of the silicon nitride ceramics must be solved.

Says Ziegler, 'To enhance the toughness of a silicon nitride ceramic, it is often necessary to engineer a thin (nanoscale) film in the ceramic's grain boundaries, which cracks when the ceramic begins to fracture. This promotes the formation of grain bridges which span across the crack, making it more difficult for the crack to propagate'.

Understanding the nature and properties of these nano-sized intergranular films is crucial to enhancing ceramic toughness, according to Ritchie and Ziegler. However, critical information about the chemical composition, atomic structure and bonding characteristics of such films has long been missing. 'The problem was the nanometer dimensions of the intergranular films', Ziegler says. 'To gain information on the local atomic structure and bonding characteristics requires characterization at Angstrom (single-atom) to sub-Angstrom scales. Until recently, no microscopes or chemical analysis probes have been able to resolve such information at these length scales'.

NCEM, however, houses a scanning transmission electron microscope (STEM) which is optimized for materials applications that require the highest resolutions for both imaging and spectroscopy. With the help of NCEM staff members Kisielowski and Browning, Ritchie, Ziegler and the collaboration used this microscope, in combination with an imaging technique called 'high-angle annular dark-field STEM', and a chemical analysis technique, called electron-energy-loss-spectroscopy (EELS), to examine a silicon nitride ceramic doped with several different rare-earth elements. They specifically looked at how the atomic bonding configuration of the intergranular phase changed with a change in the rare earth sintering additive.

'We were able to determine the exact location of each rare-earth atom and to see how these atoms specifically bonded to the interface between the intergranular phase and the matrix grains of the ceramic', Ritchie says. 'We saw that each rare-earth element attaches to the interface differently, depending on its atomic size, electronic configuration, and the presence of oxygen atoms along the interface. This information can be related to the fracture toughness of the ceramic, which means we should be able to atomistic ally tailor the grain boundaries in future ceramics to give optimum mechanical properties'.

The collaboration says that its results with the silicon nitride ceramic and the rare earth glassy films should be applicable to other types of advanced ceramics as well.

Says Ritchie, 'It's interesting, but intergranular glassy films used to be thought of as an undesirable feature in ceramics, much like inclusions in steels. We now realize they are the key feature that promotes ceramic toughness'.

7.37 A CONVEYOR BELT FOR THE NANO-AGE

In a development that brings the promise of mass production to nanoscale devices, Lawrence Berkeley National Laboratory scientists have transformed carbon nanotubes into conveyor belts capable of ferrying atom-sized particles to microscopic worksites.

By applying a small electrical current to a carbon nanotube, they moved indium particles along the tube like auto parts on an assembly line. Their research, described in the April 29 issue of nature, lays the groundwork for the high-throughput construction of atomic-scale optical, electronic, and mechanical devices that will power the burgeoning field of nanotechnology.

"We're not transporting atoms one at a time anymore — it's more like a hose", says Chris Regan of Berkeley Lab's Materials Sciences Division, who co-authored the article along with fellow materials sciences researchers Shaul Aloni, Ulrich Dahmen, Robert Ritchie, and Alex Zettl. Aloni, Regan, and Zettl are also scientists in the University of California at Berkeley's Department of Physics, where much of the work was conducted. The ability to shuttle a stream of particles to precise locations fills a void that has stymied the efficient assembly of nanostructures. For years, scientists have been able to simultaneously deliver millions of atoms to millions of sites simply by mixing chemicals. Although this fast technique has grown quite sophisticated, it remains far too blunt to build atomic-scale devices. On the other end of the spectrum is the ability to manipulate individual atoms, a feat that came of age in 1990 when IBM researchers spelled out the company logo by positioning 35 xenon atoms with a scanning tunneling microscope. Although precise, this technique is painstakingly slow, with no way to swiftly deliver atoms to the work area.

"It's either all at once, or excruciatingly serial", says Regan. "So we combined incredibly precise localization with something that has higher throughput".

This middle ground is made possible by carbon nanotubes, which are hollow cylinders of pure carbon about ten thousand times smaller than the diameter of a human hair. Since their discovery in the sooty residue of vaporized carbon rods, these incredibly strong and versatile macromolecules have been engineered into frictionless bearings, telescoping rods, and the world's smallest room-temperature diodes. Now, they're poised to change the way these and other devices are constructed.

As described in their nature article, the research team thermally evaporated indium metal onto a bundle of carbon nanotubes. The amount of evaporated metal is so small it populates the tubes' surfaces as isolated indium crystals, instead of uniformly coating them. The bundle is then placed inside a transmission electron microscope, where a tungsten tip mounted on the end of a nanomanipulator approaches one nanotube. After physical contact is made between the tip and the free end of the nanotube, voltage is applied between the tip and the other end of the nanotube, creating a circuit. This sends an electrical current through the nanotube, which generates thermal energy that heats the indium particles.

Next, if the voltage and thermal energy is carefully controlled, something strange occurs. Real-time video of the nanotube's surface captures an indium particle as it disappears, while the particle to its right grows. Several seconds later, that newly enlarged particle also disappears, replaced by another even further to the right. Like squeezing the last bits of toothpaste from a tube, particles to the left become smaller while those to the right grow.

In this manner, the thermally driven indium atoms inchworm along the nanotube, momentarily occupying a reservoir where a particle is located, and then moving to the next, until all of the indium piles up at the end of the nanotube. In the future, this nano-sized conveyor belt could be aimed anywhere scientists want to deliver mass atom-by-atom — the makings of a formidable nanoassembly tool. Moreover, if the voltage is slightly increased, the indium's temperature increases, and the metal moves from left to right more quickly.

"It's the equivalent of turning a knob with my hand and taking macro scale control of nanoscale mass transport", Regan says. "And it's reversible: we can change the current's polarity and drive the indium back to its original position".

In other words, indium can be repeatedly moved back and forth along the nanotube without losing a single atom. Nothing is lost in transit. This conservation of mass occurs because the atoms don't evaporate from the system during their journey — an advantage in any process meant to deliver valuable material to a worksite. Instead, the atoms hug the nanotube's surface as they move, tethered by a process called surface diffusion.

"To build a structure we have to be able transport material to the construction site, and we're developing a better way to do that", Regan says. "Our nanoscale mass delivery system is simple and reversible. It requires only a nanotube, a voltage source, and something to transport".

"Carbon Nanotubes as Nanoscale Mass Conveyors", by Chris Regan, Shaul Aloni, Robert Ritchie, Ulrich Dahmen, and Alex Zettl, appears in the April 29, 2004 issue of Nature.

7.38 SCIENTISTS POST A LOWER SPEED LIMIT FOR MAGNETIC SWITCHING

The speed of magnetic recording —a crucial factor in a computer's power and multimedia capabilities–depends on how fast one can switch a magnet's poles. An experiment at the Stanford Synchrotron Radiation Laboratory (SSRL) found that the ultimate speed of magnetic switching is at least 1,000 times slower than previously expected. The result, which appears in the April 22 issue of the journal Nature, has implications for future hard disk computer drive technologies.

In the push toward ever-faster magnetic recording, experts expected to find a physical limit, a threshold speed beyond which materials would respond chaotically. "If you had asked me a year ago, 'How fast does one have to create a pulse that does not switch magnetization'? my answer would have been one femtosecond (one thousandth of a trillionth of a second)", said Jo Stöhr, Deputy Director of SSRL. "Chaotic behaviour was not expected in this experiment, which ran in the picoseconds (trillionth-of-a-second) range".

The SSRL is a division of the Stanford Linear Accelerator Center (SLAC), a US Department of Energy (DOE) research facility operated by Stanford University. The collaboration for the nature paper was led by SSRL scientists Hans Christ of Siegmann and Professor Joachim Stöhr, and included researchers from Seagate Technology, the world's largest manufacturer of hard disk computer drives.

"This is a fascinating experiment that has given completely new information on the limits of magnetic switching", said Raymond L. Orbach, director of the DOE office of science. "It is also a wonderful illustration of the value of very different disciplines working together: scientists from a synchrotron light source using a high energy physics linear accelerator to do an experiment on magnetism".

Seagate's Head of Media Research, Dieter Weller underlined, "This collaboration has evolved, over time, into a very fruitful exchange between academia and industry, Aligning ourselves with such high caliber people as Hans Christ of Siegmann and Jo Stöhr is a real feat for us".

In a computer hard drive, a writing head hovers over a disk that's rapidly spinning—at up to 15,000 rotations per minute, or 150 times faster than a CD player. An electric current running in the head creates a magnetic field, which records data by turning tiny areas of the disk's surface into microscopic magnets. The disk is coated with a special, grainy material that allows only two, opposite directions of the magnetization, representing the 0 or 1 of a basic unit of data, or bit. High recording speed requires the coating material to respond and switch its poles quickly enough to record each bit reliably.

The experiment relied on the unique capabilities of SLAC's 2-mile-long linear accelerator (linac). The beam of electrons produced by the linac played the role of the electric current in a hard drive's writing head, based on the fact that moving electrons carry along a magnetic field that swirls around the electrons' path. The idea came to Siegmann in the mid-1990, literally out of a lightning bolt: He realized that the linac could magnetically record the same way that a lightning leaves a magnetic signature when it strikes a rock. The linac's beam, made of tightly packed bunches of electrons traveling close to light speed, creates magnetic pulses that are some of the world's strongest – at up to 10 Tesla, or 200,000 times the strength of the Earth's magnetic field —and briefest, at 2 picoseconds (2 trillionths of a second).

The researchers shot up to seven electron bunches in a row through samples of magnetic recording media. In the photographs of the results, they expected to see dark and light areas, neatly arranged in concentric rings around the focus point of the beam. The two colours would correspond to grains magnetized in either of the two possible directions. Instead, the pictures showed all shades of grey, indicating that some grains had switched while others had not.

They observed similar results with different types of magnetic grains, or even with a continuous magnetic film. With the help of theoretical physicist Alexander Kashuba of the Landau Institute for theoretical physics in Moscow, the SSRL researchers realized that their data bore the signature of a chaotic system—one whose parts behave in a random, unpredictable way. "That's the new thing", said Siegmann. "It's like roulette. You can't tell in advance whether it will be dark or light".

The challenge now will be to understand why the maximum speed seems to be at least 1,000 times lower than expected. The explanation, Siegmann said, could lie in the way thermal motion interacts with the magnetization process.

The limit on recording speed must be somewhere between 100 billion and a trillion bits per second, but is unlikely to ever affect technology, says Seagate's Weller. State-of-the-art drives can now record about 1 billion bits per second, and long before that speed can be increased 100-fold, other physical constraints will get in the way, he says. In particular, higher speed requires smaller magnetic grains, but their size cannot go below the size of atoms.

The SSRL result could be an important step toward understanding the basic physics of data recording, leading to the development of entirely new technologies. A promising idea, Weller says, is heat-assisted recording, where a small section of the recording medium is temporarily brought to a high temperature, to speed up its magnetization reversal.

With the help of SLAC's new Linac Coherent Light Source (LCLS), scheduled to start operating in 2008, researchers will be able to gain a solid understanding of the magnetic properties of matter. The LCLS will use the linac's electron beam to produce laser-like X-ray pulses lasting just one femtosecond, enabling researchers to take snapshots of the magnetization process. "We will take images observing not only what has happened", said Stöhr. "We will be able to see those processes while they happen".

Proof of new state of matter is in the jelly.

In the community of very tiny particles that make up all matter in the universe, there are two main citizens: bosons and fermions. Bosons are socially oriented and tend to stick together, while fermions are solitary entities, preferring to go it alone.

That's why NASA-funded researchers overcame an important technical challenge when they recently persuaded reclusive fermion atoms to act like their friendly boson buddies and jiggle together in an ultra-cold, jelly-like state of matter.

The findings, published in the online version of physical review letters, represent the first direct evidence for a fermion atom super fluid—a frictionless fluid made up of a gas of fermion atoms. Such bizarre atomic soups will help physicists understand how matter behaves at its most fundamental level, and will serve as models for exotic systems in nature, such as neutron stars and new super-high-temperature superconductors that could potentially function at thousands of degrees celsius.

"When you build airplanes, you begin with small-scale models. The same idea can be applied to very high-temperature superconductors, only the models are fermionic super fluid gases", said Dr. John Thomas, a physics professor at Duke University, Durham, North Carolina, and principal investigator for the new study. His research is funded by a grant from NASA's Office of Biological and Physical Research, Washington, through the Jet Propulsion Laboratory, Pasadena, Calif.

Superconductors are materials that permit electrical currents to flow without resistance. They have potential applications in the computer, power and transportation industries. Rail system designers imagine using superconductors to build magnetically levitated trains. But current superconductors require impractically cold temperatures to work. Fermionic super fluid gases—the atomic equivalent of superconducting electrons—might serve as rudimentary blueprints for creating superconductors that work at much higher temperatures.

The latest research builds on previous work by the Duke University team outlining the creation of a possible fermion atom super fluid. In that paper, reported in Science Magazine in 2002, the researchers used an "optical bowl" made of a focused laser beam to trap lithium- 6 atoms (fermions) into a cigar-shaped cloud. They then chilled their concoction to less than a millionth of a degree above absolute zero, a temperature just above the point where no more cooling can occur, and applied a small magnetic field.

The result was a transformation in character; the normally antisocial atoms stacked up like peas in a pod and exhibited unusual behaviour. For instance, the gas expanded in one direction while standing still in another when released from the bowl. Nonetheless, the researchers could not find direct evidence that what they were seeing was an actual super fluid—until now.

This time the researchers repeated the experiment at different temperatures and then observed how long the fermionic gas jiggled, or oscillated, after they switched the optical trap off and on, essentially giving the gas a light tap. In a normal gas, collisions between atoms should decrease as the temperature goes down, producing an imperfect jelly whose oscillations die out quickly. In a super fluid, the atoms act even more unified when temperatures are lowered, and the oscillations should theoretically last forever. What the team discovered was a jelly-like gas that wobbled for increasingly longer times as the temperature decreased — the first direct signs of a fermion atom super fluid and the beginnings of a new model for exploring the possibility of extremely high- temperature superconductors.

8 Nondestructive Testing

INTRODUCTION

Nondestructive testing is the test used to examine a material, to detect imperfections, determine action of its properties or to assess its quality without changing its usefulness. Nondestructive testing is a term generally applied to certain tests that are used in industrial operations. The techniques of ultrasonics and X-ray radiography are used in the field of medicine and Industries as well. Industrial uses of nondestructive testing include monitoring the quality of products in a manufacturing process. It is also useful in the maintenance of process machinery. Nondestructive testing is commonly employed in ship building, in manufacturing of aerospace vehicles and automobiles, metals manufacturing, rail road maintenance, electric power plant construction and maintenance, etc.

8.1 THEORY AND PRACTICE OF ULTRASONIC TESTING

The term ultrasonics is applied to mechanical vibrational waves whose frequencies are higher than the human audibility limit, *i.e.* 20,000 Hz. Mechanical vibrations above human audible range was achieved more than a century ago by Rudoph Koening. It began with the construction of a high power ultrasonic generator by Paul Langevin using a quartz crystal. Next important development was made in 1927, when Wood and Loomis generated ultrasonic waves in the frequency range 100 to 700 kHz by exciting quartz disks at about 5000 volts. Application of ultrasonics for flaw detection in various substances was first suggested by Sokolov in the year 1929. This was the beginning of nondestructive testing (NDT) by ultrasonics. During the year 1945–46, Firestone in USA and Sproul in UK independently developed pulse echo test unit. The developments in the field of transducer technology has played a significant role in the development of ultrasonic test machines. The introduction of polarized ceramic, *i.e.* Barium Titanate — a piezoelectric material in 1947, has revolutionised industrial applications of ultrasonics. This has led to the generation of higher acoustic power and reduced the cost of ultrasonic machines. In recent years, the use of computers in ultrasonics another important advancement in this area.

8.2 PHYSICAL PROPERTIES

Ultrasonic waves obey laws of reflection, refraction and diffraction as in the case of optics. But unlike light they cannot travel through vacuum. These waves propagate through the material as stress or strain waves. Propagation of these waves through the material medium depends upon the elastic properties of the medium.

Ultrasonic waves are classified as longitudinal waves, transverse waves, Rayleigh waves and Lamb waves based on the particle displacement of the media. Longitudinal waves are also known as compressional waves. They are most widely used in the ultrasonic inspection of materials. The waves travel through the material media as a series of alternate compression and rarefaction, in which particles vibrate back and forth in the direction of the wave travel.

Transverse waves or shear waves are also extensively used in the ultrasonic inspection of materials. In this mode the particles of the medium vibrate up and down in a plane perpendicular to the direction of motion.

Surface waves or Rayleigh waves travel along flat or curved surface of thick solids without influencing the bulk of the medium below the surface. The depth to which these waves propagate below the surface with effective intensity is of the order of a wavelength only. Practically all of its energy is attenuated at this depth. These waves are used to detect flaws or cracks on or near the surface of test objects. During the propagation of surface waves, the particles follow elliptical orbits

Lamb waves or flexural waves are also known as plate waves and are produced in thin metal whose thickness is comparable to wavelength.

The ultrasonic waves obey the Snell's laws of reflection, *i.e.*

(*i*) the incident ray, the reflected ray and the normal at the point of incidence, all lie in one plane.

(*ii*) the angle of incidence is equal to the angle of reflection.

The laws governing the phenomenon of refraction of ultrasonic waves are similar to those applicable to light waves, *i.e.*

(*i*) The incident ray, the normal to the refracting surface at the point of incidence and the refracted ray lie in one plane.

(*ii*) The sine of the angle of incidence bears a constant ratio to the sine of the angle of refraction which is equivalent to the ratio of the sound velocities in the media concerned.

i.e. $$\frac{\sin i}{\sin r} = \text{constant} = \frac{c_1}{c_2}$$

where c_1 is the velocity in medium *I* in which sound is incident c_2 is the velocity in medium II in which sound is refracted.

8.3 ACOUSTIC IMPEDANCE

For plane harmonic waves, acoustic impedance is defined as the product of ρ the density of the medium and c the velocity of the ultrasonic waves and is given by ρc. The acoustic impedance is analogous to the refractive index of a transparent medium.

8.4 REFLECTION AND TRANSMISSION AT THE SURFACE

When ultrasonic waves are incident on a surface they are reflected and transmitted. If z_1 and z_2 are the acoustic impedances of the two media then

$$\text{Transmission coefficient} = \frac{\text{Amplitude of transmitted wave}}{\text{Amplitude of incident wave}} = \frac{2z_2}{z_2 + z_1}$$

and $$\text{Reflection coefficient} = \frac{\text{Amplitude of reflected wave}}{\text{Amplitude of incident wave}} = \frac{z_2 - z_1}{z_2 + z_1}$$

8.5 MODE CONVERSION

When ultrasonic wave strikes an interface between two materials having different acoustic impedance at oblique incidence, some of its energy is converted into modes of vibration other than the incident mode is as shown in Fig. 8.1.

For any angle of incidence, other than the normal every longitudinal wave has a reflected and refracted component of both longitudinal and shear wave. As the angle of incidence increases, the angle of refraction for longitudinal waves reaches 90°. The angle of incidence corresponding to 90° then the angle of refraction is called the first critical angle. For greater angle of incidence the longitudinal wave is totally internally reflected in medium I and no longitudinal mode exists in medium II. Further if the angle of incidence is increased till the angle of refraction for shear wave mode is 90°, the incident angle is called the second critical angle for shear wave.

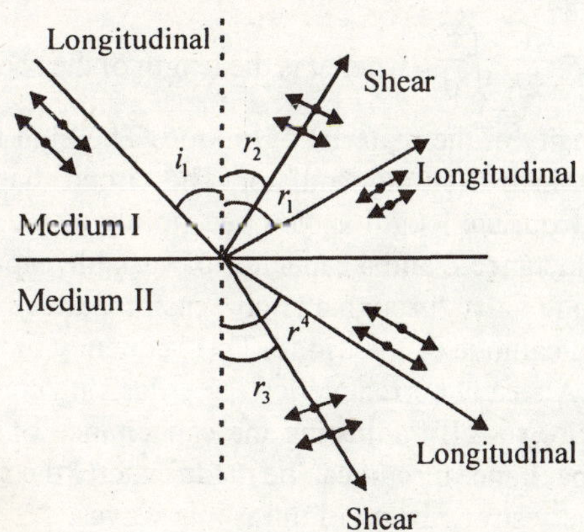

Shear and longitudinal vibrations produced for oblique incidence

Fig. 8.1

If the angle of incidence is further increased, total reflection for both longitudinal and shear modes occurs. Similar situation holds good for shear wave incident at the interface.

Thus, both shear and longitudinal waves are generated at the interface, when the angle of incidence is less than the first critical angle. Besides these it is found that Rayleigh and Lamb waves are also generated. Ultrasonic waves undergo diffraction just like light and sound waves. When the ultrasonic waves travel through the medium, they get scattered and part of the energy is absorbed. The loss of ultrasonic energy due to scattering and absorption is referred to as attenuation.

8.6 PRODUCTION OF ULTRASONIC WAVES

Ultrasonic waves can be generated by the method to be employed depends upon the power output necessary and the frequency range to be covered. Generators of mechanical type such as Galton's whistle can be used upto 10,000 CPS. For generating still higher frequencies, piezoelectric effect or magnetostrictive effect is made use of.

8.7 GALTON'S WHISTLE

Galton devised a miniature organ pipe which is in the form of a whistle to generate ultrasonic waves of low frequency. At one end of the pipe there is a movable plunger and its position can be varied by means of a micrometer screw. By this means the length of the air column within the pipe and consequently the pitch of the note can be varied.

8.8 MAGNETOSTRICTION GENERATOR

The principle of this method is that when a rod of ferromagnetic material, specially nickel, is placed parallel to a magnetic field, it suffers increase in length depending on the strength of the field. So if an alternating magnetic field is parallel to a magnetic field, it suffers an increase in length depending on the oscillating field. If, however, the rod magnetised first by a steady current and then an alternating current is superimposed on it, the rod will oscillate with the frequency of the alternating current.

But this will not generate ultrasonics, because its vibrations being a forced one amplitude of vibrations is very small. By changing the frequency of the alternating current the rod may be set to resonant vibration when the rod will vibrate with the maximum possible amplitude, giving rise to a very strong beam of ultrasonics.

This happens when the frequency of the oscillating field is equal to the natural frequency of the rod given by

$\eta = \frac{1}{2l}\sqrt{\frac{E}{\rho}}$, where l is the length of the rod, E = axial Young's modulus of the material of the rod and ρ is the density of the material of the rod. The high frequency alternating current required to generate ultrasonic is obtained from an oscillator. The circuit diagram of the generator is shown in the Fig. 8.2 A nickel rod of appropriate length is clamped rigidly at the centre. The oscillating circuit consisting of an inductive coil of inductance L and a capacitor of variable capacitance C is connected between the plate and the cathode of the triode valve through a High Tension battery. The feed back coil L of the oscillator is placed between the grid and cathode of the triode. The frequency of the oscillator is given by.

A steady current called the polarising current is passed through a separate coil wound uniformly over the entire rod. By adjusting the capacitance of the capacitor C the frequency of oscillations of the inductance capacitance circuit can be made exactly the same as that of the rod. The rod oscillating in resonance produces a very strong beam of ultrasonic waves.

8.9 PIEZOELECTRIC GENERATOR

This method is based on the piezoelectric effect. In 1880 P. Curie discovered that if some crystals such as quartz, tourmaline, rochelle salt, etc. a mechanical pressure or tension is exerted in a specific direction a maximum potential difference is developed in a direction perpendicular to it. This fact is known as piezoelectric effect. The two axes are respectively called mechanical axis and electrical axis. The effect is most prominent in quartz. Quartz crystals occur as hexagonal prisms with pyramids at each end. The axis ZZ' joining the apex of the pyramids is called optic axis of the crystal. The section of the crystal perpendicular to the optic axis is a regular hexagon. The line perpendicular to two opposite sides of the hexagon is the mechanical axis of the crystal and the line perpendicular to it is its electrical axis. A rectangular cut of the crystal is taken in such a way that its thickness t is along the electrical axis, the length l is parallel to the mechanical axis and the breadth b is parallel to the optic axis. Such a cut is called an X-cut crystal. The natural frequency of the crystal along its thickness is $\eta_t = \frac{1}{2t}\sqrt{\frac{E}{\rho}}$, where E is the Young's modulus of the crystal and ρ is its density. The natural frequency of the lengthwise vibration is $\eta_l = \frac{1}{2l}\sqrt{\frac{E}{\rho}}$.

The complete outfit for generation of ultrasonics is shown in the Fig. 8.4. The alternating emf from an oscillator is applied between two parallel plates when an alternating electric field is produced between the plates. The crystal cut Q is subjected to this alternating electric field by placing it between the plates. The resonant circuit LC is tuned to the natural frequency of the crystal. R_g is the gridleak resistance. C_1 is a capacitor which by-passes the high frequency alternating current. The circuit shown in the diagram is the well-known Hartley oscillator.

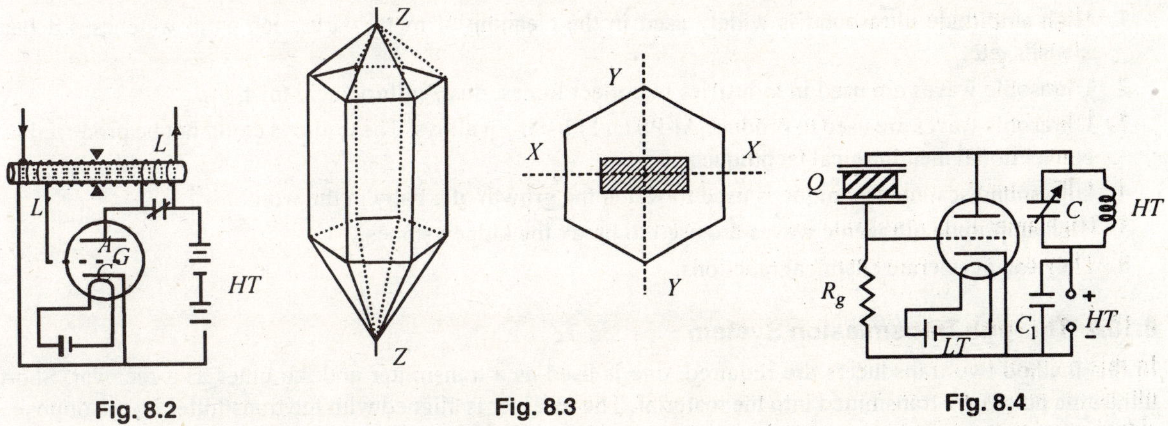

Fig. 8.2 **Fig. 8.3** **Fig. 8.4**

8.10 ULTRASONIC TESTING SYSTEMS

There are three basic ultrasonic test systems.

8.10.1 Pulse Echo System

This system is most commonly employed. Here short pulses of ultrasonic waves are transmitted into the material being tested. These pulses get reflected from discontinuities on their path or from any boundary of the material on which they strike. The received echoes are then displayed on a cathode ray tube screen as shown in Fig. 8.5. The CRT screen furnishes specific data as to the relative size of a discontinuity in terms of signal amplitude. The location of the discontinuity with respect to the scanning surface can be obtained by proper calibration of the CRT time base scale. Sometimes a single transducer is used both as a transmitter and a receiver. Otherwise two transducers are separately used for this purpose.

1. The tested specimen will not be damaged. The same specimen can be used again for the purpose for which it was prepared.

a = Transmitted echo

b = Flow echo

c = Back echo

Tr = Transducer

CRT = Cathode ray tube

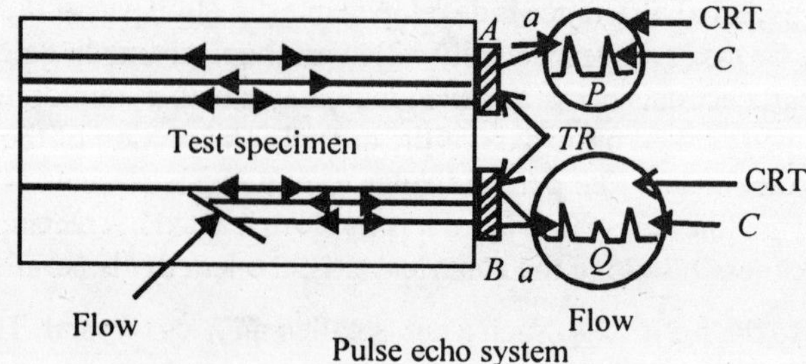

Fig. 8.5

2. This technique is cheaper and the speed of operation is high.
3. NDT is more sensitive than radiography where X-ray is used to find the defects.

Sonar: Sonar is an acronym for sound navigation and ranging. Sonar is used to determine the distance of a body remaining deep inside the sea. The working principle of sonar is similar to that of radar. By measuring the time interval between the transmitted ultrasonic pulse and the received echo signal, the distance of the body is calculated. Knowing the change in frequency of the echo signal, the velocity and the direction of a moving body can be determined by Doppler effect.

Other applications

1. High amplitude ultrasound is widely used in the cleaning of metal parts such as in watches, kitchen utensils, etc.
2. Ultrasonic waves are used in industries to collect fumes, dust, sulfuric acid mist, etc.
3. Ultrasonic waves are used to produce Al-Pb and Pb-Sn-Zn alloys. These alloys could not be produced by conventional metallurgical techniques.
4. Ultrasound scannic technique is used to watch the growth of a baby in the womb.
5. High amplitude ultrasonic waves are used to break the kidney stones.
6. They can accelerate chemical reactions.

8.10.2 Through Transmission System

In this method two transducers are required, one is used as a transmitter and the other as a receiver. Short ultrasonic pulses are transmitted into the material. The receiver is aligned with the transmitter on the opposite side to pick up the sound waves which pass through the material. The quality of the material is indicated by the energy lost by the sound beam as it transverse the material as shown in Fig. 8.6. A marked reduction in received energy amplitude indicates a discontinuity.

8.10.3 Resonance System

To detect flaws in a given object. The resonance phenomenon issued here continuous longitudinal waves are transmitted into the material. The frequency of the waves is varied till the standing waves are set up within the specimen causing the specimen to vibrate or resonate at a greater amplitude as shown in Fig. 8.7. Resonance is then sensed by a device and presented on the cathode ray tube. A change in resonant frequency which cannot be traced to a change in material thickness, is usually an indication of a discontinuity.

8.11 DETERMINATION OF VELOCITY OF ULTRASONICS IN LIQUIDS

In a longitudinal wave and hence in ultrasonics as well pressure varies periodically. The density and refractive index of the liquid of medium will also show a periodic variation because the relation of density and refractive

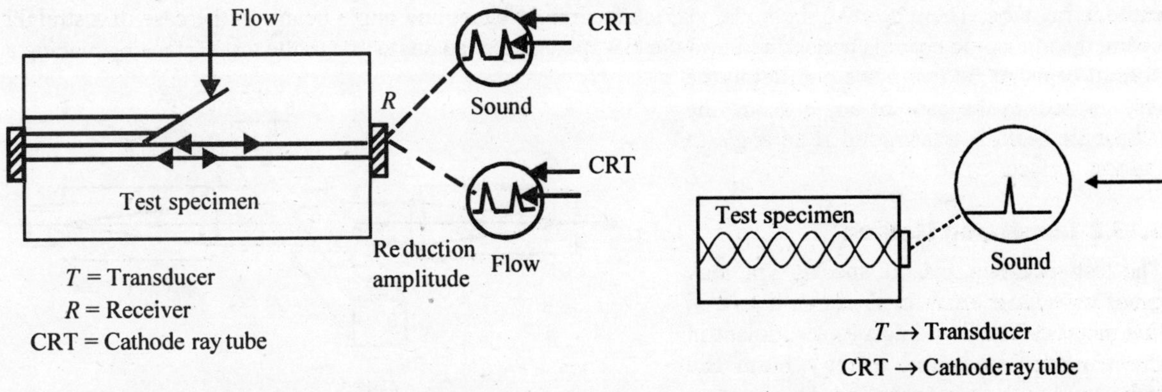

Fig. 8.6 Through transmission system

Fig. 8.7

index is $\frac{\mu - 1}{\rho} = \frac{\mu' - 1}{\rho'}$ known as Gladstone and Dale's law. If there is stationary ultrasonic waves formed in a confined liquid, the density of the liquid will be maximum at the nodal planes and minimum at the antinodal planes. Hence, the whole medium will act like a plane transmission grating.

This is called *acoustic grating*. The grating element will be equal to the wavelength of the wave. If now a strong beam of monochromatic light is passed through this 'acoustic grating' diffraction pattern corresponding to plane transmission grating will be formed. If θ is the angle of diffraction the *n*th order maxima, then $\lambda_u \sin \theta = n\lambda$ where λ_u is the wavelength of ultrasonic waves and λ is the wavelength of monochromatic light.

But c (velocity of ultrasonics) $= v\lambda_u$

where v = frequency of ultrasonic wave

= frequency of the oscillator generating ultrasonics.

$$\therefore \quad c = \frac{nv\lambda}{\sin \theta}.$$

From this, c can be calculated.

8.12 EXPERIMENTAL ARRANGEMENT

The experimental liquid is taken in a rectangular glass vessel called the cell and at one end of it is placed a quartz crystal actuated by a valve oscillator. This is placed on the table of a spectrometer set up for parallel rays. The length of the cell is set at right angles to the collimator. The waves emitted by the crystal parallel to the length of the cell are reflected back from the opposite wall of the cell and a stationary wave pattern is formed in the liquid. The light beam passing through the cell which acts as transmission grating is diffracted. Measuring the angle of diffraction with the help of telescope and the attached scale, velocity of ultrasonics in a liquid can be determined.

8.13 ULTRASONIC TESTING METHODS

There are two methods of ultrasonic testing, (*i*) Contact method, and (*ii*) Immersion method.

8.13.1 Contact Method

In this method the transducer is brought in contact with the test specimen through the use of a suitable couplant like water, oil, grease, etc. the ultrasonic wave reflected from the flaw or the discontinuity is displayed on the cathode ray tube. There are two methods, viz. using straight beam and angle beam. In the case of a straight beam, the ultrasonic beam is transmitted into the test specimen at an angle 90° to the test surface by placing a normal beam probe in a plane parallel to the test surface. In the case of angle beam, the ultrasonic beam is transmitted at an angle to the test surface.

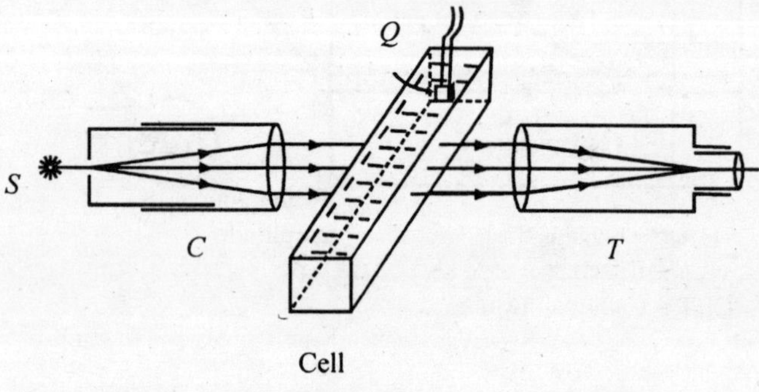

Fig. 8.8

8.13.2 Immersion Method

The test specimen and the special type leak proof transducer are immersed in a liquid, in this method. The liquid acts as a couplant in the transfer of sound energy from the transducer to the test specimen. This method provides testing flexibility since transducer can be moved under water to introduce the sound beam at any desired angle. The transducer does not touch the specimen and therefore is not subjected to wear.

In the immersion method we can have pulse-echo, and through transmission systems as shown in Figs 8.7–8.10.

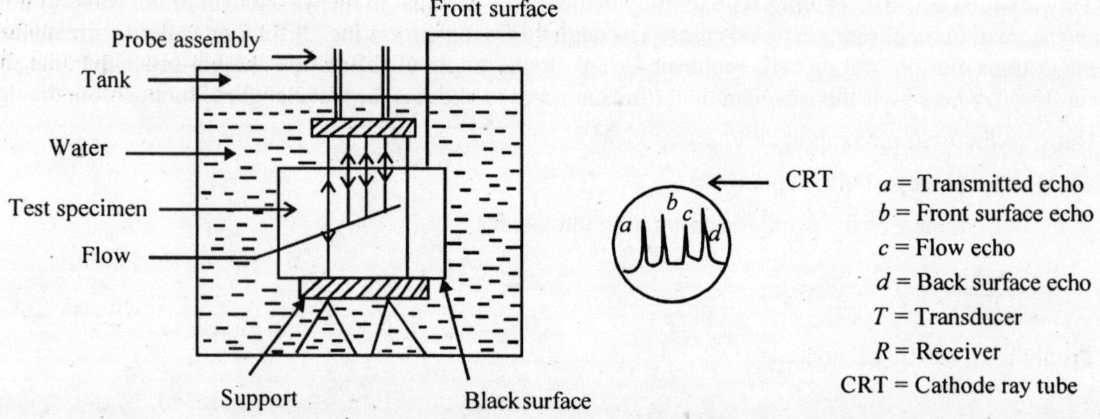

Fig. 8.9 Immersion pulse echo method

8.14 SURFACE WAVE TECHNIQUES

Surface waves are also employed for the detection of flaws near the surface such as cracks, tool marks, etc.

8.15 ULTRASONICS IN MEDICINE

The ultrasound waves have been widely used for examining the shape and movement of organs within the body. The images of ultrasonic waves that are reflected from the boundaries of the organs are obtained and is known as ultrasound scanning. The ultrasonic waves reflected from moving objects such as red blood cells exhibit a Doppler shift frequency that is used to measure the flow of blood.

8.16 PROPERTIES OF ULTRASONICS

(*i*) The energy current, that is, intensity of ultrasonic waves is very large because of their high frequency ($I \propto \eta^2$).

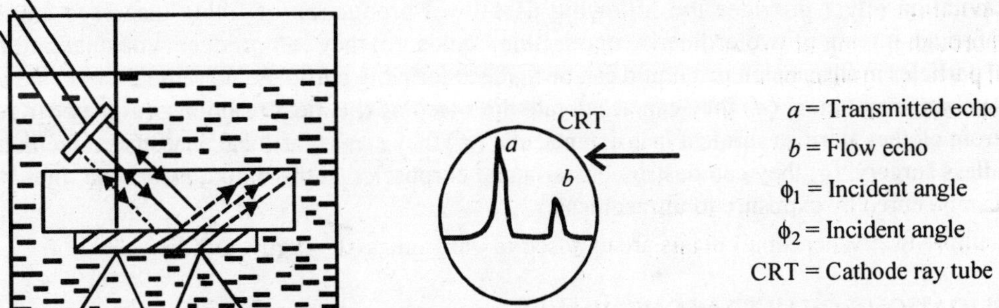

Fig. 8.10 Angle beaming using longitudinal waves

(*ii*) Because of their short wavelengths the diffraction of ultrasonic waves is very small. They can give a well collimated beam without spreading rays even after a very large distance.

(*iii*) They have got very strong disruptive effect called the cavitation effect. Let us explain this property by simple figures. If an ultrasonic wave has frequency 3×10^8 Hz and an energy current 10^3 Wm^{-2}, calculations based on sinaple harmonic motion will show that in such a wave, particles of the medium will have a maximum acceleration of 10^5 times the acceleration due to gravity and a pressure variation of ± 5 atmospheres. Such large stresses in an ultrasonic wave propagation through a liquid cause tearing apart of the liquid with production of hollow spaces filled with vapour. This hollow spaces collapse with great violence. This is what is called the cavitation effect.

(*iv*) They can propagate in all three media—liquid, gas and solid-whether conducting or nonoconducting. Electromagnetic waves are heavily damped in conducting media. Sea-water is conducting and so an electromagnetic wave cannot propagate very far through sea-water, but ultrasonics can easily do so. That is why in SONAR ultrasonics are used with greater advantage.

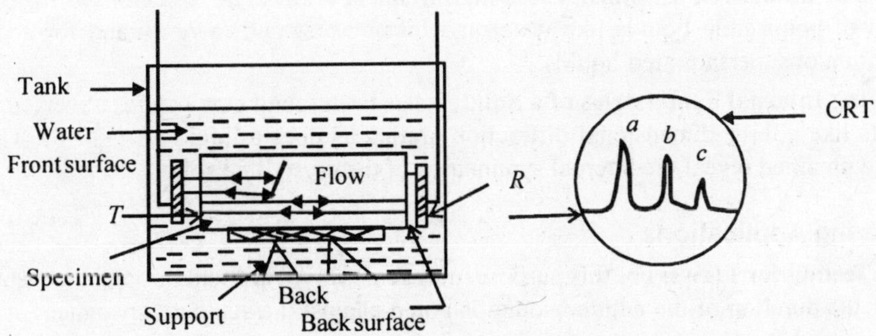

Fig. 8.11 Immersion through transmission system

8.17 APPLICATIONS

The uses of ultrasonics are numerous. Because of their short wavelengths, they are convenient in echo sounding or antisubmarine detection than ordinary sound waves recently, ultrasonics have become a new tool in industry, in medicine, in agriculture, etc.

(*i*) In industry ultrasonics are used conveniently than X-rays in flaw detection in metal structures, because production and detection can be done by a small unit, incorporated into one called a 'transducer'.

(*ii*) Its cavitation effect provides the following uses—(*a*) production of finely dispersed emulsions, (*b*) thorough mixing of two ordinarily immiscible liquids, (*c*) they can produce 'coagulation', that is, small particles in suspension in a liquid can be made to form big particles and settle down at the bottom by supersonic agitation, (*d*) they can accelerate the speed of chemical reactions, (*e*) they can remove dirt from clothes sterilise surgical instruments, etc. (*f*) they can act as local anaesthesia facilitating in bloodless surgery, (*g*) they can destroy bacteria and corpuscles of the blood. Neuralgia, rheumatism, etc. can be cured by exposure to ultrasonics.

(*iii*) Agriculture use: When small plants are exposed to ultrasonics, they grow rapidly.

8.18 APPLICATIONS OF ULTRASONIC WAVES

Ultrasonic waves have found many industrial, scientific and medical uses.

8.18.1 Scientific Applications

(*i*) **Ultrasonic Signalling:** Ultrasonic waves of 40 kHz frequency can be obtained in the form of a narrow beam which can travel many kilometres in water before being absorbed. Hence, they have been found very suitable for direction signalling in submarines and for talking from one vessel to another. The quartz vibrations are modulated by the speech frequencies in the same manner as in radio transmission.

(*ii*) **Depth Sounding and Submarine Detection:** Such applications use the echo sounding technique. For depth sounding, a beam of ultrasonic waves is directed towards the bottom of the sea from where it is reflected back to a suitable receiver. The time taken by the waves to go and come back is recorded accurately. If the velocity of ultrasonic waves in water is known, the sea depth can be calculated easily. The same method is used for finding the distance and direction of a submarine or a shoal of fish in the sea. The whole system is known as sonar which stands for sound navigation and ranging.

(*iii*) **Photographic Emulsions:** Ultrasonic waves have been used to improve the homogeneity and stability of photographic emulsions. Their colour sensitivity has also been increased by thorough dispersion of the dye in the emulsion under the influence of ultrasonic waves.

(*iv*) **Emulsions of Immiscible Liquids:** These high-frequency waves have been used for forming stable emulsions of immiscible liquids like water and oil or water and mercury and for accelerating the crystallisation of supersaturated liquids.

(*v*) **Determining Internal Symmetries of a Solid:** It has been found that a solid, traversed by ultrasonic waves acts like a three-dimensional diffraction grating as does a liquid (Art. 17-6). The diffraction patterns so obtained reveal the internal symmetries of the crystalline solid.

8.18.2 Engineering Applications

(*i*) **Material Testing for Flaws:** For this purpose, ultrasonic waves are sent through the metal to be tested and either the duration of the emitted sound is noted against the standard specimen, or the resulting reflection or a absorption of waves is measured. A noticeable reflection or absorption of the waves occurs at any blow hole or crack in the interior of the metal specimen. It is so because, in either case, there is change of medium and this results in the reflection of waves both from the blow hole and the back surface of the metal. This time interval between two reflections is recorded by an oscilloscope and is utilised for locating the position of the flaw in the metal. Cracks having thickness as small as 0.005 mm can be detected in this way. The detection of such flaws in metals prior to their failure whilel in use is of great practical importance.

(*ii*) **Soldering:** Metals have been soldered or brazed by subjecting them to ultrasonic vibrations.

(*iii*) **Drilling:** Ultrasonics have also been used for drilling and meal cutting purposes.

(*iv*) **Metallurgical Use:** Ultrasonics have been put to many metallurgical uses. Many metals such as iron-lead, aluminium-cadmium and zinc-lead, etc. which cannot be alloyed in their liquid state can be brought together when subjected to high-intensity ultrasonic waves.

8.18.3 Medical and Biological Applications

(*i*) **Biological Effect:** It has been found that smaller animals like frog and fish are either maimed or killed outright when exposed to ultrasonic waves because of the destruction of their blood vessels. In the case of microorganisms like bacteria, etc. these waves either promote or retard their agglutination and virulence.

(*ii*) **Disease Treatment:** Ultrasonic therapy has been used to treat diseases such as bursitis, abscesses and lumbago, etc.

(*iii*) **Surgical Use:** Surgical use of ultrasonic waves includes the selective cutting of the tissue during an operation.

(*iv*) **Locating Abnormal Growths:** Cerebral ventricles are explored by ultrasonic waves for locating abnormal growths in the brain. For this purpose, the waves are passed through the brain and then picked up by means of separate transmitters and receivers. A special equipment is used to examine the different acoustic losses from different parts of the head both by reflection and absorption in it.

(*v*) **Dental Cutting:** Ultrasonic waves have been found very useful for dental cutting because (*a*) they make the cutting almost painless, (*b*) they cut the hard material very easily, and (*c*) they do not require any mechanical device for cutting purpose.

(*vi*) **Cleaning and Washing:** Recently, ultrasonic waves have been used industrially for large-scale washing of clothes especially silken fabrics in hospitals, factories and textile industry. Other uses include cleaning of utensils, removing dust and soot from chimney gases, dispersing of fog at airports for enhanced visibility and removing water from paper during its manufacture.

8.19 X-RAY RADIOGRAPHY

X-Rays have been put into numerous applications in the fields of research, medicine and industry. Roentgen's experiments revealed that the X-Rays have extraordinary penetrating power. Radiography progressed little beyond laboratory curiosity for another 20–25 years. The invention of Cooledge X-ray tube in 1913, providing greater output of X-rays at higher energies made industrial radiography a practical proposition. Radiography is found to be most satisfactory for finding internal defects such as porosity and voids. It is also suitable for detecting changes in material composition, for thickness measurement and for locating unwanted components hidden from view in assembled parts.

8.20 PHYSICAL PROPERTIES

X-rays like visible light originate from the transition of an atomic electron from one orbit to another. X-rays are also produced when high speed electrons strike a solid traget. Normally metal tragets are employed. The interaction of high speed electrons with metal target produces two kinds of radiation.

(*i*) Continuous X-ray spectrum consisting of all wavelengths and

(*ii*) Characteristic radiation, which is monochromatic radiation typical of the target metal.

The continuous X-ray spectrum is due to the slowing down of high speed electrons as they pass close to the nuclei of the atoms within the target of the X-ray tube as shown in Fig. 8.12 (a). The characteristic X-ray radiation is due to the following process. Sometimes a fast electron strikes a *K*-electron in a target atom and knocks it out of its orbit and free of the atom. The vacancy in the *K*-shell is filled almost immediately when an electron from an outer shell of the atom falls into it, as shown in Fig. 8.12 (b) and in the process a characteristic X-ray photon is emitted. Continuous radiation is of interest from the view point of general industrial radiography, while characteristic radiation has limited radiographic application.

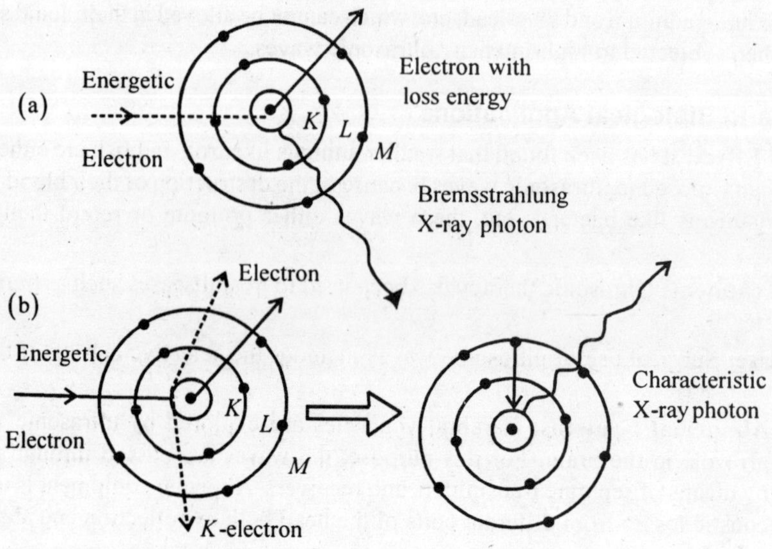

Fig. 8.12

The general arrangement for the production of X-rays is shown in Fig. 8.13. X-rays are produced in a highly evacuated glass envelope in which two electrodes are separated by a small distance. The cathode is electrically heated to produce electrons. A high voltage of several thousand volts is applied between cathode and the anode. Electrons emitted from the cathode are accelerated by the applied field and strikes the target metal giving rise to X-rays. Typical X-ray spectrum obtained is shown in Fig. 8.14.

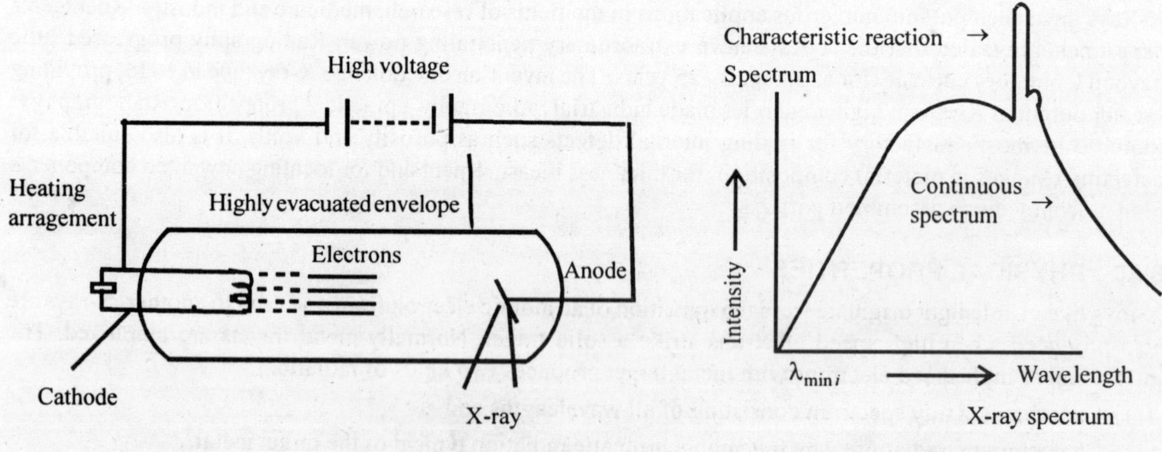

Fig. 8.13 General arrangement for generating X-rays

Fig. 8.14

Figure 8.12, generation of X-rays (*a*) a test electron is diverted near the nucleus and loses some of its energy as a bremsstrahlung X-ray photon, (*b*) a fast electron knocks a K-electron free of the atom; when an outer electron falls into the vacancy an X-ray characteristic of the target atom is emitted.

Properties of X-ray

(*i*) They are electromagnetic waves which can propagate in vacuum.

(*ii*) They are unaffected by electric and magnetic fields.

(*iii*) They propagate in straight lines.

(*iv*) They ionize gases.

(*v*) They are capable of blackening the photographic plate.

(*vi*) They produce fluorescence and phosphorescence. Fluorescence and phosphorescence are phenomena in which some materials exhibit visible light on being excited by X-rays.

(*vii*) Damage or kill living cells and produce genetic mutations.

(*viii*) Liberate photoelectrons.

In so far as industrial radiography is concerned the following properties are relevant.

(*i*) Rectilinear propagation.

(*ii*) Differential absorption in matter.

(*iii*) Ability to affect photographic plates.

(*iv*) Fluorescence.

(*v*) Liberation of photoelectrons.

8.21 INTERACTION OF X-RAYS WITH MATTER

When X-rays pass through matter, it interacts with it in a complex manner. The effect of this interaction is the attenuation of the X-ray radiation. The X-ray intensity when it passes through matter can be written as

$$I_x = I_o\, e^{-\mu x}$$

where I_o is the intensity of the incident beam

I_x is the intensity after traversing a distance x.

μ is the linear attenuation or absorption coefficient.

Major factors for the attenuation of X-rays are

(*i*) Photoelectric emission

(*ii*) Compton scattering

(*iii*) Pair production

(*i*) In photoelectric effect, the incident energy is transmitted to an orbital electron, as a result the electron is no longer bound to the atom and gets released with a particular velocity as shown in Fig. 8.15.

(*a*) in the photoelectric effect all of the photon energy is given to the photoelectron.

(*b*) in the Compton effect, some energy is given to an electron and some goes into a scattered photon.

(*c*) in pair production, a high energy photon is converted into an electron and positron.

(*ii*) In Compton scattering, the incident X-ray photon is scattered by free electron. Due to this collision process there is a change in the wavelength of the incident X-rays photon as shown in Fig. 8.15.

(*iii*) In pair production the incident X-ray photon directly interacts with the nucleus of the atom giving rise to an electron positron pair as shown in Fig. 8.15.

From the point of view of industrial radiography it is of interest to know that

(*i*) Photoelectic emission is predominant around 200 to 100 keV.

(*ii*) Compton scattering predominates in the region 100 to 200 MeV.

(*iii*) Pair production dominates above 6 MeV.

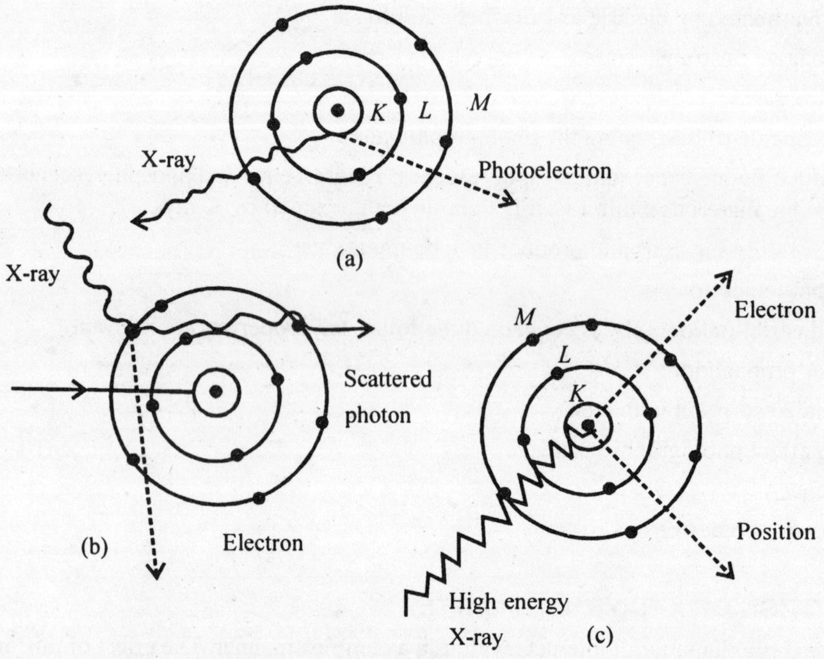

Fig. 8.15

8.22 BASIC PRINCIPLES OF RADIOGRAPHIC INSPECTION

The basic principles of radiographic inspection is shown in Fig. 8.16. The X-ray source emits radiation that travels in straight lines and penetrates the object under investigation. Once the radiation has passed through the material, its image as received on a recording plane opposite to the source is used to evaluate the condition of the part being inspected. Films or fluoroscopic screens are made use of for recording the image. If film is used it must be chemically developed in order to obtain the image.

It is clear that the image of the object will appear on the film provided there is a sufficient difference in the radiation intensities received by the film under the defect as compared to that received through the remainder of the material, *i.e.* the absorption coefficient of X-rays passing through the defect should be considerably different from those passing through the remainder of the material. X-ray radiographic technique is most helpful in locating internal defects such as inclusions, blow holes, porosity and big voids.

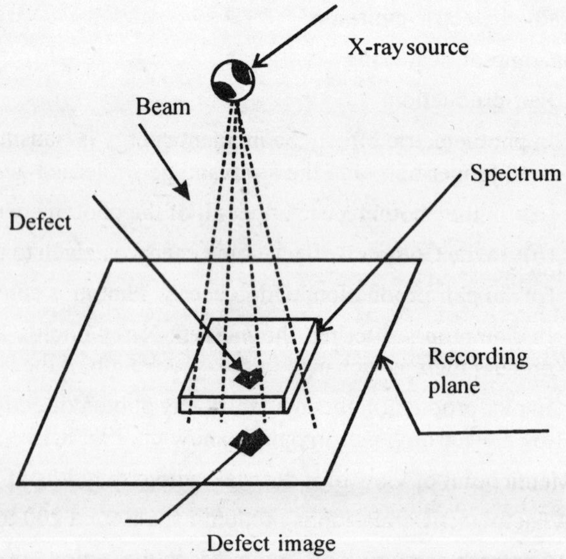

Fig. 8.16 Layout for radiographic inspection

SOLVED NUMERICAL PROBLEMS

1. Calculate the length of the nickel rod required to generate ultrasonics of frequency 40,000 Hz. Young's modulus of nickel = 2.07×10^{11} Nm^{-2}, and its density = 8900 kgm^{-3}.

   Solution: We know that, the expressions for the frequency and the young's modulars are related by the equation is given by

$$n = \frac{1}{2l}\sqrt{\frac{Y}{\rho}}$$

$$l = \frac{1}{2n}\sqrt{\frac{Y}{\rho}}$$

or $$= \frac{1}{2 \times 40{,}000}\sqrt{\frac{2.07 \times 10^{11}}{8900}}$$

$$= 060 \text{ m}$$

   This is the expected length of the nickel rod.

2. The thickness of an X-cut crystal is 20 mm. Calculate the frequency of the thickness vibrations of the crystal. Density of the quartz = 2660 kgm^{-3} and Young's modulus = 7.3×10^{10} Nm^{-2}.

   Solution: We know that, the expressions for the frequency and the young's modulars are related by the equation is given by

$$n = \frac{1}{2l}\sqrt{\frac{E}{\rho}}$$

$$= \frac{1}{2 \times 2 \times 10^{-3}}\sqrt{\frac{7.3 \times 10^{10}}{2660}}$$

or $$n = 1.3 \times 10^{7} \text{ Hz.}$$

   This is the expected frequency.

EXERCISE

1. How are ultrasonic waves produced?
2. Mention some of the properties of ultrasonic waves.
3. What is the principle involved in the ultrasonic testing of materials?
4. Describe ultrasonic testing systems and the methods involved in nondestructive testing of materials.
5. How are x-rays produced?
6. Mention the properties of X-rays as relevant to X-ray radiography?
7. What are the factors responsible for X-ray attenuation in matter?
8. What is the principle behind X-ray radiography?
9. (*a*) Distinguish between ultrasonics and supersonics.

(*b*) Explain in short, how ultrasonic waves are generated with piezoelectric crystal? How is ultrasonic frequency related to the thickness of a crystal?

(c) Give two industrial applications of ultrasonics.

10. Can we use ultrasonic equipment for long distance telecommunication? Why?
11. (*a*) What is piezo-electric effect? Explain how it may be utilized to generate ultrasonic waves.

    (*b*) Draw the circuit diagram of a piezo-electric oscillator and explain its action.

    (*c*) Mention three uses of ultrasonic waves.
12. (*a*) Explain the terms (*i*) magnetostriction, and (*ii*) piezo-electric effect.

    (*b*) Describe piezo-electric method of producing ultrasonic waves. Give three uses of piezo-electric waves.

    (*c*) Give an account of the methods used in the detection of ultrasonics.
13. Write short note on ultrasonics—their properties and methods of production.
14. What are ultrasonics? Explain at least two methods of generating them.
15. Write short note on ultrasonics—their properties and uses.
16. Write short note on ultrasonics and their use in industry.

9 Deformation and Creep in Materials

INTRODUCTION

The permanent deformation of materials on the application of a load can be either plastic deformation or creep. In crystalline materials, at lower temperatures than ~ 0.4 T_m, (where T_m is the melting point in kelvin) the permanent deformation occurs is called plastic deformation. In this temperature range, the amount of deformation that occurs after the application of load is small and is ignored. The rate at which the material is deformed, play a role in determining the deformation characteristics. At temperatures above 0.4 T_m, permanent deformation occurs which is a function of time, under the application of the load. This behaviour is termed as creep.

Table 9.1

Quantity	*Write as*	*SI units read as*	*Other units*
Shear stress τ			
Tensile stress σ	psi, ksi,		
Yield strength σ_y	MN m^{-2}	meganewton per square metre	kgf/mm^2, tons/sq. in.
Ultimate tensile strength (UTS)			
Strain ε	—	—	—
Strain rate ε	s^{-1}	per second	—
Shear modulus μ	GN m^{-2}	giganewton per square metre	psi, kgf/mm^2, dyne/cm^2
Dislocation velocity v_d	m s^{-1}	metre per second	—
Dislocation density ρ	m^{-2}	per square metre	per sq. inch
Grain diameter d	mm, μm	millimetre, micrometre	inch
Hall-Petch constant k	MN $m^{-3/2}$	meganewton per	kgf/$mm^{3/2}$ (metre)$^{3/2}$

9.1 UNITS, CONSTANTS AND CONVERSION FACTORS

Note: Yield strengths and tensile strengths are conveniently written in units of MN m^{-2} ($\equiv 10^6$ Nm^{-2}). In this unit, a low strength such as that of a linear chain polymer is about 20 MN m^{-2} (2 kgf/mm^2), while that of a high strength alloy steel is about 2000 MN m^{-2} (200 kgf/mm^2). MN m^{-2} can also be written as MPa. MN $m^{-2} \equiv$ N mm^{-2}.

Constants

Gas constant R = 8.314 J mol^{-1} K^{-1}

Boltzmann's constant k = $1.380 \cdot 10^{-23}$ J K^{-1}

Conversion factor

1 kgf/mm^2 = 9.81 MN m^{-2}

1 ksi = 6.90 MN m^{-2}

1 psi = 6.90 kN m^{-2}

1 ton/sq. in.	=	15.46 MN m^{-2}
1 dyne/cm^2	=	0.1 N m^{-2}
1 kgf/$mm^{3/2}$	=	0.310 MN $m^{-3/2}$
1 inch	=	25.4 mm

9.2 PLASTING DEFORMATION

The plastic deformation can occur under tensile, compressive or torsional loads.

9.3 STRESS–STRAIN CURVE

A typical tensile stress–strain curve is shown in Fig. 9.1. The applied load is plotted against the elongation or extension of the test specimen shown in Fig. 9.1 (a). The applied load P divided by the initial cross-sectional area A_o of the specimen gives the engineering stress:

$$\text{Engineering stress} = \frac{P}{A_o} \tag{1}$$

Engineering strain is given by the fractional increase in the gauge length l_o:

$$\text{Engineering strain} = \frac{\Delta l}{l_o} \tag{2}$$

where Δl is the increase in gauge length.

The deformation of the specimen is elastic up to the yield point, beyond which it becomes plastic. The load at the yield point divided by the initial cross-sectional area of the test specimen is called the yield stress or the yield strength of a material. Mild steel exhibits a clearly defined yield point, but a number of other materials do not have a clear demarcation between the elastic and the plastic regions.

Beyond the yield point, the linear elastic region is followed by a nonlinear plastic region. In this region, the load required to cause further deformation increases with increasing strain. This phenomenon is called work hardning. As the elongation increases the slope of the load-elongation curve decreases. It becomes zero at maximum load. The engineering stress corresponding to the maximum load is called the ultimate tensile strength (UTS) of the material. Beyond this maximum load a neck forms somewhere in the middle of the specimen, where the cross-sectional area locally decreases. The applied load decreases up to the point of fracture, where the specimen breaks into two pieces across the reduced cross-section of the neck.

The Fig. 9.1 (a) can be replotted in terms of true stress and true strain. True stress is the actual stress, which is the applied load divided by the minimum cross-sectional area A_i of the specimen at any instant is given by

$$\text{True stress} = \frac{P}{A_i} \tag{3}$$

When the increase in length dl is expressed as a fraction of the length l at any instant, then the true strain is given by

$$\text{True strain} = \int_{l_o}^{l} \frac{dl}{l} \tag{4}$$

When the deformation of the specimen becomes nonuniform after necking starts, the true strain becomes a function of the length over which it is measured. In order to avoid ambiguity, it is specified as the integral of $-dA/A$, where A is the cross-sectional area at the neck. The true stress-true strain curve is plotted and is as shown in Fig. 9.1 (b). Unlike the load-elongation curve, there is no maximum in the true stress-true strain curve. The slope in the plastic region decreases with increasing strain, but does not become zero before fracture. This indicates that there is no work softening beyond the maximum in the load-elongation curve. The material work hardens continuously till fracture, although at a decreasing rate.

A power relationship is often used to express the true stress σ as a function of true strain ε is given by

$$\sigma = K\varepsilon^n \tag{5}$$

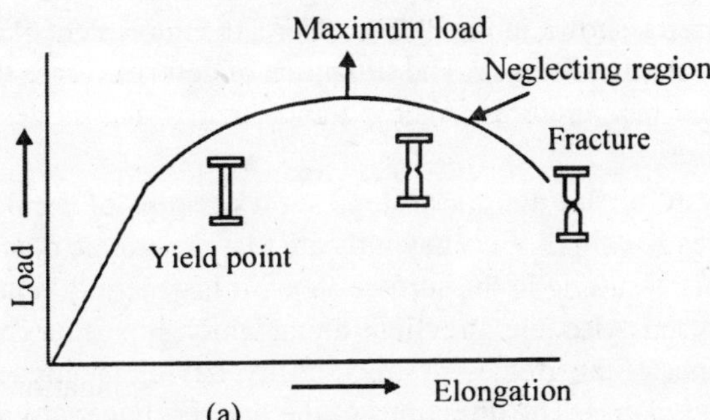

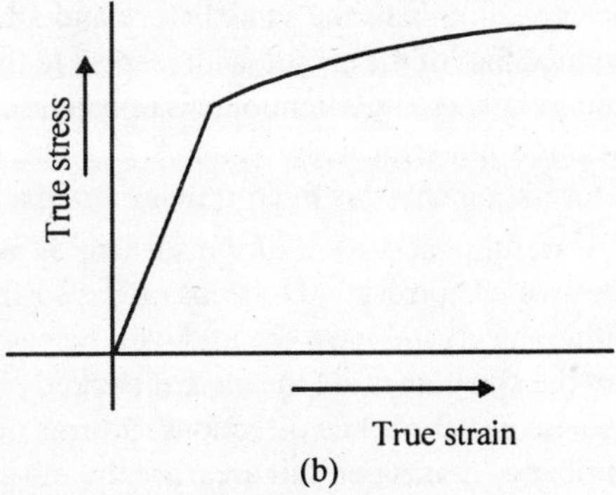

Fig. 9.1 The tensile load-elogation curve and the true stress-true strain curve for a ductile material

where K is called the strength coefficient and n is the work hardening exponent. Materials which have a high work hardening exponent such as copper and brass ($n \sim 0.5$) can be given a large plastic strain more easily than those which have a smaller n, such as heat treated steel ($n \sim 0.15$).

Time does not enter as a parameter in Fig. 9.1. Ordinarily, the deformation characteristics can be taken to be independent of time. If, however, the rate of straining is varied over several orders of magnitude, the stress–strain curve may change as a function of strain rate ε. Another power relationship can be used to express σ at a given strain ε, in terms of the strain rate ε is given by

$$\sigma = A\varepsilon^m \tag{6}$$

where A is a constant and m is the index of strain rate sensitivity. If $m = 0$, the stress is independent of the strain rate and the stress–strain curve would be the same for all strain rates, $m \sim 0.2$ for common metals. If $m = 0.4 - 0.9$, the material may exhibit superplastic behaviour, that is, deform by several hundred percent of strain without necking. The reason for this is that, as soon as necking starts in some region, the strain rate increases locally, resulting in a rapid increase of the stress required to cause further deformation in that region. The deformation then shifts to another region of the material, where there is no necking. Here, the strain rate and hence the stress to cause deformation are smaller. Some strainless steels and aluminium alloys with a very fine grain size exhibit superplastic behaviour. The glass blower is able to pull his working material to very long rods without necking, because the exponent, m for glass approaches one. If $m = 1$, the material behaves like a viscous liquid and exhibits Newtonian flow.

9.4 PLASTIC DEFORMATION BY SLIP

The X-ray diffraction studies shows that the crystalline order in the solid is not lost during plastic deformation, even though more imperfections are introduced. The atom movements are such that the crystal structure remains the same before and after plastic deformation.

There are two basic modes of plastic deformation and called are as slip and twinning. Slip is a shear deformation that moves atoms by many interatomic distances relaltive to their initial positions, as shown in Fig. 9.2. Steps are created at the surface of the crystal during slip, but the orientation of all parts of

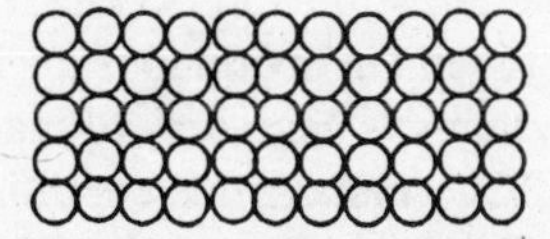

(a) Before slip or twinning

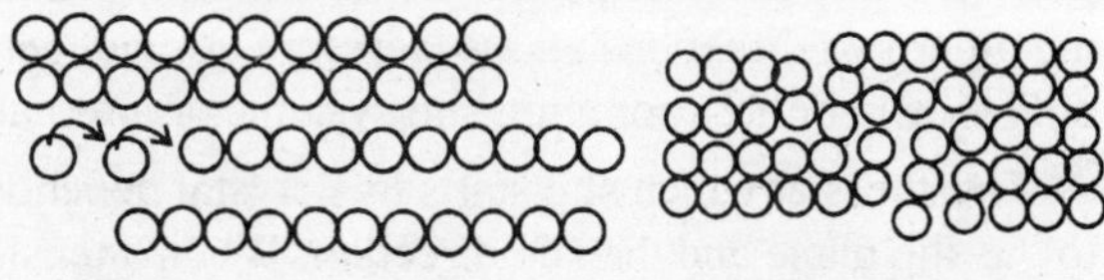

(b) After slip (c) After twinning

Fig. 9.2 The slip mode and the twinning mode of plastic deformation

the crystal remains the same before and after slip. Twinning, on the other hand, changes the orientation of the twinned part of the crystal with respect to the untwinned parts, shown in Fig. 9.2 (c). Here, the movement of an atom relative to its neighbours is only a fraction of an interatomic distance. The slip mode of deformation is the common mode in many crystals at ambient and elevated temperatures. At low temperatures, the mode of deformation changes over to twinning in a number of cases.

Careful observation of the surface of a deformed crystal under the microscope shows groups of parallel lines which correspond to steps on the surface. They are called slip lines. They indicate that the atomic planes within the crystal have sheared with respect to one another resulting in the surface steps. It is generally found that the slip planes are the closest packed planes in the crystal. Also, the directions along which slip occurs are also the closet packed directions. It turns out that the planes of the greatest atomic density, having the highest number of atoms per unit area, are the most widely spaced planes. The directions of the greatest linear atomic density have the smallest translation distance from one minimum energy position to the next. In ionic crystals, the slip planes and the slip directions are such that the ions of the same polarity do not become just opposed as nearest neighbours during shear, as this would mean a huge increase in the potential energy of the crystal. The common slip planes and slip directions for some simple crystals are given in Table 9.2.

Table 9.2

Crystal		*Slip planes*	*Slip directions*
FCC		(111)	< 110 >
BCC	more common	(110)	
	less common	(112), (123)	< 111 >
NaCl		(110)	< 110 >
HCP	more common	basal plane	
	less common	prismatic and close packed	
		pyramidal planes directions	

9.5 SLIP PLANES AND DIRECTIONS IN SOME CRYSTALS

A slip plane and a slip direction that lies on it together constitute a slip system. For example, the combination (111) and (110) forms a slip system, but not (111) and (110), as the (110) direction does not lie on the (111) plane. Counting the slip systems for the most densely packed slip planes only, there are 12 slip systems in FCC and BCC crystals, while there are only 3 in HCP crystals. The NaCl crystal has 6 slip systems.

In a polycrystalline material, slip in any crystal has to be accommodated by slip in neighbouring crystals, if the grain boundaries are to remain continuous during slip. According to the Von Mises criterion, a minimum of five independent slip systems is necessary to maintain the integrity of the grain boundaries during plastic deformation. In FCC and BCC crystals, this condition is fulfilled. In HCP crystals, for slip on the basal close packed plane, there are only two independent slip systems. These correspond to any two of the three close packed directions. A slip displacement in the third direction can be expressed as the resultant of the slip along the other two directions and is therefore not independent. In polycrystalline form, HCP materials can deform only by slip on less common slip systems or alternatively by twinning.

The stress at which slip starts in a crystal depends on the relative orientation of the stress axis with respect to the slip plane and the slip direction. When a tensile stress σ is applied to a crystal, as shown in Fig. 9.3, the shear stress τ resolved on a slip plane whose normal makes an angle of ϕ_1 with the stress axis, along a slip direction inclined at an angle of ϕ_2 to this axis, is given by

$$\tau = \sigma \cos \phi_1 \cos \phi_2 \tag{11.7}$$

This resolved shear stress should reach a critical value called the critical resolved shear stress (CRSS) for plastic deformation to start. It is evident that all slip systems in a crystal will not have the same resolved shear stress for a given tensile stress along an axis. As the applied tensile stress is increased from zero, deformation will be initiated first on that slip system for which the resolved shear stress is a maximum and so reaches the critical value first.

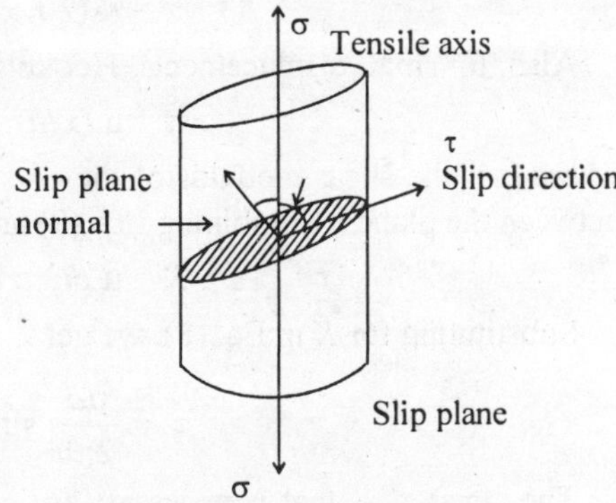

The resolved shear stress τ on a slip plane direction depends on their orientation with respect to the stress axis

Fig. 9.3

9.6 THE SHEAR STRENGTH OF PERFECT AND REAL CRYSTALS

Figure 9.4 shows that how a perfect crystal can undergo plastic deformation by the slip process. By using simplifying assumptions, we can estimate the stress required to cause slip in the perfect crystal. The Fig. 9.4 shows the stress-potential energy-displacement relationships, when a plane perpendicular to the paper is sheared towards the right by one interatomic distance with respect to another parallel plane. At position 1, no stress is applied and the potential energy is a minimum. The stress τ required to cause a displacement x will be proportional to dW/dx, where W is the potential energy. As the atoms of the top plane are displaced to position 2, the applied stress reaches its maximum value, corresponding to the inflection point on the potential energy curve.

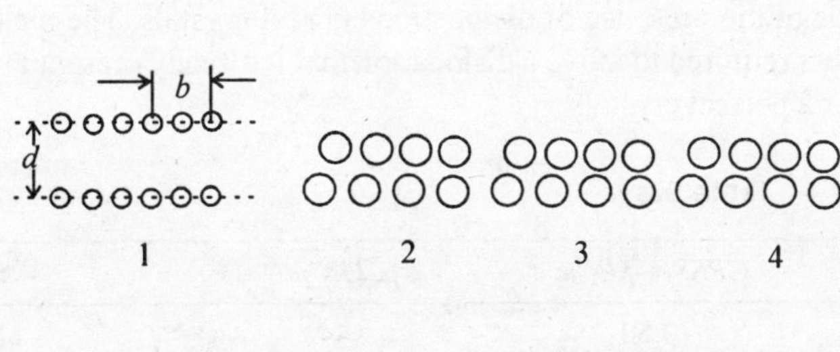

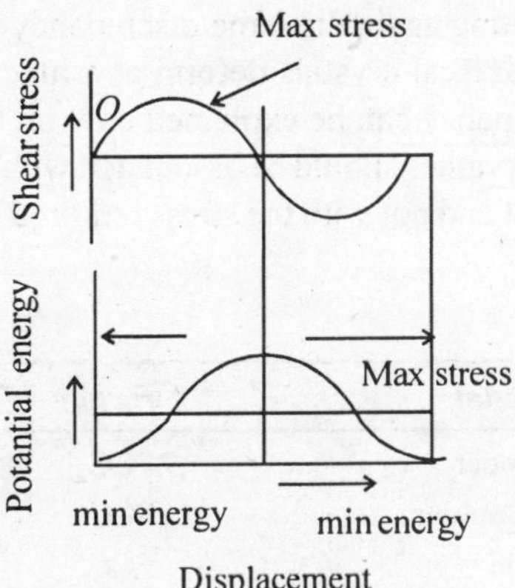

Fig. 9.4 Stress-potential energy-displacement curves during shear over one interatomic distance in a perfect crystal

Position 2 is quarter way from position 1 in going to the next equilibrium position. Beyond position 2, the slope of the potential energy-displacement curve decreases. Consequently, a progressively-decreasing amount of stress will be required to cause further displacement, until the stress falls to zero at the half way position 3. This position is in unstable equilibrium and the potential energy here is a maximum. Beyond 3, the slope of the potential energy curve becomes negative. The stress is also negative till the next equilibrium position is reached. The negative sign of the stress means that the stress in the opposite direction is necessary to stop the top plane from sliding into position 4, after it has crossed position 3. If the stress-displacement curve is taken to be sinusoidal, we can write:

$$\tau = K \sin (2\pi x/d) \tag{8}$$

where d is the atomic spacing in the direction of displacement and K is a constant. For small displacements, $2\pi x/d << 1$ and so we can write

$$\tau = K(2\pi x/d) \tag{9}$$

Also, for small displacements, Hooke's law is valid and hence,

$$\tau = \mu(x/a) \tag{10}$$

where μ is the shear modulus of the crystals and x/a is the shear strain, a being the distance of separation between the planes. Combining Eqs (9) and (10), we obtain,

$$K = \mu\, b/2\pi a \tag{11}$$

Substituting for K in Eq. (8), we get

$$\tau = \frac{\mu d}{2\pi a} \sin(2\pi x/a) \tag{12}$$

The stress τ_{max} that is necessary for causing plastic deformation in a perfect crystal corresponds to the maximum amplitude of the sinusoidal wave and is equal to $\mu b/2\pi d$. b/d is about 1 for many crystals, so that $\tau_{max} \approx \mu/6$.

The shear modulus for typical crystals that exhibit plastic deformation is in the range of 10–100 GN m^{-2}, so that the maximum stress τ_{max} should be in the range of 2–20 GN m^{-2} or 2000–20000 MN m^{-2}. The experimental values of the critical resolved shear stress (CRSS) for initiating plastic deformation in some typical crystals are given in Table 9.3 These values are very much lower than the range 2000–20000 MN m^{-2}. Table 9.3 also gives the shear moduli and the ratio μ/τ for the same crystals. The ratio μ/τ is predicted to be around 6 on the basis of the perfect crystal calculations above. But the observed ratios given in Table 9.3 fall in the range 5000–100000, indicating again the large discrepancy between the above model of plastic deformation and the experimental results. Real crystals deform at a much lower stress than predicted by the model for a perfect crystal. This discrepancy can be explained only on the basis of the presence of dislocations in real crystals. The measured CRSS values should be associated with the stress required to move a dislocation that is already present in a real crystal and not with the stress required to shear a perfect crystal.

Table 9.3

Crystal	*Structure*	*CRSS τ, MN m^{-2}*	*μ, GN m^{-2}*	*μ/τ*
Copper	FCC	0.50	44	88000
Aluminium	FCC	0.75	25	33000
Gold	FCC	0.50	26	52000
Nickel	FCC	5.0	70	14000
Silver	FCC	0.50	26	52000
Iron	BCC	15.0	70	4700
Zinc	HCP	0.30	33	110000

9.7 OBSERVED SHEAR STRENGTHS AND SHEAR MODULI OF SOME CRYSTALS

When a shear stress is applied parallel to a slip plane in a crystal, the dislocations on it move. As they sweep through the crystal, each one of them causes a displacement of the top part of the crystal with respect to the bottom by a distance equal to the magnitude of its Burgers vector. Unlike in the perfect crystal model of plastic deformation, the shear does not occur by the simultaneous displacement of all the atoms in a slip plane. The dislocation moves in a sequential fashion, occupying successive positions during its motion. Recall the atomic displacements during the motion of an edge dislocation sketched. During each step of the motion, the displacement of the atoms surrounding the dislocation is only a fraction of an intratomic distance and not a full one, as in the case of the perfect crystal. The incomplete plane of atoms is not bodily shifted but the configuration of the dislocation moves by small adjustments in the bond lengths in the dislocation region.

The analogy of the motion of a caterpillar helps in visualizing the dislocation motion. As illustrated in Fig. 9.5, the caterpillar moves by making a hump on its back and then transferring the hump from the tail to the head. It does not move forward all along its length simultaneously, but the hump moves sequentially. The frictional forces between the hump and the ground are absent, just as the bonds in the dislocation region are strained, though not broken. Therefore, the caterpillar spends much less effort in moving the hump. The magnitude of the displacement caused by the hump moving from one end to the other depends on the size of the hump. This aspect bears a direct analogy to the magnitude of the displacement caused by a moving dislocation, which is equal to its Burgers vector.

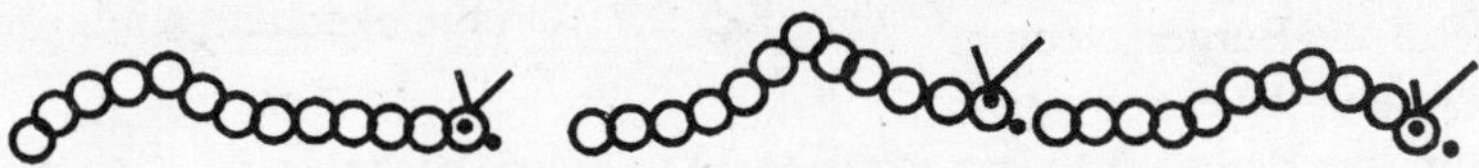

Fig. 9.5 The motion of a caterpillar

Normally, dislocations are always present in crystals. Whiskers are special crystals which are very thin and almost free of dislocations. Such crystals are crystals which are very thin and almost free of dislocations. Such crystals can withstand stresses much higher than ordinary crystals, without undergoing plastic deformation. If, however, a dislocation is introduced accidentally, for example, at the surface, the crystal abruptly loses all its strength and there is a big drop in the stress required to cause further strain, which is permanent. This is illustrated for a copper whisker in Fig. 9.6. The maximum stress the whisker withstands is 700 MN m^{-2}, but on the introduction of a dislocation, this stress falls precipitously to a much lower value and considerable plastic deformation ensues.

Stress Mnm^{-2} ↑ 600 400 200

0 2 4 6

Elongation % →

Fig. 9.6 A copper whisker deforms plastically, as soon as a dislocation is created, with a big drop in the stress required to cause further strain

9.8 THE STRESS TO MOVE A DISLOCATION

The stress required to move a dislocation in a crystal in the absence of other imperfections and impurities has been computed by Peierls and Nabarro. These calculations from first principles are not accurate enough to predict the CRSS of different crystals and to correlate it with the observed value. However, we are in a position to understand the differences in the plastic deformation behaviour of crystals in a qualitative way.

Consider the two edge dislocations sketched in Fig. 9.7. In Fig. 9.7 (a), the dislocation is stiff, that is, no relaxing displacements have taken place in the adjacent planes surrounding the dislocation region. In Fig. 9.7 (b), such displacements have occurred around the incomplete plane. In the first case, the bond lengths are the normal value around the dislocation except in the region below the incomplete plane, where the bonds are virtually broken. In the second case, the strains are distributed more evenly among the bonds around the edge, resulting in some compression above and some tension below. Here, we say that the dislocation is relaxed. The width of a dislocation is a measure of the distance on either side of the dislocation, up to which the stress-relaxing displacements are appreciable. When a dislocation is said to be wide, it is meant that the displacements are distributed over an appreciable distance. When a wide dislocation moves, the adjustments that take place in the bond lengths in the dislocation region are distributed over a number of bonds in the region, but the change in any one bond length is very small. In a narrow dislocation such as that shown in Fig. 9.7 (a), it is clear that, during the motion of the dislocation, the row of atoms below the slip plane has to move through a distance of one full interatomic distance. Therefore, narrow dislocations are more difficult to move than wide dislocations.

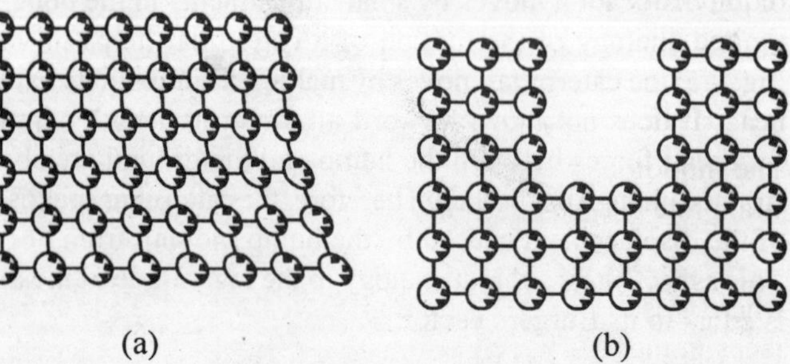

Fig 9.7 An unrelaxed and a relaxed dislocation

Peierls and Nabarro related the stress to move a dislocation $\tau_{P\text{-}N}$ to its width w

$$\tau_{P\text{-}N} = \mu \exp(-2\pi w/b) \qquad (13)$$

where μ is the shear modulus of the crystal and b is the magnitude of the Burgers vector of the dislocation. Because of the exponential form of this relationship, the Peierls-Nabarro stress (τ_{PN}) is a very sensitive function of the width and the burgers vector. The effect of width on τ_{PN} is illustrated below when the dislocation has zero width, the crystal requires a very high stress to move a dislocation, comparable to the stress to deform a perfect crystal. Even as the width increases to $10b$, the stress falls to negligible values. A very precise determination of the dislocation width is then necessary to calculate the P-N stress in any reliable manner. Unfortunately, such measurements are inherently difficult and have not been made. We have to limit our discussion to the qualitative variation of the P-N stress.

The nature of the chemical bonding in a crystal determines the extent of the relaxation and the width of a dislocation. In covalent crystals, where the bonding is strong and directional, the relaxation is small resulting in narrow dislocations. The Peierls-Nabarro stress is correspondingly high. The application of a tensile stress usually results in brittle fracture by crack propagation, before the stress required to move a dislocation is attained. Hence, covalent crystals such as diamond and silicon are brittle. They do not undergo plastic deformation. In typical metallic crystals, such as copper, the bonds are non-directional and not so strong as in a covalent crystal. So, the dislocations are wide and the Peierls-Nabarro stress is low. These crystals exhibit a considerable amount of plastic deformation and are said to be ductile. A copper wire can be cold drawn to hundred times its original length without breaking, in spite of the work hardening that occurs during drawing. In metals of the transition groups, such as iron, some covalent character persists due to d orbital bonding which is directional and, correspondingly, the transition metal crystals are harder than copper. They cannot be cold worked to the same extent as copper.

Table 9.4

Width w	0	b	$5b$	$10b$
τ_{PN}	μ	$\mu/400$	$\mu/10^{14}$	$\mu/10^{27}$

In ionic crystals, the bonds are of moderate strength and nondirectional. From this, one would have expected simple ionic crystals to have some ability to undergo plastic deformation. However, plastic deformation occurs here only under special circumstances, when the crystal surface is free of cracks that can cause brittle failure. The main reason for this behaviour seems to lie in the fact that the Burgers vector of dislocations in ionic crystals tends to be large. b is in the denominator of the exponential in Eq. (13). This tends to increase the Peierls-Nabarro stress. Only dislocations with a full lattice translation for the Burgers vector will not bring two cations as nearest neighbours during the slip process. The Burgers Vector of a full dislocation in NaCl is 3.95 Å, as compared to that in copper, which is 2.55 Å.

Intermetallic compounds and other complex crystal structures such as Fe_3C and $CuAl_2$ do not have favourable crystal planes and directions for easy slip and therefore tend to be brittle. Ordered compounds such as CuZn require dislocations to move in pairs in order to preserve the order during slip. Hence, they possess very limited ductility.

9.9 THE EFFECT OF TEMPERATURE ON THE STRESS TO MOVE-DISLOCATION

The motion of a dislocation can be assisted by thermal energy. The rate of plastic deformation, that is, the strain rate ε is proportional to a Boltzmann's probability factor and is given by

$$\varepsilon \propto \exp\left(-Q/kT\right) \tag{14}$$

where Q is the activation energy for dislocation motion. Q arises from the activation barrier that a dislocation faces in moving from one minimum energy position to the next. This barrier will be much smaller than the barrier shown for shear of a perfect crystal in Fig. 9.8. The strain rate during plastic deformation can be expressed in terms of the velocity v_d of the dislocations and is given by

$$\varepsilon = \rho b v_d \tag{15}$$

where ρ is the density of mobile dislocations and b is their Burgers vector.

Equation (14) indicates that a dislocation can change its position solely due to thermal fluctuations even without the application of an external stress. Such motion would be random, where a dislocation line will have equal probability of moving to the next minimum energy position on either side of the line. Therefore, such motion does not result in plastic deformation. A certain threshold stress is necessary for dislocations to move in a specified direction. In addition to this, the applied stress can lower the activation energy Q. Let $\tau_{P\text{-}N}$ be the stress to move a dislocation in the absence of thermal energy, that is, at 0 K. At 0 K, an applied stress τ_{app} equal to $\tau_{P\text{-}N}$ is necessary to move a dislocation, as no assistance from thermal energy is available. At higher temperatures, an applied stress lower than the P-N stress can move a dislocation, using thermal energy. The two stress terms can be multiplied by a volume term called the activation volume v to relate them to Q and is given by

$$Q = \left(\tau_{P\text{-}N} - \tau_{app}\right) v \tag{16}$$

Substituting Eq. (16) in Eq. (14), we have

$$\varepsilon \propto \exp\left[-\frac{\left(\tau_{P\text{-}N} - \tau_{app}\right) v}{kT}\right] \tag{17}$$

This indicates tells us that for a constant strain rate, the term within the exponential is constant. $\tau_{P\text{-}N}$ as given by Eq. (13) is a constant for a particular crystal. For all practical purpose we can assume the activation volume v to be constant. It follows then that, with increasing temperature, less and less applied stress is required to cause plastic deformation at a specified strain rate. Alternatively, to cause plastic deformation at an increasing strain rate at a constant temperature, the applied stress would be increased. By taking logarithms, Eq. (17) can be rewritten as:

$$\ln \varepsilon = -\frac{\left(\tau_{P\text{-}N} - \tau_{app}\right)}{kT} v + A' = A + B\,\tau_{app} \tag{18}$$

where A, A' and B are constants at constant temperature.

The yield stress of a number of crystalline materials is shown in Fig. 9.8 as a function of temperature. The P-N stress for a ductile crystal such as copper is so low that the thermal energy at most temperatures even below room temperature is sufficient to overcome the P-N stress.

The temperature dependence of the yield stress is small, as only the threshold stress and the stress required to overcome thermal obstacles, if any, are needed at most temperatures. In a transition metal such as iron, the P-N barrier is comparatively large and a high stress is needed to move the dislocation near absolute zero.

This stress decreases more or less linearly with increasing temperature. Eventually, around room temperature, the stress levels off to a nearly constant value, that is characteristic of the threshold stress and other thermal obstacles in the crystal. In covalent crystals such as diamond and silicon, the P-N stress is very high, so that even at room temperature, a high stress is needed to move a dislocation. These crystals show no plastic deformation at room temperature. However, as the stress to move a dislocation is decreasing with increasing temperature, these crystals may exhibit some plastic deformation at elevated temperatures.

9.10 MULTIPLICATION OF DISLOCATIONS DURING DEFORMATION

In an annealed crystal, the dislocation density is about 10^{10} m^{-2}. On application of a stress, if all of these were to move out and disappear at the surface of the crystal, we would be left with a nearly perfect crystal with very few dislocations in it. Experimental observations are quite contrary to this picture. During plastic deformation, the dislocation density in the crystal increases by two to six orders of magnitude, depending on the amount of deformation undergone by the crystal. A lightly cold worked rystal may have a dislocation density of 10^{12} m^{-2}, while a very heavily cold worked crystal may have 10^{16} m^{-2} of dislocations. This indicates that there are sources within the crystal which generate new dislocations during plastic deformation. One such source is called the Frank-Read source.

In Fig. 9.8, a segment *AB* of a dislocation line lies on a slip plane that is the most common one for this crystal. *A* and *B* are dislocation nodes or points beyond which the dislocation does not lie on the slip plane. Only the segment *AB* is capable of moving easily in response to an applied stress. Let a shear stress τ be applied parallel to the slip plane on which *AB* is lying. The segment cannot pull itself away from the anchoring points, because of the geometrical requirement that a dislocation cannot end abruptly within the crystal. It responds to the applied stress, by bending itself to position 2 about the fixed points *A* and *B* and thereby increasing its length. The total line energy of the segment increases in the process. The extra energy comes from the work done by the externally applied stress in bending the dislocation. Between positions 1 and 2, the slip displacement equal in magnitude to the Burgers vector b of the dislocation occurs over the shaded area of the slip plane, Fig. 9.9.

As the stress is increased, the dislocation bends progressively to a semicircular shape as illustrated. The *t* vector indicated in Fig. 9.9 follows the bent dislocation line, going from *B* to *A*. On further increase of the

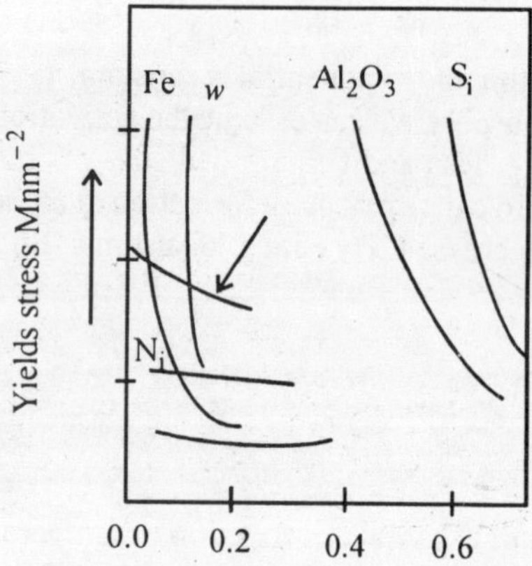

The yield stress of same crystalline material as a function of temperature

Fig. 9.8

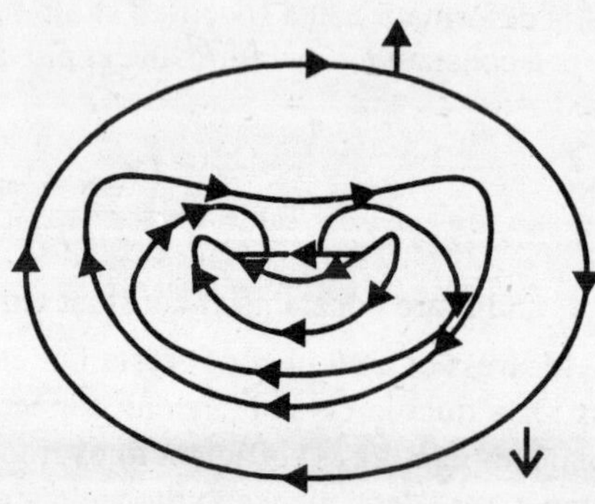

Fig. 9.9 Successive stages in the operation of a Frank-Read source

stress, the dislocation bends back on itself and the slip displacement spreads to regions on the other side of the line AB. In one of these configurations, the direction of the dislocation line at point C is opposite in sense to the direction at point C. The t vectors are in opposite directions at these two points. As the Burgers vector is invariant, this means that the segment of the dislocation line near C is opposite in sign to that near C. The two segments attract and annihilate each other. This produces a full dislocation loop and a left over piece, which springs back to the initial position AB. Note that, during this entire cycle, the direction of dislocation motion at any point on the line is perpendicular to the line at that point except at A and B.

The cycle can now be repeated to produce another full loop and so on. A Frank-Read source can operate continuously and produce an indefinite number of loops, provided the loops produced move out and disappear at the surface of the crystal. If, however, the loops are piled up against an obstacle such as a grain boundary, back stress will begin to build up at the source, the operation of the source eventually coming to a halt.

The stress required to operate a Frank-Read source is given by

$$\tau = \frac{\mu b}{l} \tag{19}$$

where μ is the shear modulus of the crystal and l is the length of the segment AB. The longer is the segment, the lower is the stress required to operate it.

9.11 WORK HARDENING AND DYNAMIC RECOVERY

In HCP crystals, the close packed basal planes are the common slip planes. They are all parallel to one another. In such cases, in a single crystal that is favourably oriented with respect to the tensile axis, the Fran-Read sources in the basal planes can operate continuously for an unlimited period, as the loops produced by the sources can move out and disappear at the surface of the crystal. There is very little work hardening here, as the same source operates at the same resolved shear stress. The stress to cause plastic deformation remains virtually independent of the plastic strain. If, however, the tensile stress axis happens to be such that there is only a small amount of resolved shear stress on the basal planes, then slip on less common slip planes can also occur. These less common slip planes, *i.e.* prismatic and pyramidal planes are nonparallel to the basal planes. Dislocations moving on nonparallel slip planes can intersect each other. In BCC and FCC crystals, the intersection of dislocations moving on nonparallel slip planes is very common. This intersection generally increases the stress required to cause plastic deformation with increasing amount of plastic strain, resulting in work hardening.

For example, consider the dislocations moving on nonparallel planes of an FCC crystal. The slip planes in FCC belong to the (111) family. Let a dislocation of burgers vector $\frac{1}{2}(0\bar{1}1)$ moving on $(\bar{1}11)$ plane meet another dislocation of Burgers vector $\frac{1}{2}(0\bar{1}1)$ moving on $(1\bar{1}1)$ plane along the line of intersection of these two planes, Fig. 9.10. These two dislocations can interact and produce a third dislocation with a decrease in the potential energy:

$$\tfrac{1}{2}(0\bar{1}1) + \tfrac{1}{2}(10\bar{1}) \rightarrow \tfrac{1}{2}(1\bar{1}0) \tag{20}$$

The strain energy decreases by 50% during this dislocation reaction. As the Burgers vector of the product dislocation lies in neither of these two slip planes, it becomes immobile (or sessile). Once the immobile dislocation forms by the dislocation reaction given above, it acts as an obstacle to the oncoming dislocations from the respective sources, which get piled up against the obstacle. A back stress builds up and the dislocation sources stop operating.

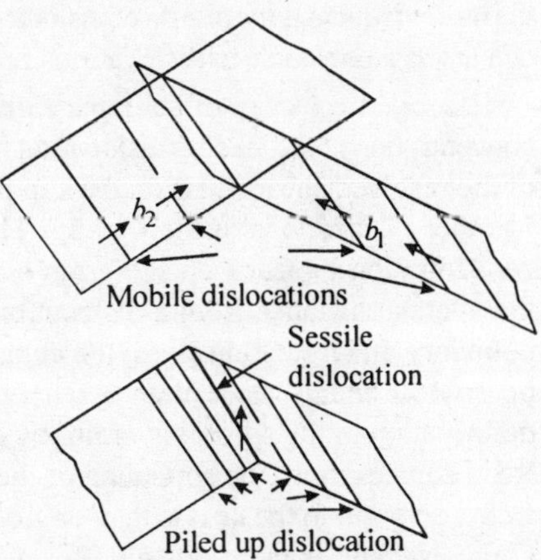

Fig. 9.10 Formation of a sessile dislocation in an FCC crystal by a dislocation reaction

The most effective Frank-Read source which is the longest is expected to operate first. Once this gets immobilized, less effective sources, consisting of shorter segments, have to operate. This requires an increase in the stress as given by Eq. (19). These sources will in turn get immobilized by the formation of the sessile dislocation. The plastic deformation can continue only with the activation of still shorter sources. Thus, a progressively increasing amount of stress is required to continue the plastic deformation process, resulting in work hardening even in single crystals. The formation of the immobile dislocation is at best one possible explanation for work hardening.

Empirically, it is found that the shear stress τ to move a dislocation increases with increasing dislocation density ρ according to the following equation.

$$\tau = \tau_0 + A\sqrt{\rho} \tag{21}$$

where τ_0 is a base stress to move the dislocation in the crystal in the absence of other dislocations and A is a constant. This equation described equally well the work hardening behaviour, as the dislocation density increases with increasing plastic strain.

For a soft crystal, the CRSS for initiation of plastic deformation is typically 0.5 MN m^{-2}. If the constant $A = 10$ N m^{-1}, a low dislocation density of 10^{10} m^{-2} corresponding to an annealed crystal gives a CRSS of 1.5 MN m^{-2}. If the crystal is now heavily cold worked, the dislocation density can increase to 10^{14} m^{-2} and the flow stress τ will now be 100 MN m^{-2}, a large increase.

The stress–strain curves for specimens of mild steel with increasing amount of cold work are shown in Fig. 9.11. Cold working increases the yield stress, but decreases the percent elongation. Cold working is a common strengthening method used in engineering practice. Under suitable conditions, the dislocation density can be increased to very high values, with the strength approaching that of a perfect crystal.

9.12 THE EFFECT OF GRAIN SIZE ON DISLOCATION MOTION

Grain boundaries provide obstacles to dislocation motion. As the orientation of the crystals on either side of a grain boundary is different and random, a dislocation moving on a common slip plane in one crystal can rarely move onto a similar slip plane in the adjacent crystal. In addition, the crystals are separated by a thin noncrystalline region, which is the characteristic structure of a large angle grain boundary. Hence, dislocations are stopped by a grain boundary and pile up against it. The smaller is the grain size, the more frequent is the pile up of dislocations. A twin boundary can act as an obstacle to dislocations in a similar manner.

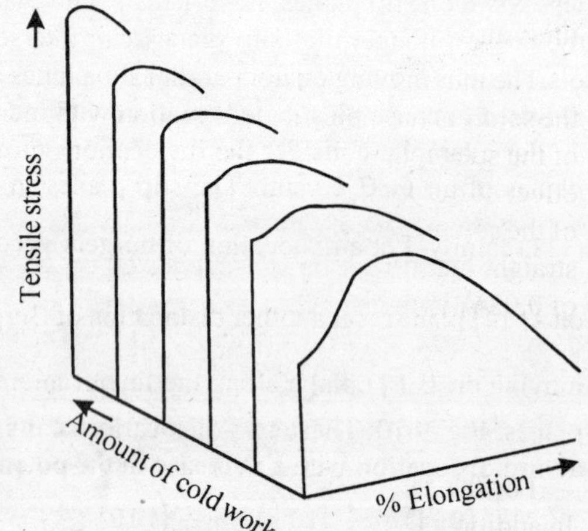

Fig. 9.11 The stress–strain curves for specimens of mild steel cold worked to different degrees.

In a piled up array of stationary dislocations, the stress concentration at the leading dislocation is equal to the product of the resolved shear stress on the slip plane and the number of other dislocations in the pile-up. The magnitude of this concentration depends on the grain size which determines the average distance from a dislocation source to the grain boundary. In large grain sizes, the concentration factor may be as 100 and is more than sufficient to initiate plastic deformation in the adjoining grain, by activating the Frank-Read sources in it. The spreading of the plastic deformation from one grain to the next is then obviously a function of the grain size, and is more effective for a large grain size. The yield strength of a polycrystalline material as observed macroscopically in a tensile test requires the initiation of the plastic deformation in a large number of grains by the spreading process. Hence, the observed yield stress increases with decreasing grain size.

Hall and Petch have derived the following equation for the yield stress σ_y of a polycrystalline material:

$$\sigma_y = \sigma_i + kd^{-1/2} \tag{22}$$

where σ_i is the yield stress for a crystal of the same material where there are no grain boundaries, k is a constant and d is the average grain diameter. The yield strength increases with decreasing grain diameter.

The grain diameter can be approximately calculated from the ASTM specification for grain size. For an ASTM number n, the number of grains per square inch (per 645 mm^2) at a magnification of 100 × is equal to 2^{n-1}. Grain size number ASTM 1 corresponds to 1 grain per square inch at a magnification of 100 ×, or $10^4/645 = 15.5$ grains per mm^2 without any magnification. This can be approximated to a grain diameter of $1/\sqrt{15.5} = 0.25$ mm. The values of σ_i and k for some polycrystalline materials are listed in Table 9.5

Table 9.5 Hall-Petch Constants for Some Materials

Material	*Fe*	*Mo*	*Nb*	*Cu*	*Al*	*Zn*
σ_i, MN m^{-2}	48	110	126	26	16	33
k, MN $m^{-3/2}$	0.71	1.77	0.034	0.11	0.068	0.22

9.13 THE EFFECT OF SOLUTE ATOMS ON DISLOCATION MOTION

Solid solutions in general offer a greater resistance to dislocation motion than pure crystals. The stress fields around solute atoms interact with the stress field of a moving dislocation, thereby increasing the stress required for plastic deformation. The solute strengthening effect depends mainly on two factors.

(*i*) the size difference between the solute and the solvent atom, and

(*ii*) the concentration of the solute atom

This size difference is small and accordingly, the strengthening effect of zinc and nickel is the least among the solutes shown in Fig. 9.12. The size difference between copper and tin is large and the slope of the line for tin is also large.

The more is the size difference between the solute and the solvent, the more intense is the stress field around the solute and its interaction with the moving dislocation is correspondingly stronger. With a large concentration of the solute, the moving dislocation interacts with the solute stress fields at many points along its length. The effect of different solutes on the yield stress of copper is shown in Fig. 9.13 as a function of the concentration of the solute. The yield stress is approximately proportional to the concentration of the solute. The slope of the straight lines in the figure depends on the size difference. For example, zinc and nickel have a size difference of 0.03 Å with respect to copper

$$| r_{Zn} - r_{Cu} | = 0.03 \text{ Å}$$

$$| r_{Ni} - r_{Cu} | = 0.03 \text{ Å}$$

For the same size difference, the smaller atom, nickel, produces a greater strengthening effect than zinc. In addition to the size difference, the intensity of the stress field around a solute atom is also dependent on the elastic modulus of the solute. Nickel having a higher elastic modulus than zinc produces a more intense stress field and a greater strengthening effect.

To improve the strength of a crystal by solute strengthening, it would appear that the maximum size difference coupled with the maximum concentration of the solute would give the best resuslts. However, these two factors are mutually exclusive. Recall from Hume-Rothery's rule that the more is the size difference between the solute and the solvent, the smaller is the equilibrium solubility. The solubility can be increased by producing a super-saturated metastable solid solution by quenching from an elevated temperature.

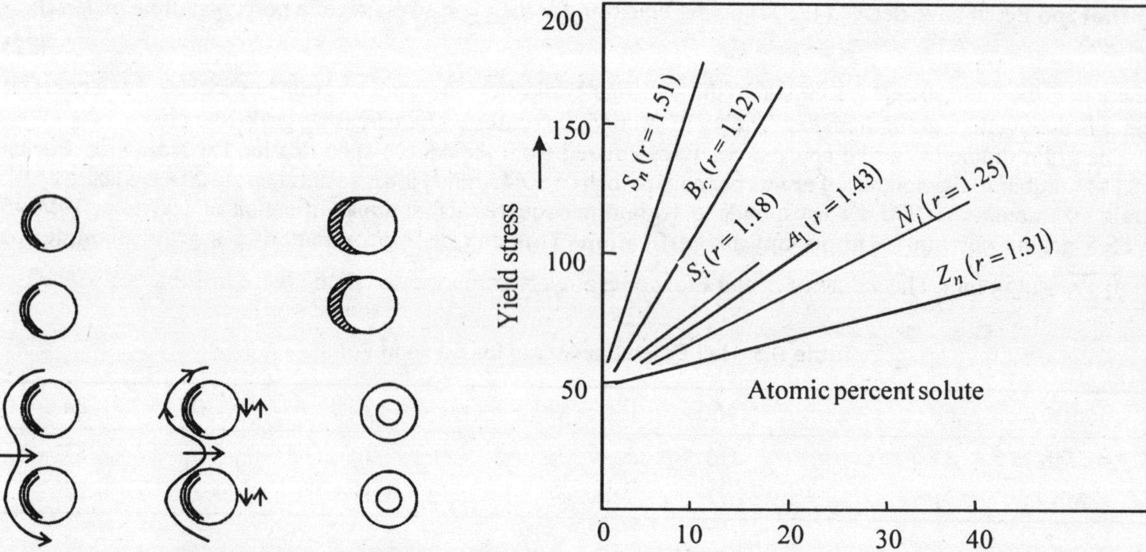

Fig. 9.12 A moving dislocation either (a) cuts through the precipitate particles or (b) by-passed them

Fig. 9.13 Solute strenghening of a copper crystal by solutes of different sizes

The effect of the size difference and the concentration of the solute described above for substitutional solutes is also valid for interstitial solutes. The inerstitial atoms are usually larger than the interstitial voids they occupy. Here, the strengthening effect can be very strong, except that the equilibrium solubility tends to be small. Nature has provided mankind with a unique reaction in steel, which permits an unusually large solubility of interstitial carbon in iron in spite of a very unfavourable size effect. Martensite in steels is a supersaturated solution of carbon in iron, obtained by quenching the steel and not allowing the carbon to diffuse out of the iron lattice. For example, an eutectoid steel contains 0.8% carbon in martensite, which is some 40 times more than the equilibrium solubility in ferrite (~ 0.02%) at the eutectoid temperature. On quenching, the carbon gets trapped in the interstitial positions, as the austenite shears over to form the martensitic structure. The stress field produced by the oversized carbon atoms is so intense that the dislocation motion is very effectively hindered. Indeed, it is so effective that it becomes necessary to temper the martensite to restore some ductility, at the expense of some hardness.

In addition to solute strengthening, the martensitic plates may contain either a high dislocation density or very fine transformation twins, depending on the type of martensite obtained. These features also aid in increasing the strength of martensite.

The carbon atoms in the ferrite phase are responsible for the occurrence of a sharp yield point in mild steel. They segregate around the dislocation cores and reduce the total distortional energy. The segregation is known as a Cottrell atmosphere and the pinning effect of the atmosphere on the dislocations raises the yield stress of the crystal. The dislocations are strongly locked by the atmosphere and do not get freed for motion during plastic deformation. Very few free dislocations are available at the start of the plastic deformation near the upper yield point of the mild steel. Once the deformation starts, the rapdi multiplication of new dislocations and the stress dependence of their velocity are responsible for a fall in stress to the lower yield point at the beginning of plastic region. Thereafter, the stress–strain curve shows the usual work hardening characteristics.

9.14 THE EFFECT OF PRECIPITATE PARTICLES ON DISLOCATION MOTION

Precipitate particles in a matrix provide obstacles to the motion of dislocations. The dislocations moving through the matrix phase have two alternatives. They can either cut through the precipitate particles or bend around and bypass them, Fig. 9.13. The first alternative will be possible, only when the slip plane is continuous from the matrix through the precipitate particle and when the stress to move a dislocation in the crystal structure of the precipitate particle is comparable to that in the matrix. Cutting through occurs for very small precipitate particles, which can be considered to be groups of segregated solute atoms. Cutting through will not be possible, if there is an interface between the precipitate and the matrix and if the orientation changes abruptly at the interface. Under such circumstances, the dislocations bend around and bypass the precipitate particles, as illustrated in Fig. 9.13 and *d*. The mechanism of by-passing is very similar to the operation of a Frank-Read source. The stress required to bend a dislocation is inversely proportional to the average interparticle spacing. Effective strengthening is achieved in the bending process, when the precipitate particles are submicroscopic in size. The spacing between them should be typically a few hundred angstroms. Optimum ageing results in the right interparticle spacing. Overageing allows the fine particles to coalesce into larger and fewer particles. The interparticle spacing is increased thereby and the yield stress decreases.

Table 9.6 lists of the yield strength and percent elongation of some precipitation hardened alloys.

Table 9.6 Yield Strength and Percent Elongation of Some Precipitation Hardened Alloys

Alloy	*Approximate composition*	*Heat treated condition*	*Yield strength, elongation MN* m^{-2}	*Percent*
Al-Zn-Mg	5.6% Zn, 2.6% Mg	annealed	103	16
		aged	503	11
Cu-Be	2% Be	annealed	221	47
		aged	1069	7
Stainless	17% Cr, 7% Ni,	annealed	324	39
steel	0.07% C	aged	1276	9
Maraging	8% Co, 5% Mo,	annealed	648	18
steel	0.8% Nb, 0.4% Ti	aged	1530	11

Real crystals are weak due to the presence of dislocations in them. Perfect crystals (whiskers) which are free of dislocations have very high strength, but they are so thin that they cannot be used as structural and machine components. The key to increase the strength of crystalline materials against plastic yield is to increase the stress required to move a dislocation.

As a review, the four main methods of strengthening crystalline materials against plastic yield are summarized as follows.

9.15 STRAIN HARDENING

When metals are cold worked, their strength and hardness increase. Annealed crystals have a dislocation density of about 10^8 m^{-2}. This can be increased to 10^{10} – 10^{12} m^{-2} by moderate cold working and to 10^{14} – 10^{16} m^{-2} by heavy cold working. As the dislocation density increases, the stress required to move any one dislocation increases due to the interfering effect of the stress fields of the surrounding dislocations. This phenomenon is the basis of work hardening and is described by means of the following equation:

$$\tau = \tau_0 + A\sqrt{\rho}$$

where τ is the stress to move a dislocation in a matrix of dislocation density ρ, τ_0 is the stress to move the dislocation in the same matrix with zero dislocation density and A is a constant.

9.16 GRAIN REFINEMENT

As the orientation changes at a grain boundary, the slip plane in a grain does not continue in the same direction beyond the boundary. So, dislocations gliding on a slip plane are unable to cross the boundary but get piled up against it. This happens more often in a fine grained material. The yield strength σ_y of a polycrystalline material is a function of its grain size as given by the Hall-Petch equation:

$$\sigma_y = \sigma_i + kd^{-1/2}$$

where σ_i is the yield strength at 'infinite' grain size (single crystal), k is the Hall-Petch constant, and d is the mean grain diameter, which can be derived from the ASTM grain size number.

The strengthening effect for a given grain size depends on the magnitude of the constant k. For example, k for BCC iron is 0.71 MN $m^{-3/2}$, whereas it is only 0.11 and 0.07 for the FCC metals, copper and aluminium. So, a given amount of grain refinement produces a greater strengthening effect in iron.

9.17 SOLID SOLUTION STRENGTHENING

The strengthening effect of solute atoms in a crystal depends on the size difference between the solute and the solvent atoms and the concentration of the solute. The more is the difference in the atomic sizes, the more is the strengthening effect. Higher concentrations of the solute also increase the strength proportionately. It is not possible to dissolve a large concentration of a solute, which differs substantially in size from the solvent, as this would be contradicting the Hume Rothery rules. However, this difficulty could be partly overcome by quenching from an elevated temperature and retaining the larger solubility at the higher temperature in the form of a supersaturated solution at the lower temperature. Quenching austenite retains all the carbon in solution in the product phase martensite, producing very hard steels.

9.18 PRECIPITATION STRENGTHENING

A microstructure that consists of a very fine, submicroscopic distribution of precipitates in a matrix has a good strength. The dislocations moving in the matrix are effectively hindered by the closely-spaced precipitate particles, as they have to bend around and bypass the particles. The strengthening effect is inversely proportional to the particle spacing, the minimum attainable spacing being of the order of 100 Å. Such a microstructure tends to become unstable at elevated temperatures, as the particles undergo coarsening which increases the inter-particle spacing and decreases the strength correspondingly. This happens during overageing in aluminium alloys.

9.19 CREEP

Creep is the permanent deformation of a material under load as a function of time. It is appreciable only at temperatures above 0.4 T_m, room temperature for iron is 0.1 6 T_m and for copper it is 0.22 T_m, so that creep at room tempeature is negligible in these materials. On the other hand, room temperature is about half of the melting point (in kelvin) for lead and so it undergoes creep at room temperature. Creep of lead under its own weight is seen in old time roofs, where a thick rim of lead is found at the edge of the roofs.

9.20 MECHANISMS OF CREEP

The Fig. 9.14 shows typical creep curves. The strain is plotted as a function of time under constant load or constant stress. The creep curve exhibits three stages. In I stage, the creep rate decreases with time the effect of work hardening is more than that of recovery in this stage. In II stage, the creep rate is a minimum and is constant with time; here, the work hardening and recovery processes are exactly balanced. In III stage, the creep rate increases with time until fracture occurs. In this stage, necking of the specimen starts and other processes that ultimately result in failure set in.

The temperature and time dependence of creep deformation indicates that it is a thermally activated process. Several atomic processes are known to be responsible for creep in crystalline materials, Fig. 9.15 (b).

When considering work hardening, it was pointed out that dynamic recovery occurs in the later stages of the plastic deformation process. Dynamic recovery was attributed to the cross-slip of screw dislocations. In the low temperature region of creep, the cross slip continues with the aid of thermal energy and causes further plastic strain as a function of time.

The Burgers vector and t vector of edge dislocations are nonparallel. Therefore, they cannot cross-slip like screw dislocations. However, if the temperature is high enough for an appreciable diffusion rate of vacancies, these may diffuse to the edge dislocations or away from them, making them to climb up or down, Fig. 9.15 (a). Thus edge dislocations piled up against an immobile dislocation can move to other parallel slip planes by climb and continue their motion in response to the stress. The rate controlling step in the climp process is the diffusion of vacancies. The measured activation energy for creep agrees with the activation energy for self diffusion by the vacancy mechanism in a number of materials.

Another mechanism of creep is called diffusional creep. Here also the diffusion of vacancies controls the creep rate, but the mechanism does not involve the climb of edge dislocations. Vacancies move in response to the applied strees from surfaces of the specimen transverse to the stress axis to the surfaces that are parallel to the stress axis. Over a period of time, this movement would elongate the specimen in the direction of the stress axis and contract it in the transverse direction resulting in creep, Fig. 9.15 (b).

The third mechanism of creep is the sliding of neighbouring grains with respect to the boundary that separates them, Fig. 9.15 (c) . Grain boundaries lose their strength at a lower temperature than the grains themselves. This effect arises from the noncrystalline structure of the grain boundaries. As illustrated in Fig. 9.15, the viscosity of a noncrystalline substance maintains its viscosity at a high value right up to the melting point. At temperatures above 0.5 T_m, the viscosity of the grain boundaries is small enough for them to behave like a very viscous liquid separating the neighbouring grains and allowing them to slide against each other. This behaviour of the grain boundaries is in contrast to their low temperature behaviour. At low temperatures, they do not flow viscously, but provide effective obstacles to dislocation motion. At high temperatures, the grain boundaries facilitate the deformation process by sliding, whereas at low temperatures, they increase the yield strength by stopping the dislocations.

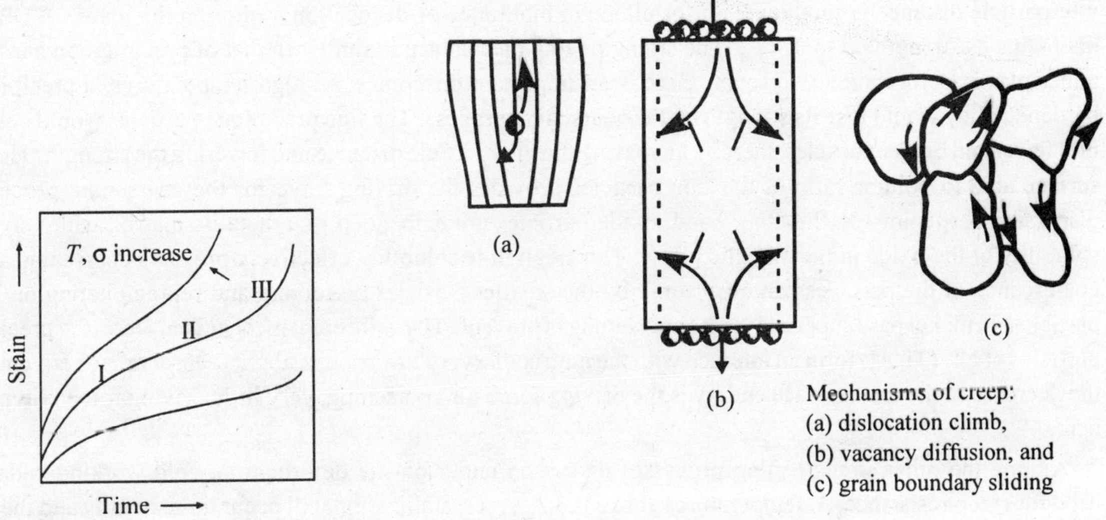

Fig. 9.14 Creep curves showing the increase in plastic strain as a function of time

Fig. 9.15

9.21 CREEP RESISTANT MATERIALS

Materials for machine parts and structural components used at elevated temperatures must be creep resistant. A number of engineering processes require the use of elevated temperatures. Cracking stills in the petroleum industry need high temperatures to accelerate the reaction rates (recall the Arrhenius law) and to control the reaction products. In energy conversion, the efficiency η of a heat cycle such as the Carnot cycle is dependent on the difference between the operating temperature T_2 and the sink temperature T_1 and is given by

$$\eta = \frac{T_2 - T_1}{T_2} \tag{23}$$

A thermodynamic efficiency of less than 30% with the use of steam at atmospheric pressure can be increased to more than 75% for a gas turbine with an operating temperature of 900° C. Materials used for turbine blades must be creep resistant. They should not deform by creep during service. Creep can result in seizing of the blades with the turbine casing.

As a first requirement, the materials to be used at high temperatures must have a high melting point, as creep becomes significant at temperatures greater than 0.4 T_m. Refractory oxides such as MgO and Al_2O_3 have high melting points and, as such, are very suitable for high temperature use. Their brittleness limits their use to applications where only compressive stresses are encountered. Recent developments of tougher ceramics such as silicon nitride (Si_3N_4) indicate that selected parts of a heat engine such as piston rings and cylinder heads can be produced from ceramic materials.

Metals and alloys can be used under more versatile conditions. Most creep resistant alloys consist of a base metal of a fairly high melting point. Since the ability to fabricate into required shapes of the components, the relative cost of production and the density are the other considerations in the choice of a material, metals such as tungsten with a melting point of well over 3000° C have not found widespread use. The commonly used high temperature alloys are iron-base, nickel-base and cobalt-base alloys. All the three base metals have moderately high melting points around 1500° C. The creep resistance of these alloys in the temperature range from 0.5 T_m up to the melting point is considerably improved by a special strengthening process called dispersion hardening. In thoria dispersed (TD) nickel, fine particles of thoria (ThO_2) are dispersed in the nickel matrix and the interparticle distance is small enough for effective hindrance of dislocation motion in the matrix. TD nickel maintains its strength up to 0.9 T_m. The strengthening mechanism is similar to that of precipitation hardening under plastic deformation. However, there is an important difference. At high temperatures, a precipitation hardened alloy would lose its strength by the coarsening process. The fine precipitate particles would coalesce into fewer and bigger particles, thereby increasing the interparticle distance and lowering the strength. The high surface area to volume ratio of the fine particles provides the driving force for the coarsening process. In dispersion hardening, on the other hand, oxide particles are embedded in a metallic matrix, with very little solubility of the oxide in the metallic phase. This negligible solubility effectively prevents coarsening, as the coalescence of the particles can occur only by the smaller particles dissolving and reprecipitating on larger particles. In nickel base superalloys, the coarsening is prevented by a different mechanism. Here, the precipitate particles of Ni_3 (Ti, Al) form an interace with the matrix of a very low energy, about 0.005 J m^{-2} (5 erg/cm^2). As the decrease in the total surface energy is the driving force for coarsening, very little driving force is available here.

Among the other strengthening processes discussed under plastic deformation, cold working cannot be used for creep resistance. At temperatures above 0.4 T_m, recrystallization will occur quite readily and the cold-worked strength will be lost on recrystallization. Solid solution strengthening can be used for better creep resistance, in the same way as in plastic deformation.

A fine grained material is desirable for better mechanical properties in a low temperataure application, where creep is not important. On the other hand, for high temperature applications, fine grained materials are to be avoided, as grain boundary sliding can add to creep deformation. There are no grain boundaries in a single crystal and, therefore, grain boundary sliding can add to creep deformation. There are no grain boundaries in a single crystal and, therefore, grain boundary sliding is not a problem here. Single crystal titanium turbine blades have been tried out, even though the cost is an inhibiting factor here. A less expensive way of minimizing grain boundary sliding is to orient the grains in such a way that the boundary sliding does not result in cavities at grain junctions. These cavities, if present, can coalesce together and cause creep fracture.

SOLVED NUMERICAL PROBLEMS

1. The axis of a cylindrical crystal of copper lies in the (111) plane. Show that, when a tensile stress σ is applied along the axis of the crystal, the resolved shear stress τ in any direction on this plane is zero. Does this mean that this crystal will not plastically deform however large is the applied stress ?

   Solution: The angle ϕ_1 is 90° for the plane (111) so that $\cos \phi_1 = 0$. So, from Eq. (7), the resolved shear stress τ for any value of ϕ_2 is zero. This does not mean that the crystal will not plastically deform, as all planes in the family of (111) are not parallel to one another. There are planes other than (111), which will have a finite resolved shear stress. As soon as this reaches the critical value, plastic deformation will be initiated.

2. The activation volume for dislocation motion in a crystal is 20 b^3, where b is the Burgers vector of the moving dislocation. $b = 2$ Å. The P-N stress for this crystal is 1000 MN m^{-2}. For a specified rate of dislocation motion, the activation energy $Q = 40$ kT. Calculate the stress required to move the dislocation at (*i*) 0 K, (*ii*) 100 K, (*iii*) 300 K, and (*iv*) 500 K.

   Solution:

   (*i*) at 0 K, kT = 0. The stress required to move the dislocation is equal to the P-N stress, 1000 MN m^{-2}.

   (*ii*) at 100 K,

   $$(\tau_{P\text{-}N} - \tau_{app})\ \ 20b^3 = Q = 40\ \text{kT} = 5.52 \times 10^{-20}\ \text{J}$$

   $$\tau_{app} = 1000 - \frac{5.52}{20} \times \frac{10^{-20} \times 10^{-6}}{2^3 \times 10^{-30}}\ \text{MNm}^{-2}$$

   $= 655$ MN m^{-2}, which is about two-thirds of the P-N stress.

   (*iii*) at 300 K, the thermal energy kT is three times larger than at 100 K. Hence, the stress to be applied externally falls to zero. That is, the barrier to dislocation motion is entirely overcome by thermal fluctuations. However, a threshold stress to move the dislocations in a specified direction will be necessary.

   (*iv*) The thermal energy available at 500 K is more than that needed to overcome the barrier to dislocation motion.

   The activation volume denotes the volume over which the thermal energy is to concentrate in order to achieve the activation. In the above example, the activation volume was relatively small, $20b^3$. Dislocations face different kinds of obstacles during their motion through the crystal. In the Peierls-Nabarro model, the obstacle is due to lattice friction, as the barrier depends on the basic crystal structure. For other kinds of obstacles, the activation volume may be large. In such cases, the thermal energy may not be of much help in overcoming the obstacles. These obstacles are called athermal obstacles to dislocation motion. For example, the stress fields of the other dislocations in a crystal may pose an obstacle to the motion of a dislocation. Such obstacles have a large activation volume (~ 10000 b^3). Overcoming them would require virtually the entire stress to be applied externally, irrespective of the available thermal energy.

3. The length of a dislocation line between two pinning points is on an average equal to the reciprocal of the square root of the dislocation density in a crystal. Calculate the dislocation density in copper, work hardened to a stage where slip occurs at a shear stress of 35 MN m^{-2}.

Solution: The shear modulus of copper is 44 GN m^{-2}. The Burgers vector of dislocations in copper is $3.61/\sqrt{2} = 2.55$ Å. From Eq. (19), we know that,

$$l = \frac{\mu b}{\tau}, \text{ on substituting the values}$$

$$\frac{44 \times 10^9 \times 2.55 \times 10^{-10}}{35 \times 10^6} = 0.32 \ \mu\text{m}$$

$$\frac{l}{l^2} = \frac{1}{0.32^2 \times 10^{-12}} = 1.0 \times 10^{13} \ \text{m}^{-2}$$

Single crystal experiments show that, after a certain stress level is reached, the rate of strain hardening decreases with further strain. This phenomenon is called dynamic recovery. This stress level is sufficient to activate screw dislocations to cross-slip. As the Burgers vector of the screw dislocations is parallel to the dislocation line, they are free to move on any out of several equivalent slip planes. When deformation is initiated, they move on those planes where the resolved shear stress is a maximum. If they happen to pile up against an immobile dislocation, they can cross-slip provided the applied shear stress is sufficiently large to give the necessary resolved shear stress on the new slip plane.

4. The yield strength of a polycrystalline material increases from 120 MN m^{-2} to 220 MN m^{-2}, on decreasing the grain diameter from 0.04 to 0.01 mm. Find the yield stress for a grain size of ASTM 9.

Solution: By substituting the two yield stresses and the two grain diameters in Eq. (22), we obtain

$$120 = \sigma_i + k\,(0.04 \times 10^{-3})^{-1/2}$$
$$220 = \sigma_i + k\,(0.01 \times 10^{-3})^{-1/2}$$

Solving for σ_i and k, we get

$$\sigma_i = 20 \text{ MN m}^{-2}$$

and $$k = 0.633 \text{ MN m}^{-3/2}$$

Grain diameter for ASTM 9 is

$$d = 1/\sqrt{10^4 \times 256/645} = 0.0159 \text{ mm}$$

Yield stress for this grain size is

$$\sigma_y = 20 + 0.633 \times \left(0.0159 \times 10^{-3}\right)^{-1/2}$$
$$= 179 \text{ MN m}^{-2}$$

The normal range of grain sizes in metals varies in ASTM specification from ASTM 1 to ASTM 8. In recent years, ultrafine grain sizes have been produced in microalloyed steels, which contain small quantities of strong carbide forming elements such as Nb, V or Ti. These elements go into solution in austenite on reheating the steel billet to about 1300° C. When the steel is subsequently hot rolled, the temperature falls gradually and the solubility of the alloying elements decreases. Very fine particles of alloy carbides precipitate from austenite. These precipitates effectively pin down the migrating grain boundaries during repeated recrystallization of the deformed austenite between passes at successive stands of the rolling mill. This decreased growth rate due to pinning yields a fine-grained recrystallized austenite. A fine-grained austenite provides more potential sites at the grain boundaries for the nucleation of ferrite as the temperature further drops. The end result is thus a very fine grain size of ferrite in the steel of about 2-3 μm (ASTM 14-15). The yield strength increases as per the Hall-Petch equation. The increase is about 50%. The other advantages are that the excellent weldability of mild steel remains unaffected and the micro alloyed steel is produced without a heat-treatment step, which adds to the cost.

5. One million precipitate particles per unit volume of a precipitation hardened alloy coalesce into 1000 particles due to overageing. Assuming the particles to be uniformly distributed and their volume to be small compared to the volume of the matrix, estimate by how much the yield strength would have changed.
Solution: When 10^6 particles coalesce to form 10^3 particles, the linear distance of separation between them increases to $\sqrt[3]{10^6/10^3} = 10$ times the initial value. The stress required to bend the dislocation around precipitate particles is inversely proportional to the spacing between the particles. So, the yield strength would have decreased to 10% of the initial value, due to the coarsening.

UNSOLVED NUMERICAL PROBLEMS

1. The yields stress, the strength coefficient K and the work hardening exponent n for a quenched and tempered steel and for annealed copper are listed below.

Material	σ_y, MN m^{-2}	K, MN m^{-2}	n
Steel	520	1270	0.15
Copper	55	317	0.54

Calculate the difference between the flow stress at 0.1 plastic strain and the yield stress, for the two materials. **Ans.** Steel: 379 and coppeer: 36 MN m^{-2}

2. Find all the slip systems that have the following slip planes: (*i*) (111) in an *FCC* crystal, and (*ii*) (110) in a BCC crystal.
3. Show that the resolved shear stress reaches a maximum value, when $\phi_1 = \phi_2 = 45°$.
4. Sometimes, slip occurs on planes other than the close packed planes in HCP crystals. If the c/a ratio of Zn is 1.86 and that of Co is 1.62, which one is likely to slip more often on close packed planes and why?
5. A tensile stress of 10 MN m^{-2} is applied along the (112) direction of an iron (BCC) crystal. What is the shear stress in the (010) direction lying on the (001) plane? **Ans.** 3.33 MN m^{-2}
6. A zinc crystal (HCP) is oriented with normal to the basal plane making an angle of 60° with the tensile axis and the three slip direction x_1, x_2 and x_3 lying on its plane making angles of 38°, 45° and 84° respectively with the tensile axis. If the plastic deformation is first observed at a stress of 2.3 MN m^{-2}, find which of the three slip directions has initiated slip and at what value of the resolved shear stress? **Ans.** x_1 at 0.9 MN m^{-2}
7. An FCC crystal has a CRSS of 0.7 MN m^{-2}. What tensile stress must be applied along the (100) direction of the crystal to initiate plastic deformation? **Ans.** 1.7 MN m^{-2}
8. Estimate the width of the dislocations in copper. The shear stress to initiate plastic deformation is $\mu/10^5$. **Ans.** 4.7 Å
9. The activation volume for the intersection of a moving dislocation with other dislocations in the matrix can be as high as $5000b^3$. Show that the dislocation intersection is essentially an athermal process at room temperature.
10. A cube of copper of 10 × 10 × 10 mm is sheared at a rate of 10 mm per min. Estimate in order of magnitude the minimum number of dislocations in motion in this cube at a given instant. Assume a dislocation velocity of 1 km s^{-1}. **Ans.** 10
11. Make a series of sketches to show how a Frank-Read source operates. How many loops are created from a F-R source in a single crystal of copper, when the corresponding slip displacement is 1 mm along the slip plane? **Ans.** 4×10^6
12. The lengths of Frank-Read sources in a copper crystal vary from 10^{-8} to 10^{-6} m. Estimate the shear stress at which plastic deformation is initiated. **Ans.** 11 MN m^{-2}

13. Compute the stored energy of copper cold worked from the annealed state to a dislocation density of 10^{14} m^{-2}. **Ans.** 0.14 MJ m^{-3}
14. In a cold worked copper crystal, having a dislocation density of 3×10^{14} m^{-2}, the shear stress to move a dislocation is found to be 100 MN m^{-2}. This value reduces to 1 MN m^{-2} on annealing. Find the shear stress to move a dislocation in a copper crystal cold worked to a dislocation density of 2×10^{12} m^{-2}. **Ans.** 9.08 MN m^{-2}
15. Show that the average grain diameter decreases by half, for every increase of 2 in the ASTM number for grain size.
16. Estimate the yield stress of a polycrystalline Fe-3% Si alloy, when the grain size is ASTM 1, 4, and 8 respectively. Assume $\sigma_i = 80$ MN m^{-2} and $k = 0.63$ MN m$^{-3/2}$. **Ans.** 120, 147 and 214 MN m^{-2}
17. What difficulty would you expect in increasing the concentration of the solute tin in bronze to 40% as in brass, for increasing the solid solution strengthening?
18. In a duralumin alloy, the precipitates are 1 μm diameter and the average distance between centres of neighbouring particles is 10 μm. Estimate the yield stress of the alloy. Is this alloy aged to the optimum extent? Explain. **Ans.** 0.79 MN m^{-2}
19. Supersonic aircraft with a speed greater than Mach 2 attain a skin temperature of over 250° C. Explain why a duralumin alloy cannot be used for such an application.
20. Compare the methods available for increasing the low tempeature strength and the creep strength of a crystalline solid.
21. In an FCC crystal, the tensile axis is along the (121) direction. Determine the slip system that is most favourably oriented for slip to start at the lowest applied tensile stress.

    Hint: $\cos\phi = \dfrac{h_1h_2 + k_1k_2 + hl_2}{\sqrt{\left(h_1^2 + k_2^1 + l_1^2\right)\left(h_2^2 + k_2^2 + l_2^2\right)}}$, where ϕ is the angle between two planes or two directions. **Ans.** (111) (011) or (111) (110)
22. Compute the mean grain diameters in mm, corresponding to ASTM grain size numbers (*i*) 3, (*ii*) 0, and (*iii*) – 2.5. **Ans.** 0.125 mm, 0.35 mm, and 0.84 m.
23. A grain size of ASTM 7 in a mild steel is refined to ASTM 14 by the addition of microalloying elements. Estimate the increase in the yield strength of the steel. **Ans.** 300 MN^{-2}
24. Two samples of an Fe-3% Si alloy have grain sizes of ASTM 1 and 8 respectively. They have yield strengths of 118 and 207 MN m^{-2}. Calculate the yield strength of a single crystal of this alloy. **Ans.** 81 MN m^{-2}.
25. Estimate the yield strength of a cold-worked microalloyed steel from the following data:

Peierls-Nabarro (tensile) stress for Fe	:	35 MN m^{-2}
Solid solution strengthening by Mn and Si	:	135 MN m^{-2}
Dislocation density due to cold working	:	10^{14} m^{-2}
ASTM grain size number	:	13

    Assume that the consant A in the work hardening equation is 10 N m^{-1}. **Ans.** 630 MN m^{-2}
26. Describe briefly the main strengthening methods against plastic yield at low temperatures.
27. Reexamine critically the above strengthening methods against plastic yield for temperature above 0.4 T_m.
28. Discuss the role of (*i*) grain bounaries, and (*ii*) precipitate particles, in strengthening crystalline materials against yield at low as well as high temperatures of deformation.

Index

Reader's Notes